AF684885

Conservation Methodologies and Practices for Built Heritage

Conservation Methodologies and Practices for Built Heritage

Guest Editors

Daniela Fico
Daniela Rizzo

MDPI

Basel • Beijing • Wuhan • Barcelona • Belgrade • Novi Sad • Cluj • Manchester

Guest Editors
Daniela Fico
National Research Council-Institute of Heritage Science (CNR-ISPC)
Lecce
Italy

Daniela Rizzo
Department of Cultural Heritage
University of Salento
Lecce
Italy

Editorial Office
MDPI AG
Grosspeteranlage 5
4052 Basel, Switzerland

This is a reprint of the Special Issue, published open access by the journal *Heritage* (ISSN 2571-9408), freely accessible at: https://www.mdpi.com/journal/heritage/special_issues/BuiltHeritage_Conservation.

For citation purposes, cite each article independently as indicated on the article page online and as indicated below:

Lastname, A.A.; Lastname, B.B. Article Title. *Journal Name* **Year**, *Volume Number*, Page Range.

ISBN 978-3-7258-2743-5 (Hbk)
ISBN 978-3-7258-2744-2 (PDF)
https://doi.org/10.3390/books978-3-7258-2744-2

Cover image courtesy of Daniela Fico and Daniela Rizzo

© 2025 by the authors. Articles in this book are Open Access and distributed under the Creative Commons Attribution (CC BY) license. The book as a whole is distributed by MDPI under the terms and conditions of the Creative Commons Attribution-NonCommercial-NoDerivs (CC BY-NC-ND) license (https://creativecommons.org/licenses/by-nc-nd/4.0/).

Contents

About the Editors

Daniela Fico

Daniela Fico obtained her PhD in Analytical Chemistry in 2015 from the Department of Cultural Heritage of the University of Salento (Lecce, Italy). Until 2020, she worked at the analytical chemistry laboratories of the Department of Cultural Heritage (University of Salento, Lecce, Italy), focusing on the morphological and chemical characterization of organic and inorganic materials and the development of innovative non-toxic and green protective materials for stone and innovative analytical protocols for the analysis of organic substances, using spectroscopic (μFTIR, μRaman) and chromatographic (LC-MS, Py-GC/MS, GC-MS, GC-C-IRMS) techniques, provenance studies, and statistical data processing. From 2021 to 2022, Daniela Fico worked at the Department of Engineering for Innovation of the University of Salento (Lecce, Italy), carrying out a range of research activities, from material processing through complete characterization (morphological–structural, chemical, physical–mechanical, rheological analysis) to the development of innovative materials by means of a multidisciplinary approach (differential scanning calorimetry, mechanical analysis, permeability, colorimetry, modeling and 3D printing, etc.), with a focus on sustainability and the circular economy. Daniela Fico is currently working at the National Research Council—Institute of Heritage Science (CNR-ISPC, Lecce, Italy), where she is mainly involved in diagnostic and chemical analyses of cultural heritage and the study of manufacturing processes, alterations, conservation, and preservation through advanced technologies (ATR-FTIR spectroscopy, mapping, imaging, and high-resolution mass spectrometry, including LA-ICP-MS, ICP-MS, and GC-HRMS). Daniela Fico has participated in national and international projects and supervised several theses. She is the author of papers in international Scopus-indexed journals and a reviewer and Guest Editor for several important international journals.

Daniela Rizzo

Daniela Rizzo is the technical manager of the Laboratory of Art Techniques and Restoration for Art History and a contract lecturer for the "Laboratory of artistic techniques - L-ART/04, degree course in Cultural Heritage", in the Department of Cultural Heritage, Unisalento, Italy. Daniela Rizzo holds a degree in "Cultural Heritage" from the University of Lecce, graduating with 110/110 cum laude. She also holds a Master's Degree Level II in "Management of Environmental Resources. Water, Soil, Waste and Energy", with a final mark of 30/30 cum laude, from the Department of Legal Studies, University of Salento; a Master's Degree Level II in "Territorial Sustainability and Eco-Management (STEM)" from the Department of Economics, University of Salento; a Level I Master's degree in "Health, Safety and Environmental Management" (MAHSEM) from the School of Management, LUM University; and an RSPP qualification (Module A-B-B SP4- C) from the School of Management, LUM University. Daniela Rizzo has more than 20 years of professional experience in the field of diagnostics applied to cultural heritage using analytical, non-destructive, and micro-destructive techniques for the chemical characterization of binders, natural and synthetic polymers, pigments, and dyes present in works of art and materials of historical–artistic interest in the BBCC sector and the determination of analytes in environmental matrices. Currently, she works in the field of historical knowledge of traditional art techniques through a theoretical–practical approach and a critical analysis of sources, in addition to using technologies such as 3D printing and experimentation with innovative materials.

She has participated in various training courses, conferences, events, seminars, and schools and is a member of working groups and scientific projects. She is also the author and co-author of articles in books and international scientific journals, as well as posters and conference proceedings.

Preface

The planning of conservation interventions for historic buildings requires an understanding of their properties, the site, the intervention procedures, and the unique features of the architectural asset. In recent years, numerous advances have been made in the field of conservation methodologies and practices for architectural cultural heritage. The use of innovative technologies and multidisciplinary approaches can be considered valuable tools to support the conservation, fruition, rehabilitation, and valorization of historical–artistic buildings, which are usually subject to protective restrictions and therefore require non-invasive or micro-destructive diagnostic analyses.

This reprint collects original research articles on architectural heritage conservation, focusing on material science, diagnostics and monitoring, chemical analysis, intervention, civil and materials engineering, green materials, and the use of new digital technologies.

The topics include, but are not limited to, the following conservation methodologies and practices for built heritage:

- Chemical and engineering materials and technologies in the field of the diagnosis, conservation, fruition and valorization, and rehabilitation of historical buildings;
- Environmental monitoring of monuments;
- Multivariate statistical treatments and mathematical models;
- Virtual and augmented reality, additive manufacturing, ICT, and digital heritage technologies;
- Sustainability, green chemistry, and ecofriendly materials.

Daniela Fico and Daniela Rizzo
Guest Editors

heritage

MDPI

Article

Comprehensive Overview and New Research on Carbonate Rocks of the Sé Velha Cathedral in Coimbra, Portugal

Alice Tavares [1,*], Fernando Rocha [2], Ana Fragata [2], Aníbal Costa [3] and Matilde Oliveira [1]

[1] CICECO, Department of Materials and Ceramics Engineering, University of Aveiro, Campus Universitário de Santiago, 3810-193 Aveiro, Portugal; matilde.c.oliveira@ua.pt
[2] GeoBioTec, Department of Geociences, University of Aveiro, Campus Universitário de Santiago, 3810-193 Aveiro, Portugal; tavares.rocha@ua.pt (F.R.); afragata@ua.pt (A.F.)
[3] RISCO, Department of Civil Engineering, University of Aveiro, Campus Universitário de Santiago, 3810-193 Aveiro, Portugal; agc@ua.pt
* Correspondence: tavares.c.alice@ua.pt

Abstract: In addition to the surrounding climatic environment, the intrinsic characteristics of the stones used in construction are a critical factor to understand a building's conservation state and define the necessary planning and conservation management activities. The use of environmentally susceptible stones, such as carbonate stones, in heritage buildings can be especially problematic. The present investigation presents an overview of past research and contributes to identifying the types of carbonate stones used over time in Coimbra's Old Cathedral in Portugal, which was classified as a National Monument and integrated into the Coimbra World Heritage site by UNESCO. Our mineralogical and chemical analyses revealed the use of carbonate stones from different quarries (Coimbra region), including Ançã limestone and dolostone (Porta Especiosa portal), Portunhos limestone and Outil limestone (in the perimeter of windows), and marlstones and Outil limestone (side facades). These stones, which were installed at various times during the construction and alteration of the monument, represent a great challenge for conservation planning due to their intrinsic vulnerability to adverse environmental conditions and pollution.

Keywords: built heritage; carbonate stone; limestone; dolostone; marlstone; Porta Especiosa; Coimbra UNESCO site

Citation: Tavares, A.; Rocha, F.; Fragata, A.; Costa, A.; Oliveira, M. Comprehensive Overview and New Research on Carbonate Rocks of the Sé Velha Cathedral in Coimbra, Portugal. *Heritage* **2024**, *7*, 5569–5592. https://doi.org/10.3390/heritage7100263

Academic Editor: Daniela Fico

Received: 13 August 2024
Revised: 30 September 2024
Accepted: 4 October 2024
Published: 8 October 2024

Copyright: © 2024 by the authors. Licensee MDPI, Basel, Switzerland. This article is an open access article distributed under the terms and conditions of the Creative Commons Attribution (CC BY) license (https://creativecommons.org/licenses/by/4.0/).

1. Introduction

Carbonate rocks applied in heritage buildings present considerable challenges in urban contexts. To support the long-term preservation of cultural heritage, we must improve our understanding of the nature of the stones used in monuments and the different processes, factors, and characteristics that lead to their degradation [1].

Many external decay agents operate in combination with the inherent petrographical and petrophysical characteristics of stone to accelerate its deterioration. The principal catalysts for the deterioration of building stones are widely acknowledged to be water, salt, atmospheric pollutants, biological colonization, solar irradiation, wind stress, and fluctuations in temperature. Human activity also plays a major role in the deterioration of geological heritage [2–6].

However, the nature of the stone itself and its suitability as a construction material are not always considered as sources of problems. This is a critical factor to understand the real state of a monument's conservation and plan conservation management activities throughout the building or monument's lifetime. A heritage building's artistic characteristics are unlikely to be lost due to selecting inappropriate stone for its construction. Problems arise when carbonate stone with intrinsic problems is applied to a monument. This problem occurred at the iconic Old Cathedral of Coimbra (Sé Velha, Portugal, Figure 1a), a Romanesque structure that is part of the area classified by UNESCO as a World Heritage Site at the University of Coimbra Alta and Sofia [7–9]. This monument, which was classified as a National Monument in 1910 [7], features one of the most beautiful entrances in Coimbra, the renowned Renaissance Porta Especiosa portal (Figure 1b). The stone of this portal has continuously undergone an accelerated degradation process over time, currently impairing the readability of the sculptures. Some areas of the Cathedral walls also show damage. For two decades, researchers have questioned whether the problem stemmed from the origin of the stone itself. Moreover, the risk of completely losing the sculptural elements has motivated ongoing research [10–17] to support monument management decisions.

(a) (b)

Figure 1. Different carbonate rocks in (**a**) Old Cathedral of Coimbra and (**b**) Porta Especiosa. Credits: Alice Tavares.

This paper analyses the carbonate rocks of Coimbra's Old Cathedral (primarily limestone and dolostone), which are characterized by their chemical homogeneity (one with a high percentage of carbonate and the other with dolomite), albeit with considerable variation in terms of texture, hardness, porosity, and fossil content. These differences have a significant impact on how these building stones behave and how easily they deteriorate [18–20]. Other researchers noted that the composition of limestone varies according to the sample's specific characteristics, geological conditions, and place of origin [7,11,12,14,21,22]. Nonetheless, limestone typically consists primarily of calcium carbonate minerals such as calcite and aragonite, as well as secondary minerals. Previous research also revealed the common presence of additional minerals, including quartz, dolomite, muscovite, organic compounds, and clay minerals, depending on the limestone's formation circumstances [23–25].

In addition, stones expand and contract in response to periodic temperature changes, which causes a constant redistribution of stress in the area where the heat front advances and retreats. This phenomenon has the potential to initiate and spread new discontinuities or reactivate pre-existing ones [26]. Stones placed outdoors are particularly susceptible to biodeterioration processes aided by the growth of different microorganisms in different types of stone (such as marble, sandstone, and limestone), impacted by the natural bioreceptivity of the stone [27].

1.1. The Old Cathedral of Coimbra Sé Velha and Porta Especiosa: A Case Study

The Old Cathedral (Sé Velha) of Coimbra (Figure 1a), one of Portugal's most outstanding examples of Romanesque architecture, was chosen for this case study. Built in the 12th century, this cathedral exhibits characteristics typical of the period, including robust stone construction, arched entryways, and ornate sculptures [28]. In the second half of the 12th century, the original Old Cathedral of Coimbra was rebuilt in place of another church. Construction of the current structure began in 1160, during the episcopate of Bishop Miguel Salomão [9,16]. This church was labelled a Romanesque building by Master Roberto, a French architect [9].

The Porta Especiosa portal (Figure 1b) was commissioned by the Bishop of the Cathedral of Coimbra, Jorge de Almeida, and is one of the most representative works of the Renaissance in Portugal [9].

However, no documentary evidence of its construction, the material conditions under which it was built, the precise chronology, the labour involved, or the relevant reference sources were found in the literature [9,16,28]. However, it is known that this structure was designed by João do Ruão and built in the 1530s [9,27]. Therefore, analyses of the sculptural language and materials themselves are used to enable an in-depth reading of the monument. Porta Especiosa is a medieval term given to the doors of cathedrals through which processions enter to recite the psalm Speciosa est Maria [9].

The Sé Velha Cathedral of Coimbra, including the Porta Especiosa, has been subject to restoration interventions since the 19th century along with successive attempts to maintain and reverse the severe degradation of the stone [9]. The detected pathologies are mainly related to the construction itself, the variety of carbonate rocks used, and the lithological characteristics leading to structural disintegration [10,12–14,16,27,28]. For this reason, a considerable portion of the Porta Especiosa was irretrievably lost; damage to the stone of the facade walls and internal vaults is also visible. This situation has led researchers to question the type of stone used and whether its characteristics make it suitable as a construction material.

1.2. Overview of Background Research

One common and significant obstacle to researching stone characterization in monuments is the different difficulty levels of direct physical contact, sample collection for lab testing, and in situ investigations. Consequently, less-destructive techniques, such as optical microscopy, X-ray diffraction (XRD), X-ray fluorescence (XRF), and Fourier-transform infrared spectroscopy (FTIR) are considered the most appropriate methods for assessing the quality of stones that possess historical or artistic value [29–31].

The origin of the stone used in this monument has been extensively debated within the scientific community due to the lack of clear documentation about the building's construction and stone provenance [10,12–14,16]. Consequently, in situ and laboratory analyses, along with comparisons to regional quarries, have been the primary sources of information in this area [12,15]. Other analyses have attempted to correlate the characteristics of the stones with their levels of decay susceptibility [13,14,16,29].

Throughout the 20th century, several analyses were conducted on the building's stone, primarily focusing on the Porta Especiosa due to its significant artistic value, with less attention given to the interior spaces of the Cathedral. The present study focuses on analysing these spaces to fill the aforementioned research gap. The following sections provide a summary of the research data collected prior to the current investigation.

1.2.1. Overview of Methods and Samples Collected

Most samples from Sé Velha Cathedral are loose fragments taken from damaged areas of the monument, mainly from the Porta Especiosa portal. In addition, some samples were collected in selected outcrops to compare their characteristics and composition with those of the carbonate rocks already identified in the monument [22]. Results from previous studies on these materials in both situations were also used. Thin sections were prepared and analysed using hand lenses, whenever necessary, and a petrographic microscope, combined with chemical analyses (for several samples) [16,22]. Special attention was given to the Renaissance Porta Especiosa portal located at the north façade, where the material distribution was mapped using a photogrammetric map provided by IPPC [16]. Analyses of limestone from regional quarries were used to study the altarpiece. These tests explored the macroscopic characteristics of the Ançã and Coimbra limestone based on thin sections, physical analyses, chemical analyses, the bulk density of limestone, diffractometric analyses, suction curves, porosimetry obtained via suction tests, porosimetry histograms, capillarity tests (for the mass of absorbed water), and the height of capillary saturation [16,22].

1.2.2. Debate over Geological Origin

Since the beginning of its construction in the 12th century, the Sé Velha Cathedral used carbonate rocks as the construction material for its walls and vaults [16]. To determine their origins, previous researchers compared stones from the cathedral, mainly from the Porta Especiosa portal, with stones from other quarries.

Figure 2 shows the different geological layers presented by the LNEC, where the area of Sé Velha Cathedral belongs to the beds of Coimbra, which form the main body of the hill. The beds of Coimbra s.l. are of Lias age (inferior Sinemurian–Carixian) and composed of two units: the beds of Coimbra ss and the beds of S. Miguel [16].

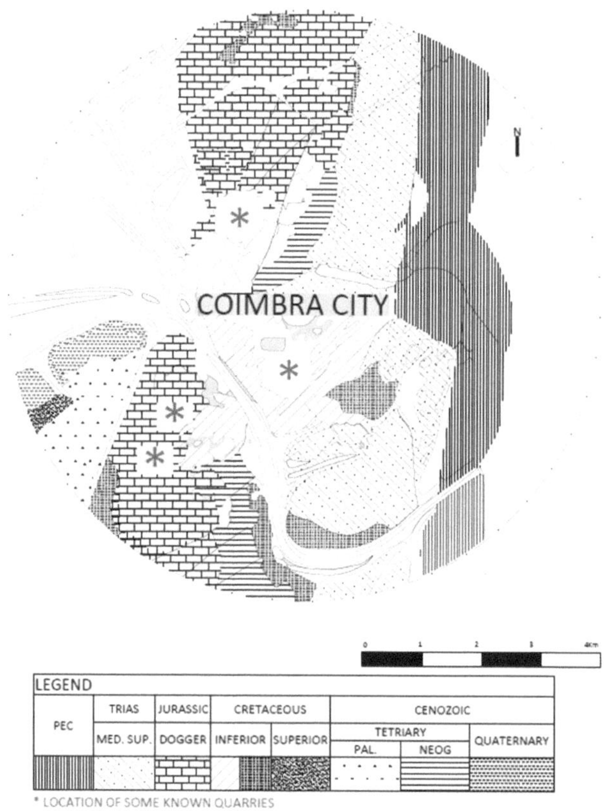

Figure 2. Geological map of the Coimbra region based on the LNEC map [16].

Coimbra is located over an area of transition between Precambrian phyllites in the east and the Meso–Cenozoic west border composed of sandstone and limestone units [16]. The correspondence between the geological time and the type of carbonate stones in Coimbra and its surroundings was also recorded [16], as follows:

- Bajocian age—Ançã limestone;
- Lias age (is Rhaetianto Toarcian in age)—dolostone and calcitic dolostone (beds of Coimbra) and limestone to dolostone (beds of S. Miguel);
- Toarcian domerian medium—marls and marly limestone of Pedrulha;
- Carixian domerian superior—marls and marly limestone of Eiras.

Table 1 presents the results from background research on the geological origin of the quarried stone.

Table 1. Results from background research on the geological origin and probable uses of the stone.

Quarries/Region	Identified Stone	Possible Use in Sé Velha Cathedral	Reference
Beds of Coimbra and hill of Cathedral	Dolostone and calcitic dolostone (with 75–98% dolomite)	Cathedral and cloister	[16]
Ançã region (10 km northwest of Coimbra), Cantanhede	Ançã limestone, milky white rocks with a very pure appearance, oolitic limestone	Porta Especiosa portal Altarpiece of Baptistery Chapel	[11,21,22]
Santo António da Pedreira (250 m south of the cathedral	White-to-grey limestone, greyish limestone, slightly marly, yellowish to white	Porta Especiosa portal	[16]
Outcrop in the south of Coimbra	White créme and greyish limestone with a residual silicate fraction similar to Cantanhede but not Ançã limestone	Porta Especiosa portal	[12,14]
D'El Rei quarry in Portunhos (Ançã region)	Dolomitic limestone	Cathedral	[21]
Santa Clara quarry in Coimbra	Dolomitic limestone	Altarpiece of the Baptistery Chapel	[22]
Ançã, Portunhos, Pena, Outil	Marlstone (probable Lower Jurassic) and Ançã limestone (with 97% calcium carbonate)	Lower part of the Porta Especiosa portal	[17]

1.2.3. Material Mapping of the North Wall (Including the Porta Especiosa Portal)

From the early 18th century until 2023, restoration work was carried out at Sé Velha Cathedral, but to the best of our knowledge, no documentation is available on the interventions prior to 1930 [16,30]. For this reason, mapping the different types of carbonate rocks is helpful to better understand the different moments of construction and repairs as well as the monument's state of conservation and to prepare future conservation and maintenance plans.

Previously, detailed mapping was conducted exclusively on the North façade, as this wall was presumed to exhibit the most severe deterioration issues. A collection of carbonate rock types was also gathered for identification purposes [16], as shown in Figures 3 and 4, corresponding to different construction phases and alterations over time. Dolostone and calcitic dolostone from the Coimbra beds were used in the main body of the cathedral (1162–1181), while limestone from Ançã was employed in the construction of the Porta Especiosa (1530). The limestone exhibits a white to grey, marly colouration, whereas the dolostone, when exposed to outdoor conditions, presents a dark patina that obscures its natural yellowish hue [16]. Although the dolostone components are generally in fair condition, certain areas show significant weathering. In contrast, the limestones are extensively deteriorated and require urgent restoration (Figure 3). However, the differences in colour and level of decay combined with the need to identify the stone types of the portal led to an ongoing debate without a consensus position. Near the windows, replacement materials were used to restore the original architectural design of the cathedral, close the aperture of the windows to their original size, and replace deteriorated blocks. These materials were also used to increase the sizes of windows in different epochs, to increase the light inside the Cathedral, and for repair interventions [16]. As shown in Figure 4, white-to-yellowish Ançã limestone and marly and grey Ançã limestone were identified in the Porta Especiosa portal, light yellow dolostone was used in several arrangements of windows, a brownish dolostone was used in the secondary lateral entrance, and dolostone was used in the rest of the cathedral walls [16].

(a) (b)

Figure 3. Details of different carbonate rock types and decay in the Porta Especiosa portal (**a**) credits: [15]; (**b**) credits: Alice Tavares.

Figure 4. Mapping of stones from North facade of Coimbra's Old cathedral (based on the LNEC drawing) [16].

Table 2 shows some of the physical properties of those stones in the literature.

Table 2. Range of some physical properties of the Sé Velha materials [16].

	Dolostone to Calcitic Dolostone	White Yellow Ançã Limestone	Marly Ançã Limestone
Porosity %	8.5–24.2	23.2–29.0	11.3–11.4
Dry Unit weight (KN/m^3)	21.13–23.97	18.87–19.52	23.51–23.53
Grain Unit Weight (KN/m^3)	26.61–28.03	25.88–26.62	26.52–26.54

1.3. Chemical and Mineralogical Analysis Background

The characteristics and properties of stone materials from the first-floor construction elements, bases of the facades, and Porta Especiosa were assessed. The provenance of the stone was assumed based on similarities between rocks and outcrop rocks known around Coimbra. Based on these studies, the researchers concluded that more than 90% of the materials used in the Cathedral are from beds in Coimbra of the Lias age [16]. The remaining stones are generally Ançã limestone of the Dogger age.

Table 3 presents the range of variation among the main chemical components of the samples analysed from the cathedral and regional quarries (Portunhos—Limestone Ançã: CA1, CA2, CA3; CC4 Coimbra limestone—calcitic dolostone).

Table 3. Range of values in the literature for the chemical analysis of carbonate materials from Sé Velha [16] and regional quarry stones [21].

Classification	% Dolomite	CaO	MgO	Fe_2O_3	Na_2O	K_2O	Al_2O_3
Dolostone	>90	29.11–30.70	19.87–21.50	0.99–2.11	0.01	0.06	0.89
Calcitic dolostone	50–90	23.42–31.43	16.42–19.55	0.83–3.77	0.01–0.94	0.06–1.42	0.89–3.35
Dolomitic limestone	10–50	39.06–43.32	8.02–8.90	0.94–1.47	-	-	-
Limestone:							
White and yellowish	<10	53.09–55.48	0.32–0.60	0.05–0.20	0.00–0.02	0.00–0.05	0.50–0.70
Grey		54.71–54.98	0.47–0.60	0.28–0.49	0.01–0.03	0.07–0.09	0.66–0.93
Marly		49.79	0.91	0.83	0.01	0.11	1.33
Regional quarries:	SiO_2	CaO	MgO	FeO_3	SO_3	CO_2 (read heat loss)	Al_2O_3
CA1—limestone	2.29	53.90	0.60	0.07	0.0	42.55	0.70
CA2—limestone	1.80	54.33	0.60	0.05	0.0	42.77	0.50
CA3—limestone	1.76	54.33	0.40	0.08	0.0	42.81	0.55
CC4—calcitic dolostone	2.06	29.35	19.30	0.85	remains	45.46	3.35

Ançã limestone is heterogeneous in its colour and composition, with the white-to-yellowish limestone commonly used for cornerstones and sculptures. Macroscopically, two groups can be distinguished: white-to-yellow limestone and marly and grey limestone [16]. The white-to-yellow limestone has a compact appearance, sometimes with a chalky quality. Researchers identified some marly limestone in the Porta Especiosa [16]. However, this discovery was surprising since this limestone is a poor construction material due to its fine grains and light-to-dark grey colouration. The authors emphasized that when compared to the white limestone, the marly limestone presented a larger amount of Al_2O_3 and Fe_2O_3 and a smaller amount of CaO due to the presence of clay minerals [16]. The high Fe_2O_3 content is due to the impregnation of clay minerals with colloidal iron such as $Fe(OH)_3$ [16]. On the other hand, there is a significant amount of Fe_2O_3 in the dolostone. Consequently, weathering of these rocks typically results in colour changes due to iron oxidation [16].

Comparative analyses of quarry stones were also performed. The researchers concluded that Coimbra limestone is macroscopically a very fine-grained rock composed primarily of calcite and dolomite [22]. These stones present a higher proportion of very fine pores, much less than that in Ançã limestone, with half the porosity (14%) and a significantly higher apparent density [22].

Analyses were carried out on four quarry samples—CA1, CA2, and CA3 from the Portunhos quarry (Ançã) and CC4 from the Santa Clara quarry (Coimbra limestone)—with the results shown in Table 4. The following conclusions can be drawn from the chemical analysis. The CA samples were mainly composed of calcite with traces of quartz, while the CC sample was mainly composed of dolomite with slight traces of quartz and calcite [22]. At that time, Ançã limestone was classified as semi-compact limestone, deemed unsuitable for construction and exposure to atmospheric agents if no preliminary tests for its approval were performed [10,22], whereas Coimbra limestone was designated as hard limestone, with some application in construction, as outlined by M. Mamillan [29].

Table 4. Analysis of stone samples from the Porta Especiosa, LNEC 1985 [11].

Sample Location	Infrared Spectroscopy	X.R.D	Thermo-Differential Analysis AID	Microscopy
Sample 1 Grey limestone slabs from the right threshold 2 metres from the staircase landing	Calcite, clay product. In addition to the above, there is an area with dark spots showing traces of sulphates. Phosphates confirmed via chemical tests.	Calcite Quartz (low content)	Calcite	Fine-grained limestone with quartz clasts scattered throughout the matrix.
Sample 2 Gray crusts collected from the recess at the base of a bas relief on the right side of the portico	Calcite and a great deal of gypsum-type sulphate. Traces of nitrates. Phosphates confirmed via chemical tests.	Calcite Gypsum	Calcite Gypsum	--
Sample 3 Yellowish crusts on the right doorframe	Calcite and a great deal of gypsum-type sulphate. It contains nitrates and clay products and phosphates confirmed by chemical tests.	Calcite Gypsum	Calcite Gypsum	Fine-grained limestone with a profusion of quartz granules and a rich matrix of isotropic clay-type products.

The authors also noted that the apparent density of Ançã limestone ranged from 1924 to 1990 kg/m^3 with a porosity of approximately 28% and an absorption coefficient of about 0.91. In contrast, Coimbra limestone had an apparent density of 2444 kg/m^3, a porosity of 14%, and an absorption coefficient of 0.77 [22].

The comparison between sound stones (from the quarries) and stones from the altarpiece showed that dolomitic limestone (Coimbra limestone) was unlikely to be used in the altarpiece due to its magnesium oxide content. However, the use of Ançã limestone, which has a very similar composition, was considered a possibility [22].

In 1985, new analyses were conducted on three damaged stone samples from the Porta Especiosa [11], and the conclusions are summarized in Table 4.

Between 1998 and 2000, several studies were conducted on the various qualities of stone used in the Porta Especiosa [10,12–14,17,28]. These studies resulted in several reports, journal articles, and scientific papers; these publications formed the basis for the specifications outlined in the intervention dossier of 2002. In the preliminary phase before the 2002 restoration [12–14,17,28], the limestone known as "brando de Ançã" was noted to vary in its composition and exhibit different colorations such as white, dirty white, yellowish, and grey [17]. One of the factors influencing this limestone's weathering behaviour is the percentage of clay content [17]. Grey Ançã limestone is predominant in the Porta Especiosa, where the highest levels of degradation are observed.

2. New Research Results—Materials and Methods

Samples from different locations within the monument (Figure 5) were analysed to provide an updated contribution to characterizing the stone, particularly in areas where the decay is most significant. To this end, only small samples were collected from the collapsed sections of the wall surface, both inside and outside the monument, without employing intrusive sampling methods (Figure 6). The collection of larger samples was not authorized due to the high loss of material in some areas. However, samples were sought from different locations to facilitate a more general approach and not merely concentrate on the Porta Especiosa portal. These small samples were analysed to determine their chemical and mineralogical composition, and the corresponding results are presented in the following sections.

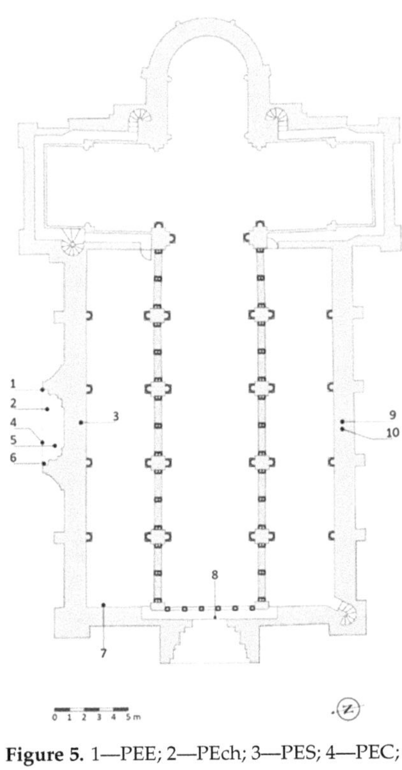

Figure 5. 1—PEE; 2—PEch; 3—PES; 4—PEC; 5—PE1; 6—PER; 7—PLE; 8—PPch; 9—PEt; 10—P3P.

Figure 6. (**a**) Samples collected from windowsills (PEt; P3P); (**b**) sample collected from the windowsill of the front façade (PPch); (**c**) sample collected from the balcony guard (PEC); (**d**) sample collected the wall at the entrance (PES); (**e**) sample collected from the wall near the floor and colonnade (PE1, PER); (**f**) sample collected from the base guardrail (PEch); (**g**) sample collected from the sill guard (PEE); (**h**) sample collected from the base of the wall (PPch). Credit: Alice Tavares.

2.1. Stone Materials

The new investigation focused on the stone of the interior faces on the side facades and the main facade, as well as that on the Porta Especiosa portal. As it was not possible to collect samples of "healthy" stone, which is considered an intrusive operation, we instead used very small splinters of stone that collapsed from the construction elements and were retained on the interior windowsills, at the top of the balcony, or on the floor of the Porta Especiosa. All samples were collected at the level of the first floor, which is not accessible for public visitation. Only very small samples that could not be used again for replacement as part of conservation or restoration activities were used. This factor informed the type and number of analyses carried out.

The following plan for the first floor of the cathedral identifies the sample collection locations. Table 5 presents the identification of the samples.

Table 5. Basic description of the samples studied.

Specimen Descriptions	Abbreviation
Collapsed stone close to the entrance of the Porta Especiosa	PES
Collapsed stone on the floor of the Porta Especiosa	PEch
Collapsed stone on the floor of the Porta Especiosa (another one)	PE1
Stone on the inside of the façade wall on the right-hand side	PEt
Colonnaded stone from the Porta Especiosa (on the guardrail)	PER
Collapsed stone on the sill of the Porta Especiosa guard (left side)	PEE
Stone from the colonnade base of the Porta Especiosa (on the balcony guard)	PEC
Stone section of the inner face of the main façade, left side	PLE
Stone in the passageway adjacent to the main façade (close to the floor)	PPch
Collapsed stone next to the third section on the right-hand side	P3P

The following figures show the sample collection locations (Figure 6).

2.2. Methods

Observations were conducted using a Zeiss Jenapshot 2000 Optical Microscope (Jena, Germany) to investigate the surfaces of the samples. Additionally, a Konica Minolta Sensing Chroma Meter CR-400 (Ramsey, NJ, USA) colorimeter was used to analyse the colour properties of the stones. The colour analysis was based on the CIELAB colour space, utilizing the parameters a*, b*, and L*. For each sample, three measurements were performed using both sides of the samples.

The mineralogical composition of the samples was determined via X-ray diffraction (XRD) with a Philips Panalytical X'Pert-Pro MPD diffractometer (Malvern, UK), Cu Kα (λ = 1.5405 Å) radiation, and a goniometer speed of 0.02° 2θ s^{-1}. The parameters for this analysis were set to 50 kV and 30 mA, with a 2θ range from 4.01° to 64.99° and a scan step time of 1 s. Analysed sample was random (unoriented) powder, about 0.500 g. For qualitative and semi-quantitative identification of the minerals, we followed the criteria provided by [30,31].

The chemical composition of the major, minor, and trace elements in the samples was determined via X-ray fluorescence (XRF) using a Panalytical Axios spectrometer PW4400/40 X-ray (Marvel Panalytical, Almelo, The Netherlands) equipped with an Rh tube under argon/methane. Reference samples were analysed using Omnian37 and Pro-Trace2021 softwares for major and minor elements analyses, respectively; Lost on Ignition was assessed on furnace at 1000 °C, for 4 h. Monument analysed samples were powders, about 0.400 g since there was very limited quantity; therefore, the used program was Malvern Panalytical OmnioanHelium (Powder Compounds Marvel Panalytical, Almelo, The Netherlands), which is semi-quantifying.

SEM/EDS analyses were also conducted using an Analytical UHR Schotty emission scanning electron microscope SU-70 (Hitachi, Bruker, Tokyo, Japan) with a voltage of 15 kV.

3. Results

3.1. Morphological and Colourimetric Analysis

Past researchers considered stone colour as a guiding parameter for stone characterization and a method for distinguishing between different types of stone. However, this information was mainly based on visual observations, with no quantitative data. In the present investigation, the colour and texture of the stones (on both sides) were recorded using a colourimeter to enable a possible comparison with future studies. A colourimetric analysis of the samples was then conducted, with the results corresponding to the obtained colour coordinates presented in Table 6. This type of analysis allows the specific properties of the samples to be determined, e.g., the presence of certain elements. Figure 7 shows the characteristics of the samples from this perspective.

Table 6. CIELAB colour coordinates for the samples under investigation.

Sample	L*	a*	b*
PEE	62.6 ± 1.6	1.0 ± 0.3	13.7 ± 0.8
PEt	60.8 ± 5.0	0.2 ± 0.2	8.0 ± 1.8
PPch	70.7 ± 3.2	2.5 ± 0.9	24.7 ± 2.2
PEch	84.5± 1.0	0.1 ± 0.09	7.8 ± 0.4
PEC	67.7 ± 0.8	1.0 ± 0.2	12.6 ± 2.6
PER	64.8 ± 8.0	1.0 ± 0.6	11.6 ± 2.2
PE1	65.9 ± 1.5	2.9 ± 0.2	19.4 ± 0.5
PLE	73.6 ± 6.4	2.6 ± 1.5	19.8 ± 5.2
P3P	60.6 ± 3.2	0.4± 0.3	6.6 ± 3.5
PES	76.8 ± 0.3	0.4 ± 0.2	8.4 ± 0.6

(a)

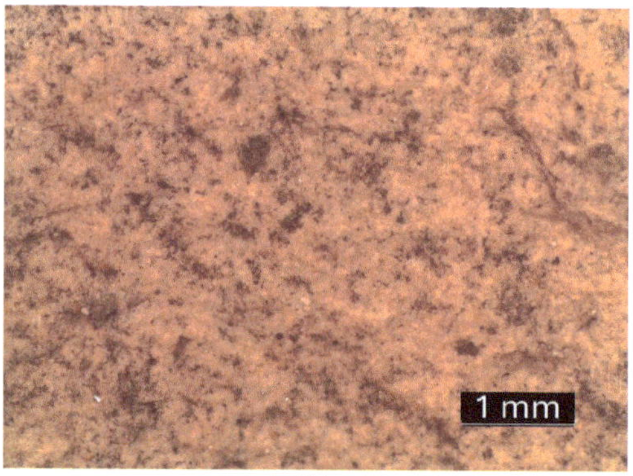

(b)

(c)

(d)

Figure 7. *Cont.*

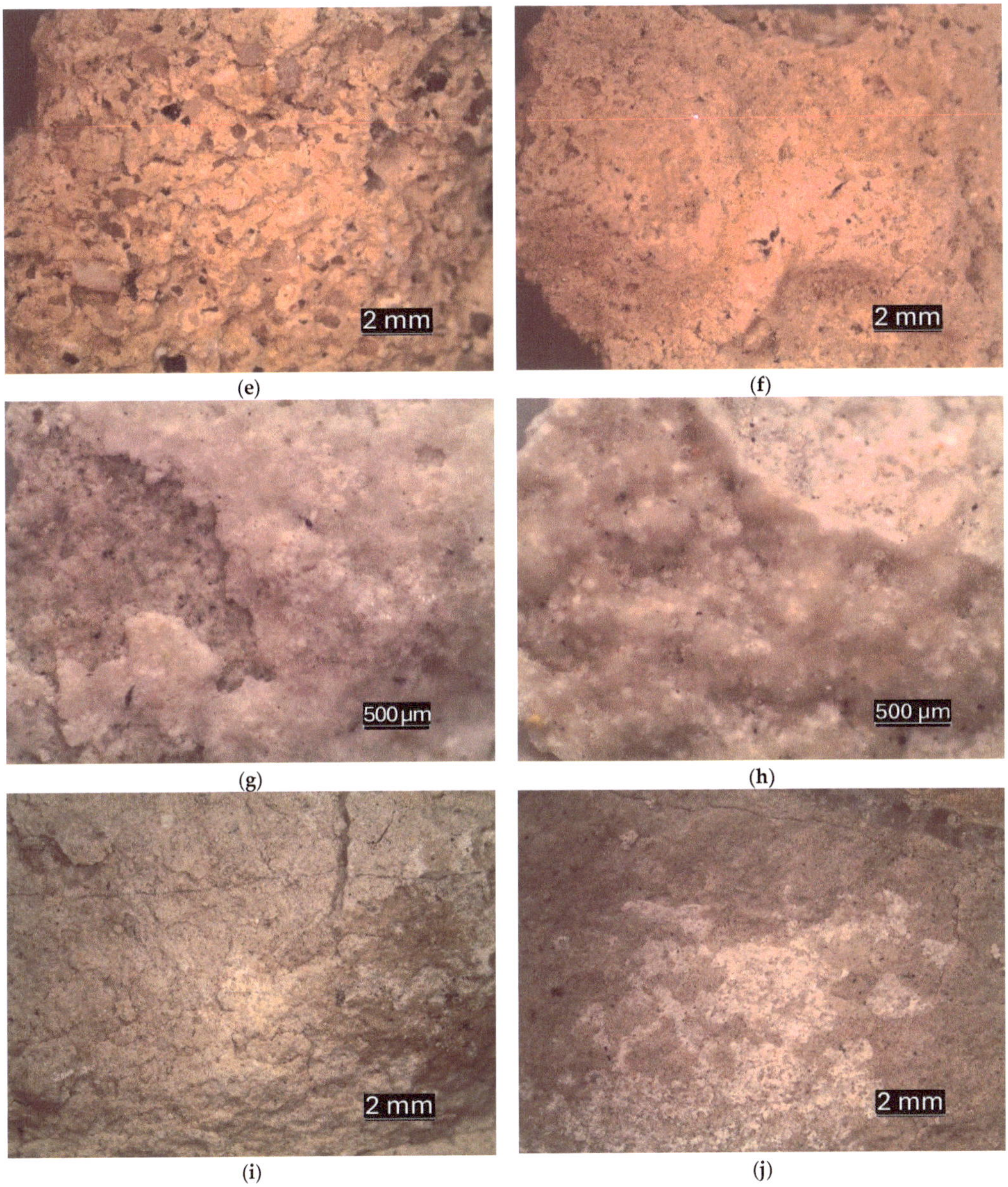

Figure 7. *Cont.*

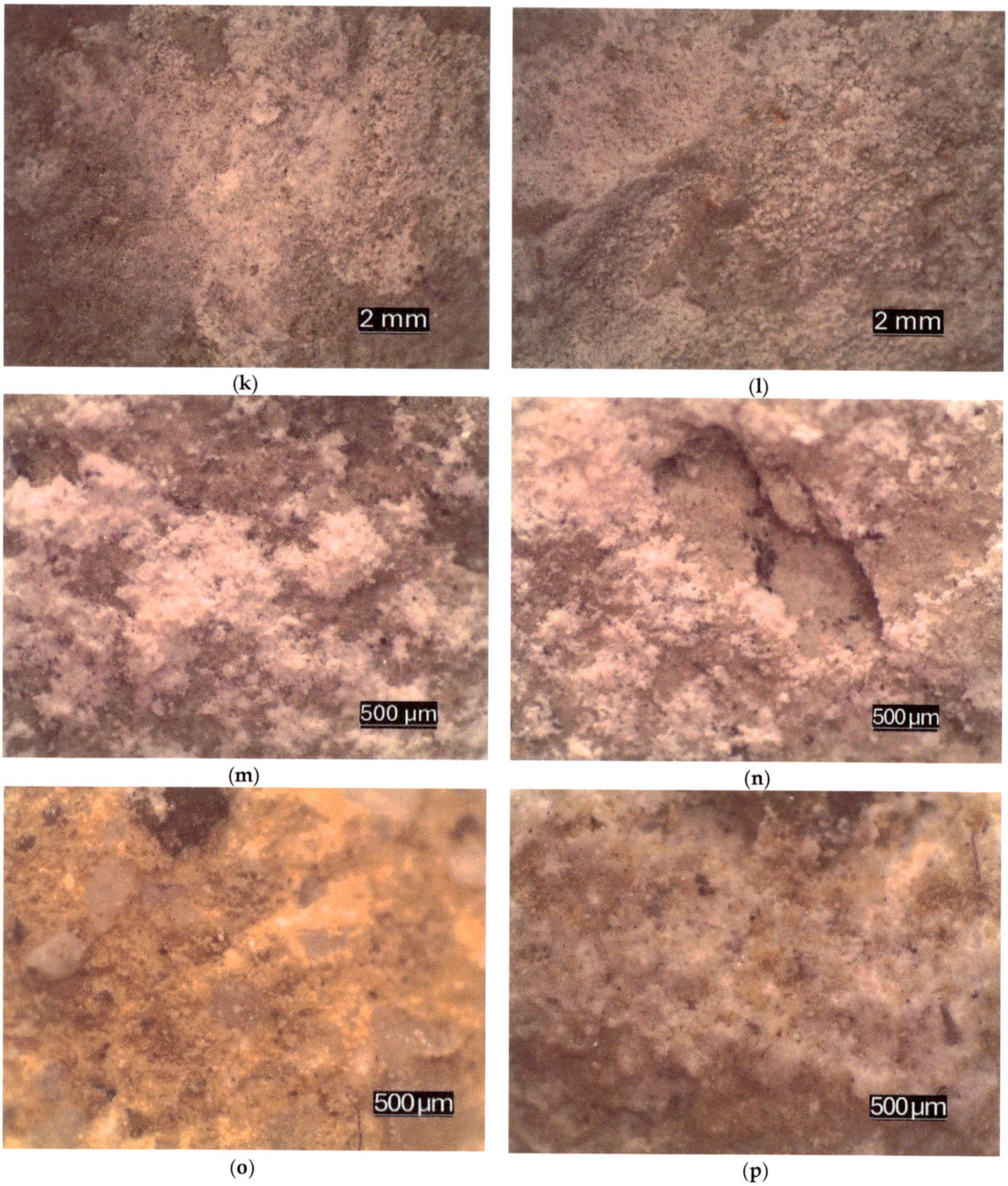

Figure 7. *Cont.*

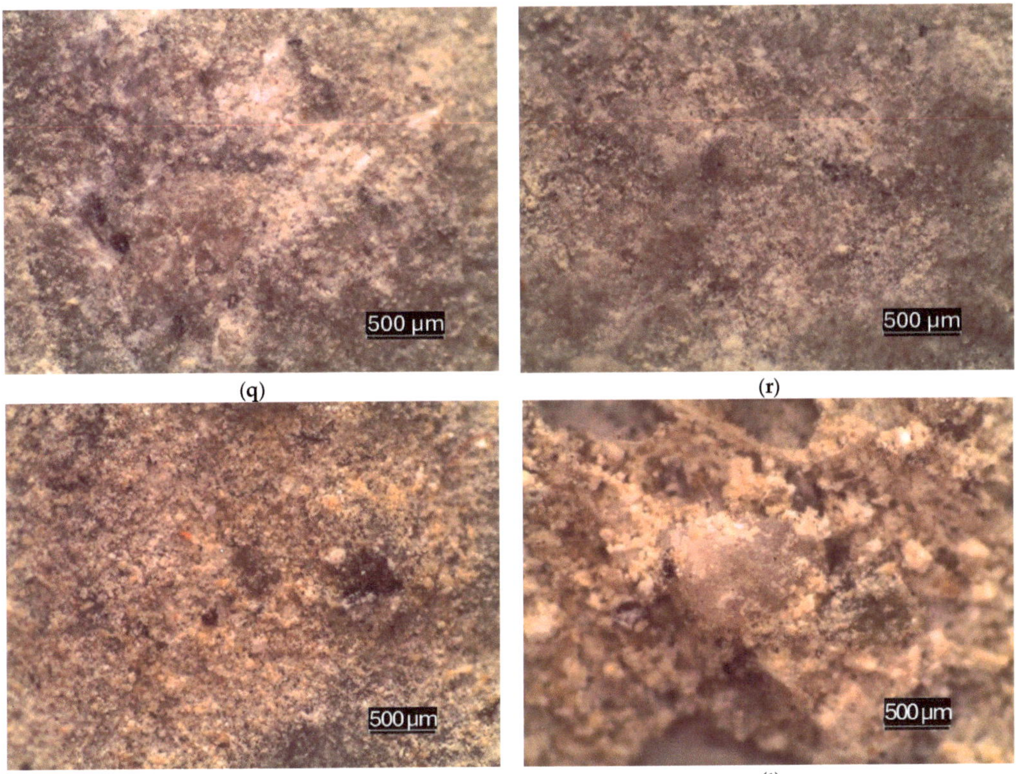

(q) (r) (s) (t)

Figure 7. Visual analysis of the topography of the samples: (**a**) PEE—top face; (**b**) PEE—bottom face; (**c**) PEt—top face; (**d**) PEt—bottom face; (**e**) PPch—top face; (**f**) PPch—bottom face; (**g**) PEch—top face; (**h**) PEch—bottom face; (**i**) PEC—top face; (**j**) PEC—bottom face; (**k**) PER—top face; (**l**) PER—bottom face; (**m**) PES—top face; (**n**) PES—bottom face; (**o**) PLE—top face; (**p**) PLE—bottom face; (**q**) PE1—top face; (**r**) PE1—bottom face; (**s**) P3P—top face; (**t**) P3P—bottom face.

The chromatic properties of the materials are primarily influenced by their composition, particularly the presence of trace elements such as iron or manganese oxide. However, these properties are also impacted by textural variations, including pore size and the presence of banding [32].

We determined some associated issues using the results obtained from the micrographs and the analysis performed with the colorimeter, particularly between the PER and PEC samples. These samples not only exhibit a very comparable appearance but also have a low ΔL (4.9), indicating that they are visually similar. The same result applies to the PEt and P3P samples—in this case, with a ΔL of 1.0. The results obtained for the colourimetric coordinates are consistent with those of a study conducted by Occhipinti et al. [33], which examined the colour properties of carbonate rocks from historic buildings in the city of Catania.

3.2. Mineralogical Analysis

The results obtained from the X-ray diffraction (XRD) analysis are summarized in Table 7 and Figure 8. These results provide information regarding the peaks associated with the mineralogical phases present in each sample, enabling categorization based on their specific characteristics.

Table 7. XRD results.

Samples	Mineralogical Peaks Identified	Categorization
PEch	Calcite	Efflorescence on limestone
	Gypsum	
	Nitre	
PES	Calcite	
	Nitre	
	Coquimbite	
	Rozenite	
	Gypsum	
	Huntite	
	Vaterite	
	Dolomite	
PE1	Calcite	Dolomitic limestone
	Dolomite	
	Quartz	
PLE	Calcite	
	Dolomite	
	Quartz (trace amounts)	
PEt	Quartz	Marlstone
	Calcite	
	Potassium feldspar	
	Plagioclase	
	Mica	
P3P	Quartz	
	Calcite	
	Potassium feldspar	
	Plagioclase	
	Dolomite	
	Mica	
PER	Calcite	Typical limestone
	Quartz	
	Kaolinite (trace amounts)	
PEE	Calcite	
	Quartz	
PEC	Calcite	
	Quartz	
PPch	Dolomite	Dolostone/Magnesian limestone
	Calcite	
	Quartz	
	Potassium feldspar	

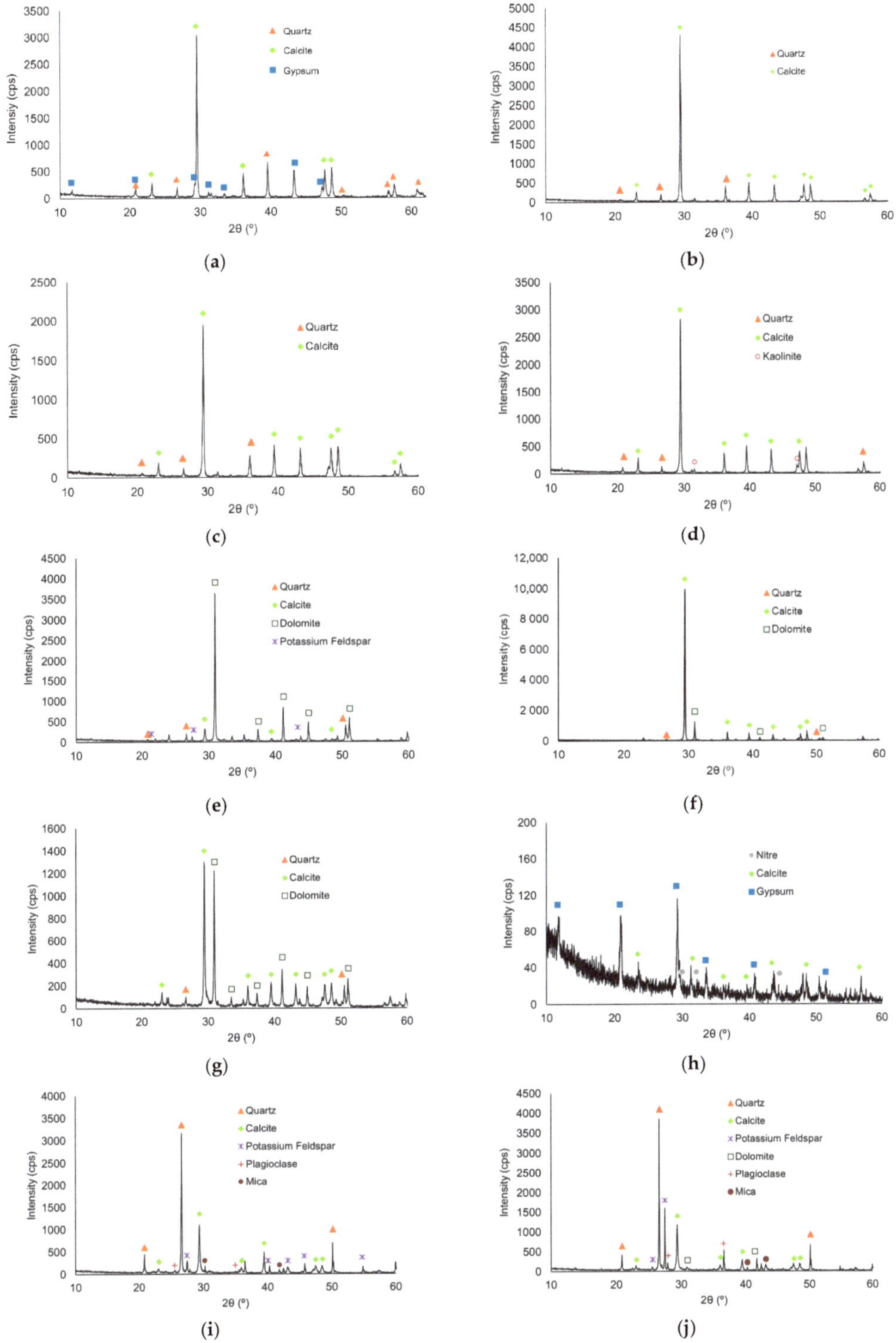

Figure 8. XRD graphs obtained: (**a**) PES; (**b**) PEC; (**c**) PEE; (**d**) PER; (**e**) PPch; (**f**) PLE; (**g**) PE1; (**h**) PEch (**i**) PEt; (**j**) P3P.

The mineralogical analysis (Table 7 and Figure 8) yielded the typical composition of regional calcitic limestone (i.e., samples PER, PEE, and PEC) and dolomitic limestone (PE1 and PLE); however, one sample was more similar to a dolostone (sample PPch). As expected, the sampled efflorescence indicated the presence of sulphates such as gypsum (Ca sulphate), coquimbite (Fe and Al sulphate), and rozenite (Fe sulphate); nitrates such as nitre (K nitrate); and secondary carbonates such as vaterite (Ca carbonate) and huntite (Fe and Mg carbonate).

3.3. Chemical Analysis

A preliminary analysis was carried out on Ançã stone from the region for comparison with the new samples and those from the literature, followed by an analysis of the samples chosen from the Sé Velha cathedral. The results are presented in Table 8.

Table 8. Chemical composition data for the original Ançã stone supplied by the stone museum (LOI: Lost on Ignition).

	%Na_2O	MgO	Al_2O_3	SiO_2	P_2O_5	SO_3	Cl	K_2O	CaO	TiO_2	MnO	Fe_2O_3	Sr	LOI
A1	0.030	0.395	0.241	2.488	0.019	0.059	0.030	0.257	54.330	0.019	0.001	0.135	0.046	41.950
A2	0.129	0.533	0.298	4.236	0.032	0.135	0.194	0.423	51.595	0.015	0.002	0.145	0.044	42.200

Additionally, the mineralogical analysis revealed that both samples (A1 and A2) were nearly monomineralic, consisting primarily of calcite, with a very minor, almost vestigial, presence of quartz. This result indicates a highly pure and, therefore, soft limestone.

The results obtained from the X-ray fluorescence (XRF) analysis for the samples investigated in this study are summarized in Table 9. These results provide information on the quantification percentage of the chemical elements present in each sample.

Table 9. XRF results.

Elements (%)	PEE	PEt	PEch	PPch	PEC	PER	PE1	PLE	P3P	PES
Cl	0.25	0.86	-	0.53	0.05	0.20	0.61	0.09	1.00	0.17
Al	0.28	1.03	0.33	0.32	0.26	0.27	0.28	0.27	0.97	0.26
Ca	86.37	62.72	29.96	75.36	88.08	85.86	81.39	87.80	67.76	88.97
Cr	-	0.04	-	0.04	-	-	-	-	0.04	0.05
Cu	-	0.03	0.86	0.11	0.03	-	-	-	0.04	0.02
Fe	3.45	6.58	2.36	9.26	2.13	2.37	6.22	2.96	6.67	2.17
K	0.89	4.84	0.40	2.66	1.05	1.19	1.59	0.92	5.78	0.81
Mg	0.10	0.24	0.23	1.88	0.23	0.09	0.81	0.51	0.30	0.09
Mn	0.07	0.14	-	1.04	0.08	0.03	0.59	0.61	0.20	0.04
Ni	0.05	0.04	0.45	0.03	0.03	0.02	0.04	0.02	-	0.02
O	6.00	9.14	53.16	5.78	5.20	7.29	6.34	5.13	5.41	4.81
P	0.30	0.47	3.77	0.30	0.25	0.21	0.24	0.20	0.38	0.22
S	0.54	0.78	6.81	0.18	0.95	1.21	0.12	0.08	0.70	0.70
Si	1.24	12.01	1.43	2.12	1.27	0.83	1.25	1.12	9.76	1.20
Sr	0.16	0.10	0.06	0.04	0.14	0.17	0.10	0.02	0.12	0.14
Ti	0.24	0.55	-	0.21	0.20	0.21	0.23	0.18	0.49	0.25
Zn	0.03	0.08	-	0.01	0.01	0.02	0.06	0.05	0.10	0.02

The results obtained from the chemical analysis (Table 9) agree with the lithological and mineralogical features of the studied samples. These limestone-type samples were found to be extremely rich in Ca, whereas the more dolomitic samples presented slightly higher Mg and Fe values. The amounts of Al remained low, as expected for carbonate-rich rocks. The efflorescence results indicated some higher amounts of S, Fe, and Cl.

3.4. SEM/EDS Analysis

Figures 9 and 10 present the results obtained from the SEM/EDS analyses of the most representative samples of the different stones.

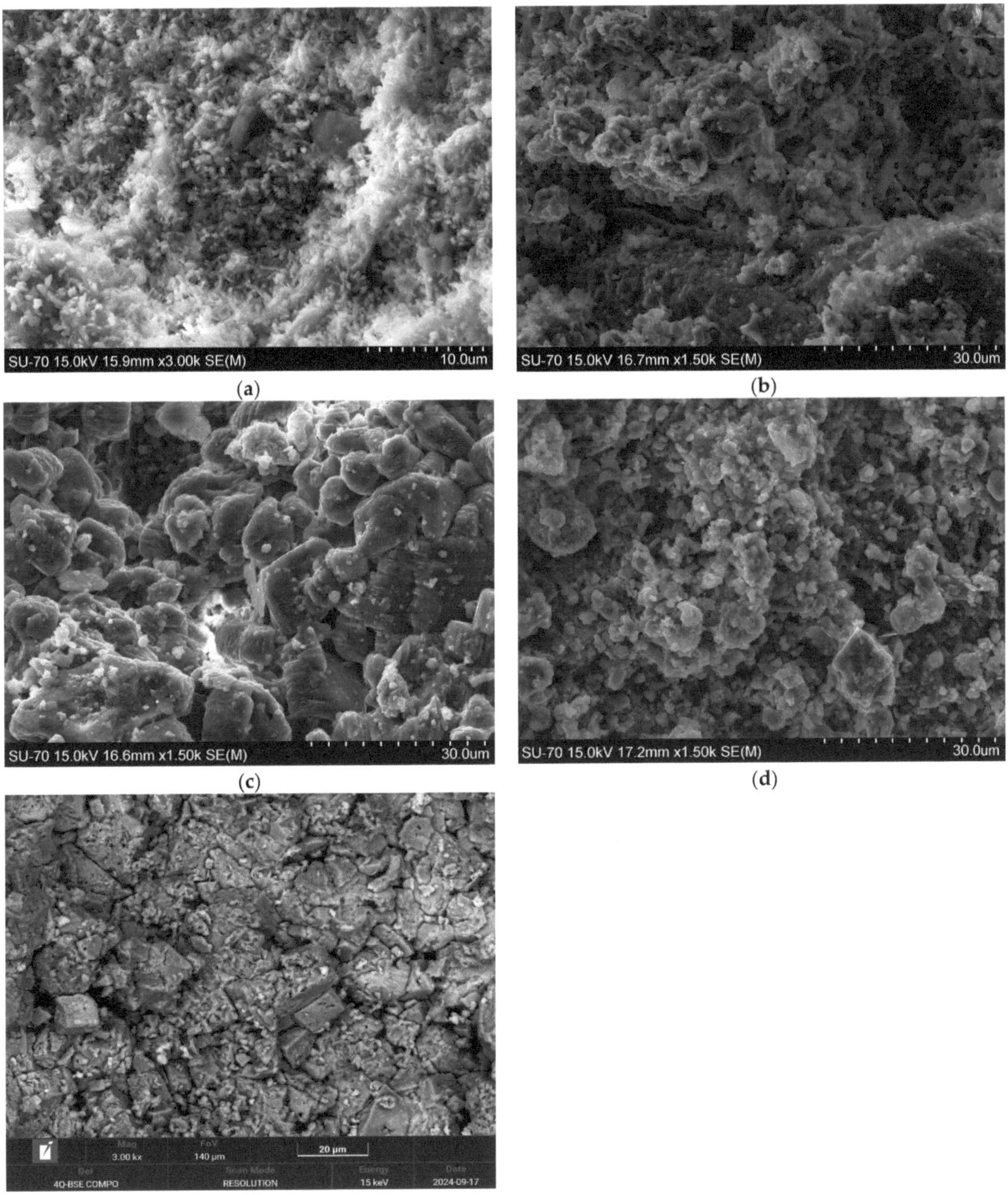

Figure 9. SEM images from the samples: (**a**) P3P; (**b**) PE1; (**c**) Pech; (**d**) PEE; (**e**) PPch.

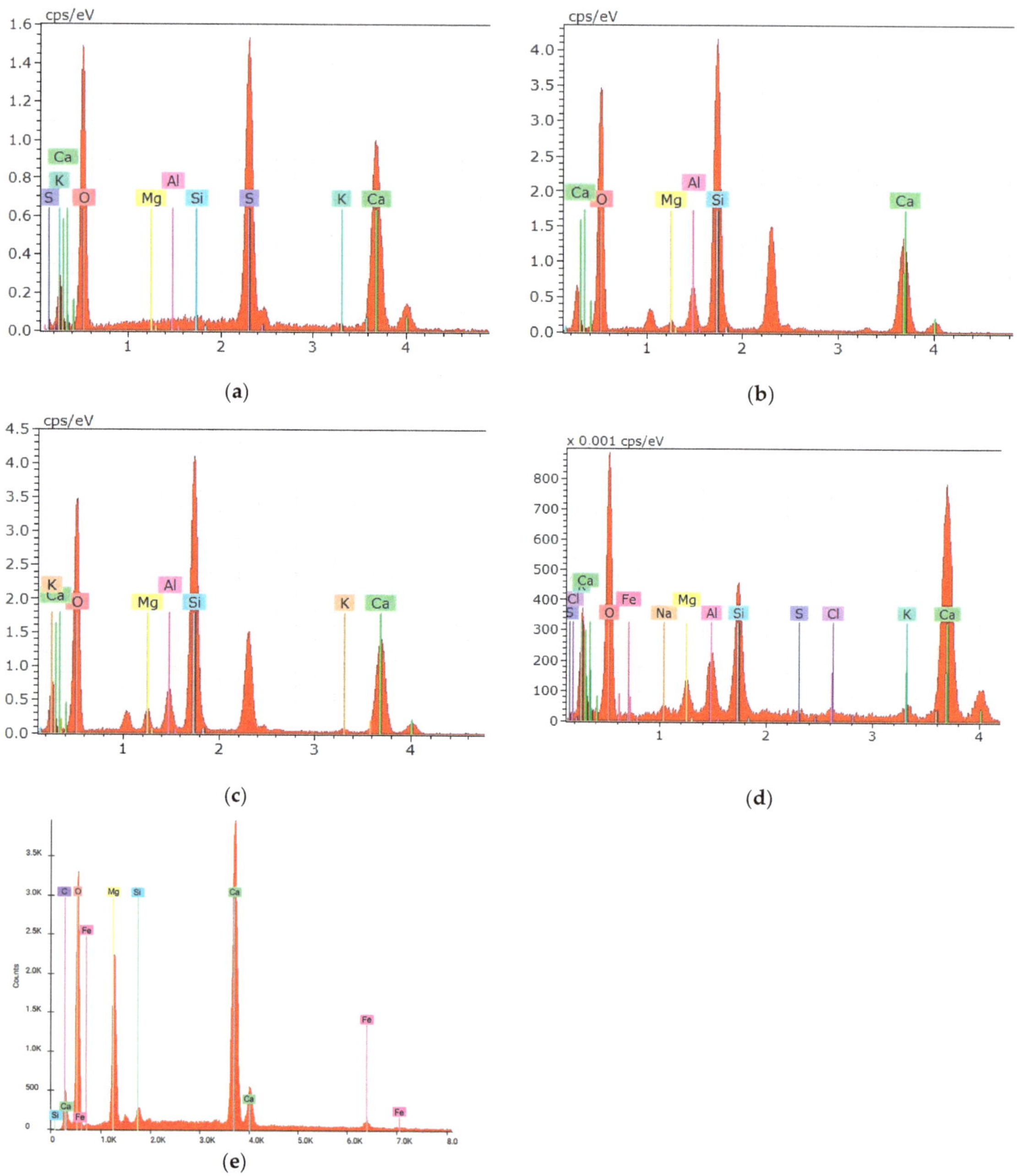

Figure 10. EDS analyses: (**a**) P3P; (**b**) PE1; (**c**) Pech; (**d**) PEE; (**e**) PPch.

SEM images show some textural differences between samples, but the main aspects are in the analysis results showing samples richer in Ca, such as PEE and PPch (the latter also in Mg), whereas others show more Si content (such as PE1 and Pech; P3P shows distinctive S presence).

4. Discussion

Previous studies focused more strongly on lower portions of the monument and the Porta Especiosa portal (separately), while analyses of the samples in the present investigation were taken from the internal faces of the wall facades and from the Porta Especiosa. These areas were defined for sampling due to their lower risk of contact with visitors and the resulting contamination. This approach enabled us to advance our understanding of this monument.

Previous studies indicated the presence of different types of carbonate rocks such as dolostone, calcitic dolostone, and Pedra de Ançã limestone (white to yellowish and marly to grey) at the Porta Especiosa portal, although this position did not achieve consensus and appears to depend upon the sampling location. Most of the results obtained from other sampling areas considered the use of dolostone, with the main body showing a dark patina that hides the yellowish colour of the dolostone used in the frames of windows.

Chemical and mineralogical analyses (Tables 7 and 9), as well as SEM/EDS results, clearly indicate the presence of different types of stone and efflorescence.

Four different types of stone were detected among the samples, based on their analysed mineralogical composition (via XRD) and chemical characteristics (via XRF and SEM/EDS, the latter just on selected samples).

The PER, PEE, and PEC samples composed primarily of calcite (calcium carbonate) with Ca contents higher than 80% corresponded to typical Ançã limestone, which agrees with the results from previous studies. PEE and PEC were both relatively pure in their mineralogical (only calcite and some quartz) and chemical composition (with around 1% of Si and predominant Ca). However, PER showed the presence of S, suggesting the additional presence of sulphates, which were likely not detected via XRD due to their amorphous structures.

The mineralogical composition of samples PE1 and PLE indicated the predominance of calcite but with a significant presence of dolomite (calcium and magnesium carbonate) and an increase in Mg content compared to that indicated via the chemical analyses. PE1 showed the presence of Fe, also suggesting the presence of iron oxides/hydroxites, which were likely not detected via XRD due to their amorphous structures. These samples corresponded to a variety of dolomitic limestone locally known as Outil Limestone.

Based on the chemical analysis, dolomite was predominant in the PPch sample, albeit with the presence of calcite due to an increase in Mg and a decrease in Ca contents. Therefore, this sample was a dolostone/magnesian limestone, which points to a carbonate rock that differs from common limestone due to being more dolomitic (instead of calcitic), which is locally designated as Portunhos Limestone.

This compositional substitution of calcite by dolomite, which is common in carbonate sedimentary deposits, usually occurs due to the effects of diagenesis over geological time, with practical consequences on the physicochemical behaviour of these rocks. Dolomite, which is a more magnesian stone, has greater chemical stability and, generally, offers higher mechanical resistance (in most of the parameters relevant to building materials, such as the compressive strength and elastic modulus), resulting in a "harder" stone [34]. Thus, in this case, the Ançã Limestone was found to be the most "vulnerable" to both chemical and physical degradation but also much more "workable" due to being "softer" (less "hard"), particularly when used for ornamental purposes.

It is possible to establish a sequence of increasing resistance (i.e., decreasing vulnerability) for Ançã Limestone (PER, PEE, and PEC)—Outil Limestone: PE1 and PLE samples—Portunhos Limestone (PPch sample), mainly containing dolomite (lower in Ançã and higher in Portunhos); in addition, dolomite generally shows higher mechanical resistance than calcite [35]. The greater vulnerability of the Ançã Limestone is well illustrated in the PES sample, which corresponds to its calcite-rich typology (Ançã Limestone) albeit with the occurrence of gypsum (hydrated calcium sulphate) and placement outdoors, leading to weathering and urban traffic.

These three types of limestone occur in the region of Coimbra and represent the main typologies of construction stone and ornamentation at the local and regional levels and are present in both historical and contemporary buildings.

The PEt and P3P samples, on the other hand, were still carbonated but presented a different chemical–mineralogical composition compared with that of Ançã, Portunhos, and Outil limestone. With the significant presence of quartz and feldspars (both silicates), the calcite content decreased, whereas the Si, Al, K, and Fe contents increased. These features indicate marlstones, which are softer and chemically more vulnerable than the other three types of limestone previously described. These two samples were located (Figure 5) in a different part of the building, relatively far from the famous Porta Especiosa on one of its more monumental components, close to all other samples. These stones likely represent a "second-class" choice or a posterior replacement using stones of lesser quality.

P3P sample SEM/EDS analysis shows a distinctive S presence, pointing to the formation of efflorescences, not detected on XRD most probably due to their amorphous state.

Considering the efflorescence, gypsum (hydrated calcium sulphate) and nitre (hydrated potassium nitrate) were found in the PES and PEch samples. In the PEch samples, we found other compounds such as coquimbite (hydrated iron and aluminium sulphate) and rozenite (hydrated iron sulphate). Sulphates (gypsum) were most common in the PES samples, while nitrate (nitre) was most common in PEch, indicating different processes/causes of alteration, and, eventually, the influence of its position in the building. Moreover, the sulphate-rich PES sample was taken from the interior of the building, whereas the nitrate-rich PEch sample was taken from the exterior and exposed to much more traffic and bird droppings.

These results reveal the use of different types of carbonate stones in different parts of of the Sé Velha Cathedral, as presented in Figure 11.

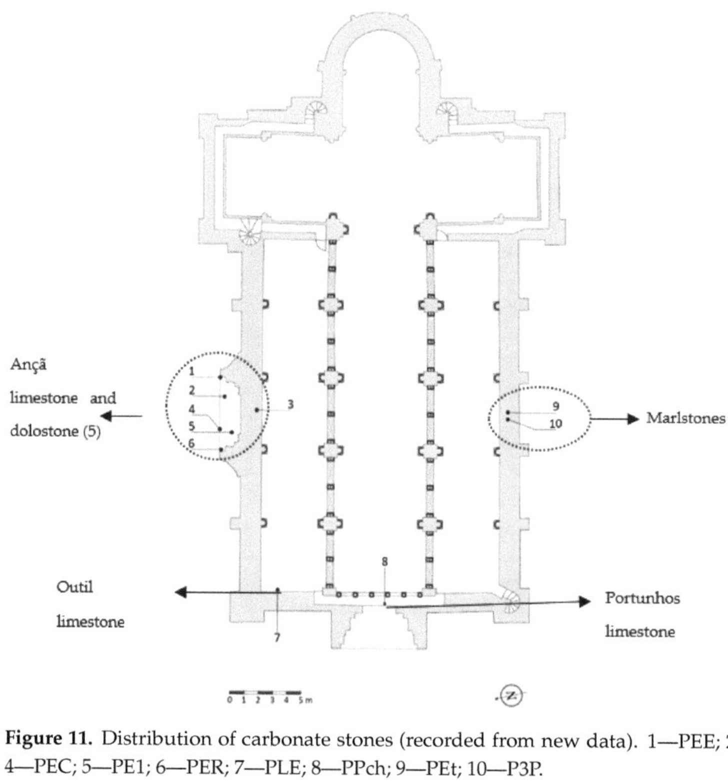

Figure 11. Distribution of carbonate stones (recorded from new data). 1—PEE; 2—PEch; 3—PES; 4—PEC; 5—PE1; 6—PER; 7—PLE; 8—PPch; 9—PEt; 10—P3P.

Lastly, the various chemical and mechanical characteristics of the different stones used in the construction of this cathedral explain the different damage patterns observed.

5. Conclusions

The present investigation analysed samples from different locations of the Sé Velha Cathedral (Coimbra) to identify the different types of limestone used in the cathedral's construction. On the upper floor of the cathedral, we identified four different types of carbonate stones. Based on the mineralogical and chemical analyses carried out, the use of Ançã limestone was apparently concentrated around the Porta Especiosa, along with dolostone. These conclusions agree with those from previous investigations by LNEC. However, other types of stone were used in other areas of the monument, namely near the large windows, which differ between the front (Portunhos limestone) and side facades (marlstones, Outil limestone). These data reveal the use of stone from different quarries in the Coimbra region for the upper floors of the cathedral, which has not yet been studied.

These results indicate that stones from different quarries were used in different stages of construction and alterations to the monument. In some cases, these stones are more vulnerable to the activities of atmospheric agents and pollution. Overall, the construction of the cathedral relied on different types of stones with different chemical and mechanical characteristics. Notably, the Porta Especiosa was constructed from more vulnerable types of stone and is already in an advanced state of degradation, making it impossible to remove it. Because this entryway faces north, it should be protected using a structure that allows one to control hygrothermal conditions.

Author Contributions: Conceptualization, A.T., F.R., A.F. and A.C.; methodology, A.T., F.R., A.F. and A.C.; validation, A.T., F.R., A.F. and A.C.; formal analysis, A.T., F.R., A.F. and M.O.; investigation, A.T., F.R., A.F. and M.O.; writing—original draft preparation, A.T., F.R., A.F. and M.O.; writing—review and editing, A.T., F.R., A.F. and A.C.; supervision, A.T. and F.R.; funding acquisition, A.T., F.R. and A.F. All authors have read and agreed to the published version of the manuscript.

Funding: This research was supported by CICECO-Aveiro Institute of Materials, UIDB/50011/2020, UIDP/50011/2020 and LA/P/0006/2020 and DOI 10.54499/2021.03830.CEECIND/CP1659/CT0032 financed by national funds through the FCT/MCTES (PIDDAC) and ENGINEER—European Union's Horizon Europe Framework Programme under the grant agreement No 101079377 and the UKRI [project number 10050486] and GeoBioTec Research Centre (UIDB/04035/2020), funded by FCT.

Data Availability Statement: Data are contained within the article.

Acknowledgments: The authors would like to thank the Regional Directorate of Culture of the Centre of Portugal Suzana Menezes for the support and opportunity to carry out this research, and Filipe Oliveira for the support in the SEM/EDS analyses.

Conflicts of Interest: The authors declare no conflicts of interest. The funders had no role in the design of the study; in the collection, analyses, or interpretation of data; in the writing of the manuscript; or in the decision to publish the results.

References

1. Winkler, E.; Hudec, P.P.; Harvey, R.D.; Baxter, J.W.; Fraser, G.S.; Smith, C.B.; Norman, G.E.; Renninger, F.A.; Nichols, F.P., Jr.; Ozol, M.A.; et al. Decay and Preservation of Stone. *Nature* **1978**, *115*, 779. [CrossRef]
2. Touloum, N.R.; Brara, A. Advanced decay of the Fontvieille porous limestone used in the construction of Pouillon's housing estates in Algiers (Algeria). *J. Archaeol. Sci. Rep.* **2023**, *47*, 103804. [CrossRef]
3. Cooke, R.U. Salt weathering in deserts. *Proc. Geol. Assoc.* **1981**, *92*, 1–16. [CrossRef]
4. Bonazza, A. Sustainable heritage and climate change. In *Routledge Handbook of Sustainable Heritage*; Routledge: London, UK, 2022; pp. 263–271. [CrossRef]
5. Winkler, E. *Stone in Architecture: Properties, Durability*, 3rd ed.; Springer Science & Business Media: Berlin/Heidelberg, Germany, 1997.
6. Rosado, T.; Silva, M.; Galvão, A.; Mirão, J.; Candeias, A.; Caldeira, A.T. A first insight on the biodegradation of limestone: The case of the World Heritage Convent of Christ. *Appl. Phys. A Mater. Sci. Process.* **2016**, *122*, 1012. [CrossRef]
7. Património Cultural-Instituto Público, "Igreja da Sé Velha". Available online: https://servicos.dgpc.gov.pt/pesquisapatrimonioimovel/detalhes.php?code=70529 (accessed on 17 July 2024).

8. University of Coimbra–Alta and Sofia. Available online: https://whc.unesco.org/en/list/1387/ (accessed on 3 October 2024).
9. DGEMN. *'A Sé Velha de Coimbra', Boletim da Direção Geral dos Edifícios e Monumentos Nacionais, no 109*; Ministério das Obras Públicas: Porto, Portugal, 1962.
10. Gomes Coelho, A.; Delgado Rodrigues, J.; Ferreira Pinto, A.P. Conservação da Porta Especiosa: Caracterização da Degradação por Ensaios In Situ; NEC, Lisbon, Portugal, 2000.
11. Aires de Barros, L.A.; Ferreira Pinto, A.P. *Relatório de Caracterização da Pedra da Porta Especiosa Report*; IST: Lisbon, Portugal, 1985.
12. Aires Barbosa, L.; Dionísio, M.A. *Relatório Preliminar Sobre a Litologia e as Patologias da Porta Especiosa—Sé Velha Coimbra e Proposta de Salvaguarda*; IST: Lisbon, Portugal, 1998.
13. Marques, F.; Lopes, C.S. *Levantamento do estado de degradação global da Porta Especiosa*; IPPAR: Lisbon, Portugal, 1999.
14. Delgado Rodrigues, J.; Ferreira Pinto, A.P. *Conservação da Porta Especiosa. Caracterização por Ensaios In Situ*; LNEC: Lisbon, Portugal, 2000.
15. Freitas Tavares, A. *Caracterização da pedra da Porta Especiosa*; Museu Nacional Machado de Castro: Coimbra, Portugal, 1989.
16. Quinta Ferreira, M.; Soares, A.F.; Delgado Rodrigues, J.; Monteiro, B. Carbonate rocks from Sé Velha Cathedral. In Proceedings of the 7th International Congress on Deterioration Conservation of Stones, Lisbon, Portugal, 15–18 June 1992.
17. Henriques, F.; Delgado Rodrigues, J.; Aires de Barros, L.; Craveiro, L. *Conservação e Restauro de Coberturas, Fachadas e 2ªfase Porta Especiosa*; LNEC: Lisbon, Portugal, 2002.
18. Bonomo, A.E.; Minervino Amodio, A.; Prosser, G.; Sileo, M.; Rizzo, G. Evaluation of soft limestone degradation in the Sassi UNESCO site (Matera, Southern Italy): Loss of material measurement and classification. *J. Cult. Herit.* **2020**, *42*, 191–201. [CrossRef]
19. Colston, B.J.; Watt, D.S.; Munro, H.L. Environmentally-induced stone decay: The cumulative effects of crystallization-hydration cycles on a Lincolnshire oopelsparite limestone. *J. Cult. Herit.* **2001**, *2*, 297–307. [CrossRef]
20. Hashim, M.S.; Kaczmarek, S.E. A review of the nature and origin of limestone microporosity. *Mar. Pet. Geol.* **2019**, *107*, 527–554. [CrossRef]
21. Dilaria, S.; Bonetto, J.; Germinario, L.; Previato, C.; Girotto, C.; Mazzoli, C. The stone artifacts of the National Archaeological Museum of Adria (Rovigo, Italy): A noteworthy example of heterogeneity. *Archaeol. Anthropol. Sci.* **2024**, *16*, 14. [CrossRef]
22. Castro, E.; do Rosario, M.; Delgado Rodrigues, J. *Conservação de Pedras e Edifícios de Interesse Histórico ou Artístico, 6º Report, Estudo do Calcário dum Retábulo da Sé Velha de Coimbra e de Calcários de Pedreira da 635 Região*; LNEC: Lisboa, Portugal, 1975.
23. Varas-Muriel, M.J.; Pérez-Monserrat, E.M.; Vázquez-Calvo, C.; Fort, R. Effect of conservation treatments on heritage stone. Characterisation of decay processes in a case study. *Constr Build Mater* **2015**, *95*, 611–622. [CrossRef]
24. Ding, Y.; Salvador, C.S.C.; Caldeira, A.T.; Angelini, E.; Schiavon, N. Biodegradation and Microbial Contamination of Limestone Surfaces: An Experimental Study from Batalha Monastery, Portugal. *Corros. Mater. Degrad.* **2021**, *2*, 31–45. [CrossRef]
25. Šiler, P.; Kolářová, I.; Bednárek, J.; Janča, M.; Musil, P.; Opravil, T. The possibilities of analysis of limestone chemical composition. *IOP Conf. Ser. Mater. Sci. Eng.* **2018**, *379*, 012033. [CrossRef]
26. Villarraga, C.J.; Gasc-Barbier, M.; Vaunat, J.; Darrozes, J. The effect of thermal cycles on limestone mechanical degradation. *Int. J. Rock Mech. Mining Sci.* **2018**, *109*, 115–123. [CrossRef]
27. Pinheiro, A.C.; Mesquita, N.; Trovão, J.; Soares, F.; Tiago, I.; Coelho, C.; de Carvalho, H.P.; Gil, F.; Catarino, L.; Piñar, G.; et al. Limestone biodeterioration: A review on the Portuguese cultural heritage scenario. *J. Cult. Herit.* **2019**, *36*, 275–285. [CrossRef]
28. IPPAR. *Concurso Público Internacional–Igreja Sé Velha, Coimbra–Conservação e Restauro das Coberturas, Fachadas e 2afase da Porta Especiosa*; IPPAR: Lisboa, Portugal, 2005.
29. Mamillan, M. *Pathologie et Restauration des Constructions en Pierre*; Faculté D'Architecture de l'Université de Rome: Rome, Italy, 1972.
30. Galhano, C.; Rocha, F.; Gomes, C. Geostatistical analysis of the influence of textural, mineralogical and geochemical parameters on the geotechnical behaviour of the 'Argilas de Aveiro' Formation (Portugal). *Clay Miner* **1999**, *34*, 109–116. [CrossRef]
31. Oliveira, A.; Rocha, F.; Rodrigues, A.; Jouanneau, J.; Dias, A.; Weber, O.; Gomes, C. Clay minerals from the sedimentary cover from the Northwest Iberian shelf. *Prog. Oceanogr.* **2002**, *52*, 233–247. [CrossRef]
32. Urosevic, M.; Pardo, E.M.S.; Agudo, E.R.; Fernández, C.C. Evaluación de las propiedades físicas de dos rocas carbonáticas usadas como material de construcción actual e histórico en Andalucía Oriental, España. *Mater. Construcción* **2011**, *61*, 465–2746. [CrossRef]
33. Occhipinti, R.; Stroscio, A.; Belfiore, C.M.; Barone, G.; Mazzoleni, P. Chemical and colorimetric analysis for the characterization of degradation forms and surface colour modification of building stone materials. *Constr. Build. Mater.* **2021**, *302*, 124356. [CrossRef]
34. Cai, W.K.; Liu, J.H.; Zhou, C.H.; Keeling, J.; Glasmacher, U.A. Structure, genesis and resources efficiency of dolomite: New insights and remaining enigmas. *Chem. Geol.* **2021**, *573*, 120191. [CrossRef]
35. Maričić, A.; Starčević, K.; Barudžija, U. Physical and mechanical properties of dolomites related to sedimentary and diagenetic features—Case study of the upper triassic dolomites from the medvednica and samobor mts., NW Croatia. *Rud. Geol. Naftni Zbornik* **2018**, *33*, 33–44. [CrossRef]

Disclaimer/Publisher's Note: The statements, opinions and data contained in all publications are solely those of the individual author(s) and contributor(s) and not of MDPI and/or the editor(s). MDPI and/or the editor(s) disclaim responsibility for any injury to people or property resulting from any ideas, methods, instructions or products referred to in the content.

heritage

Article

Sustainable Conservation of Architectural Heritage to Reduce Environmental Impact: The Morocco Pavilion on Cartuja Island in Seville

Manuel V. Castilla *,† **and Francisco Lopez** †

Higher Polytechnic School, University of Seville, 41011 Sevilla, Spain
* Correspondence: mviggo@us.es
† These authors contributed equally to this work.

Abstract: The architectural heritage of a particular place, in most cases, is characterised by vernacular and unique constructions that have been adapted to local climate conditions. For this purpose, specific materials and construction methods have traditionally been used that, in addition to the durability of the construction, also allow for the consideration of the energy efficiency of the building itself. The present intersection of climate change and architecture has led to new exposure to the external agents for which constructions were designed, forcing, in most cases, a review of building envelopes and very costly proposals. From the point of view of efficiency, intervention strategies with passive measures are proposed that not only improve the energy performance and maintenance of buildings themselves, but also lower the overall energy consumption. Using a heritage case study of the city of Seville, the Moroccan Pavilion, at Expo 92, this work includes an analysis and proposal of effective action through a methodological study of energy efficiency. The problem of high energy consumption during the summer months in Seville is tested in the Pavilion. The results indicate an urgent need for renovation, and among different options, new intervention measures are recommended as an alternative to consumption based on knowledge and tradition; moreover, passive construction elements are proposed in accordance with the climatic reality of the environment for optimal conservation in new climate scenarios.

Keywords: passive architecture; heritage construction; Expo 92; energy consumption; sustainable construction; architectural heritage; the Morocco Pavilion; Seville; Cartuja Island

Citation: Castilla, M.V.; Lopez, F. Sustainable Conservation of Architectural Heritage to Reduce Environmental Impact: The Morocco Pavilion on Cartuja Island in Seville. *Heritage* **2024**, *7*, 3851–3866. https://doi.org/10.3390/heritage7080182

Academic Editors: Daniela Fico and Daniela Rizzo

Received: 20 June 2024
Revised: 17 July 2024
Accepted: 22 July 2024
Published: 24 July 2024

Copyright: © 2024 by the authors. Licensee MDPI, Basel, Switzerland. This article is an open access article distributed under the terms and conditions of the Creative Commons Attribution (CC BY) license (https://creativecommons.org/licenses/by/4.0/).

1. Introduction

Great international events, such as world expos, Olympic Games, and international summits, provide unique opportunities for showcasing innovative architecture on a global stage [1]. This paper examines the role of singular architecture in such events, focusing on a case study and design principles that contribute to environmental success. Singular buildings built for large international events are usually designed with the intention of breaking out of the routine of local design. Often, they are designed to be models and references that show the cultural identity of a town and its architectural culture, acting as a reflection of architectural know-how.

The Expo 92 Universal Exhibition in Seville was a great showcase of architectures designed to adapt to new environments (with greater or lesser success), reflecting every country and culture that participated in it [2].

In the specific case study of the Morocco Pavilion on La Cartuja Island in Seville, we encounter a building with low energy efficiency. This pavilion, designed with curtain wall solutions, represents a clear example of a construction not adapted to its environment, with increasingly longer and more intense summer periods every year. This paper presents a time-based study of its boundary conditions and possible interventions and recommendations from a regenerative and sustainable perspective.

We understand that the countries in immediate proximity to the host city had no difficulty in implementing an efficient product adapted to the environment. Climate and implementation methods could easily be extrapolated from these countries.

The rationale behind developing this architectural design is to decrease buildings' environmental impact, improve the quality of the built environment, and increase the thermal comfort of buildings' occupants and its performance [3].

Examples from other latitudes, for which climatic rigor forced decisions to be made, should be considered, as they were not always achieved successfully. It should be possible to maintain a comfortable range of conditions without resorting to the provision of artificial cooling by using measures such as shade, thermal mass air movement, lighting controls, and low-energy lighting [4].

In this way, the need to use passive measures in architecture with vernacular elements, such as mashrabiya, cannot be overlooked.

Mashrabiya works perfectly as a protection device from direct sunlight and effectively reduces heat gain, especially during hot seasons. In the current era, various shapes derived from mashrabiya can be found on the façades of buildings in various countries around the world. Also, several studies and applications have submitted new designs and proposals for the development of mashrabiya, either using different materials instead of wood, such as aluminium, steel, ceramics, or glass-fibre-reinforced concrete (GRC), or incorporating interactive techniques for opening and closing [5]. Moreover, this knowledge can be applied to modern buildings by combining the mashrabiya concept with new solutions, improving the design according to users' needs in line with modern building systems in hot climates. Additionally, it could be effective to use this method in temperate climates, leading to thermal comfort periods [5].

An understanding of urban resilience in urban studies and planning is critical in analysing and combating the impacts of climate change in cities in the Global South [6].

2. Literature Review

Numerous attempts have been made to explain the concept of sustainability, introduced by the Brundtland Commission in their 1987 report, "Our Common Future" [7]. Consequently, various definitions have emerged to describe what constitutes sustainable architectural design. In the 1990s, the building sector began to acknowledge its significant environmental impact [8]. Since then, a crucial shift has occurred in the way buildings are designed, constructed, and operated to reduce their environmental footprint [9].

Initial ideas in construction must consider the very functioning of a building by considering its adaptation to external agents, its massiveness, and its final use. It must be understood that, until recently, structural and technical energy resources were used without sustainable limits and with the direct application of technologies. The concept of efficiency suggests that technology is not everything, and that production and consumption must be balanced [10]. It would be unfeasible to reach a balance without the use of passive measures [11].

This approach reduces the need for artificial lighting, thereby lowering energy consumption and enhancing occupant well-being. Effective passive lighting design incorporates various elements, including window placement, skylights, light shelves, reflective surfaces, and shading devices, to optimize natural light distribution while minimizing glare and heat gain [12].

2.1. Vernacular Culture/Architecture

The Islamic tradition, and consequently ours, is reflected in the architecture of the pavilion on its different fronts, although this challenges its operation on its southwest front, as shown in Figure 1a,b. The use of water, which is also a traditional and vernacular measure closely linked to Islamic tradition, requires technology for its operation [13]. The disuse of this measure in the pavilion creates a problem for its skin and façade.

(a)

(b)

Figure 1. (**a**) South elevation of Moroccan Pavilion. Isla de la Cartuja (Seville, Spain); (**b**) Southwest corner of Moroccan Pavilion. Isla de la Cartuja (Seville, Spain).

2.2. Passive Measures

Passive measures in buildings are those directly associated with the envelope of the construction, including its materiality, thickness and orientation [14]. Through these measures, vernacular architecture has achieved adaptation to the environment and local conditions without the need for extreme energy inputs. This historical wisdom has been proven to reduce the added costs of maintaining basic comfort conditions when the environment and the building are not prepared for it.

All shading devices play a crucial role in architectural design by controlling the amount of sunlight that enters a building, thereby enhancing occupant comfort, reducing glare, and minimizing energy consumption for cooling [15]. This literature review explores the various types of shading devices, their historical context, modern advancements, and the principles guiding their effective implementation.

2.3. Islamic Shading Elements

Islamic architecture is renowned for its intricate and functional designs, particularly in the use of shading devices [16]. These elements are essential in hot and arid climates where controlling sunlight and promoting natural ventilation are crucial for comfort [17]. Islamic shading devices are not only practical but also contribute to the aesthetic and cultural value of the architecture.

Shading devices in Islamic architecture were developed to address the harsh climatic conditions, primarily in the Middle East and North Africa, as well as in some countries of the Mediterranean Sea [18,19].

Key historical shading devices include:

1. Mashrabiya (مشربية): This is a traditional wooden lattice screen used in windows and balconies. The mashrabiya allows light and air to pass through while providing privacy and reducing heat gain. Often intricately carved, they add decorative value to the building façade. They can also refer to wooden or metal screens used to cover windows, similar to the concept of blinds. These can be adjusted to control the amount of light and air entering a room [20,21], as shown in Figure 2a.
2. Riwaq (رواق): A riwaq is a covered arcade or portico that surrounds courtyards and open spaces in mosques and palaces [22]. The arcades provide shaded walkways and cool resting areas, protecting occupants from direct sunlight, as shown in Figure 2b.
3. Muqarnas (مقرنص): This decorative element is a form of architectural ornamented vaulting, often used in domes, half-domes, and entrances. While primarily decorative, muqarnas structures also cast intricate shadows, contributing to the control of light and heat [23], as shown in Figure 3a.

4. Courtyards (صحن): Central courtyards in Islamic architecture are surrounded by high walls or buildings, creating shaded areas. These courtyards often contain water features, which further help to cool the air [24], as shown in Figure 3b.

(a) (b)

Figure 2. (**a**) Mashrabiya. The Hall of Comares (Alhambra, Granada, Spain); (**b**) Riwaq. Court of the Lions (Alhambra, Granada, Spain).

(a) (b)

Figure 3. (**a**) Muqarnas. Hall of the Abencerrajes (Alhambra, Granada, Spain); (**b**) Courtyards. Court of the Gilded Room (Alhambra, Granada, Spain).

Key Features and Principles:

1. Ventilation and Airflow: Islamic shading devices are designed to enhance natural ventilation. The mashrabiya, for example, promotes airflow while blocking direct sunlight, creating a cooler indoor environment [25].
2. Thermal Comfort: By reducing heat gain, these devices help maintain comfortable indoor temperatures [26]. The materials used, such as wood and stone, also have thermal mass properties that contribute to temperature regulation.
3. Aesthetic Integration: Shading devices in Islamic architecture are not just functional; they are also highly decorative. The intricate designs and patterns of mashrabiya screens and muqarnas add visual interest and cultural significance to the buildings.
4. Privacy: Islamic shading devices often provide privacy for occupants while still allowing light and air to pass through [27]. This is particularly important in residential architecture, where maintaining a private indoor environment is culturally significant.

Contemporary architects continue to draw inspiration from traditional Islamic shading devices, incorporating their principles into modern designs; these designs include:

- Al Bahar Towers (Abu Dhabi, UAE): These towers feature a dynamic façade with a responsive mashrabiya system [28]. The shading screens open and close based on the sun's position, reducing solar heat gain and glare while maintaining natural light and views, as shown in Figure 4a.

- (KAPSARC) (Riyadh, Saudi Arabia): Designed by Zaha Hadid Architects, this complex uses geometric shading devices inspired by Islamic patterns. These elements provide effective sun control and enhance the building's aesthetic [29].
- Doha Tower (Doha, Qatar): Designed by Jean Nouvel, the tower features a complex façade with aluminium mashrabiya elements [30]. These screens control sunlight and provide a unique visual identity that reflects traditional Islamic architecture, as shown in Figure 4b.

(a) (b)

Figure 4. (**a**) Mashrabiya system in Al Bahar Towers (Abu Dhabi, UAE); (**b**) Aluminium mashrabiya elements in Doha Tower (Doha, Qatar).

2.4. *Benefits of Islamic Shading Elements*

1. Energy Efficiency: By reducing the need for artificial cooling and lighting, these devices contribute to significant energy savings.
2. Enhanced Comfort: Effective shading and ventilation improve thermal comfort for occupants, making indoor environments more liveable [31].
3. Cultural Continuity: Incorporating traditional shading elements preserves cultural heritage and provides a sense of continuity in modern architecture.
4. Aesthetic Value: The intricate designs of Islamic shading devices enhance the visual appeal of buildings, adding depth and texture to façades.

3. Methodology

In the reviewed literature, various measures exist for evaluating the environmental suitability of cultural heritage sites [32]. Some methodologies concentrate on the fluctuations in the microclimate throughout the year. Nonetheless, it is important to emphasize the need for a proper study that incorporates new parameters, aspects, and indices that impact the preservation of these highly sensitive heritage structures [33].

3.1. *The Moroccan Pavilion in Seville*

Located on the Isla de la Cartuja, the Moroccan Pavilion in Seville is part of the architectural ensemble planned for the International Expo' 92 Exhibition. It is also adjacent to the most historic area of the island, next to the walls that enclose *El Real Monasterio de la Cartuja*, a significant architectural piece in Seville. This unique monastery's architecture housed Carthusian monks and later served as an iconic ceramics and porcelain factory during the 19th century and much of the 20th century; its location is shown in Figure 5.

Figure 5. Location plan of Moroccan Pavilion. Isla de la Cartuja (Seville, Spain).

This historic boundary described above, and the building's architecture, make the pavilion a bridging construction and, in some way, a "gentle transition" to the more contemporary architecture of the other pavilions around the Cartuja. On the other hand, its freestanding architecture, while making it a unique piece visible from all sides, also exposes it to external elements. Therefore, its envelope is key to its interior comfort and functionality.

The building, with 5000 m^2 of constructed area over four levels (basement, ground floor, first floor, and second floor), has a layout developed around an interior courtyard with a gallery ambulatory that organizes the space, as shown in Figure 6. Its envelope is defined as a skin separated from all the interior constructed elements, designed to withstand external weather agents.

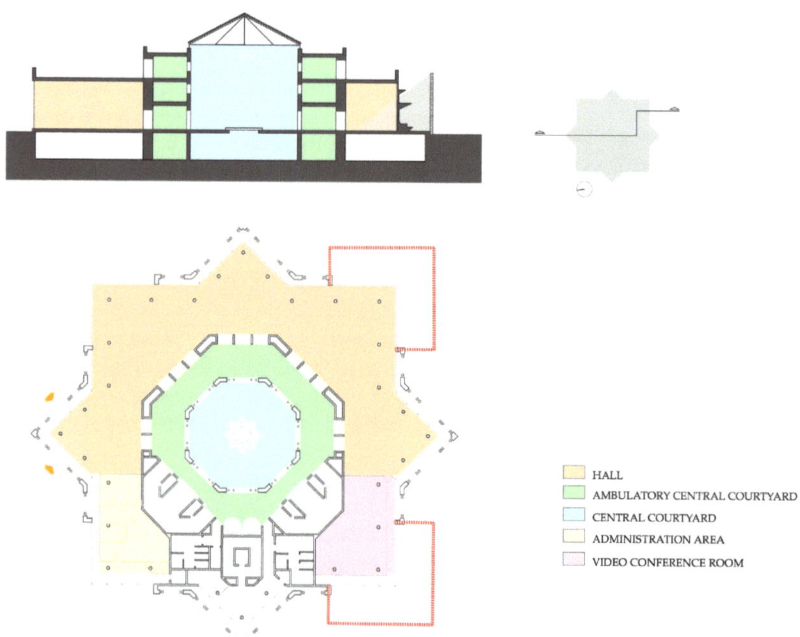

Figure 6. Transverse cross-section and ground-floor zoning plan of Moroccan Pavilion. Isla de la Cartuja (Seville, Spain).

Nowadays, this unique pavilion is part of the architectural heritage of the city of Seville and the preserved ensemble from the Exhibition Expo'92, aiming to be a reference in Seville's heritage. After numerous failed renovations, action is needed, and a new restoration project has been approved with a particular focus on energy efficiency and sustainability.

3.2. Climate Data for Seville

The city of Seville in southern Spain experiences a Mediterranean climate characterized by hot, dry summers and mild, wet winters. Here is a detailed look at the climatic data (source AEMET, Agencia Estatal de Meteorología):

3.2.1. Temperature

Seville experiences significant temperature variations throughout the year:

- Winter (December to February): Average temperatures range from 5.9 °C to 17 °C. The coldest month is January, with temperatures ranging from 5.9 °C to 15.4 °C.
- Spring (March to May): Temperatures increase from an average of 14.2 °C in March to 26.7 °C in May.
- Summer (June to August): This period is characterized by high temperatures, averaging from 25.7 °C to 35.3 °C. July and August are the hottest months, with maximum temperatures often exceeding 35 °C.
- Autumn (September to November): Temperatures gradually decrease from an average of 30.6 °C in September to around 18.8 °C in November.

3.2.2. Humidity

Humidity in Seville varies across the year:

- Winter: The highest relative humidity, around 75%, is observed in December and January.
- Summer: The lowest humidity levels occur in July, with an average of 39%.

3.2.3. Solar Radiation

Seville enjoys a high number of sunshine hours, particularly in the summer:

- Winter: Average daily sunshine hours range from 6.5 to 8 hours in December and January.
- Summer: The average daily sunshine hours peak in July, with approximately 12.7 h of sunshine per day.

3.2.4. Rainfall

Precipitation in Seville shows a marked seasonal pattern:

- Winter: The wettest months are October through December, with rainfall reaching up to 77 mm in December.
- Summer: The driest period is between July and August, with almost negligible rainfall.

3.3. Building Skin (Epidermis)

Due to the importance of implementing passive measures in this research, it is essential to know and describe the building's complete envelope. Given its heritage nature, it is crucial to maintain and complement it with non-invasive intervention measures.

The building has a star-shaped form with several entrances, although currently, only one is operational for daily use, as shown in Figure 6. There is another direct access to the basement that connects the auditorium with the exterior plaza where some events take place.

It is a building of high artistic and craftsmanship value, both in its exterior and interior. This greatly limits our possibilities for intervention without sacrificing artistic elements, whether acting on the envelope or the interior.

Regarding exterior enclosures, there are primarily two types. On the ground floor and first floor, the entire building features a curtain wall envelope with some external lattice elements adorning certain corners, interspersed with non-functional fountains. Although from the outside it may appear that a percentage of the façades are on the ground floor and first floor, from inside, we can observe that everything is external to the curtain wall, which serves as the element separating the interior from the exterior, as shown in Figure 6.

3.4. Energy Performance Evaluation

3.4.1. Energy Audit of the Moroccan Pavilion to Assess Its Current Energy Performance

A comprehensive energy assessment has been conducted, documenting various features of the building's envelope, including its walls, ceilings, floors, doors, windows, and skylights. For each of these elements, the area and thermal resistance (R-value) have been measured and estimated. Additionally, the rate of air leakage or infiltration through the building envelope has been examined, with particular attention paid to the quality of windows and doors. The aim of this study was to quantify the building's overall thermal performance. The assessment also evaluated the efficiency, physical condition, and programming of mechanical systems, such as the heating, ventilation, and air conditioning equipment [34,35].

3.4.2. Integrated Model Decision

The integrated model proposed in Figure 7 consists of two distinct phases, each with specifically selected decision factors [34–36]. The first phase is the analysis phase, followed by the simulation and materialization phase, which is supported by a decision support system. This approach can also be effectively applied to any other building requiring intervention and climate control, particularly in the context of new scenarios.

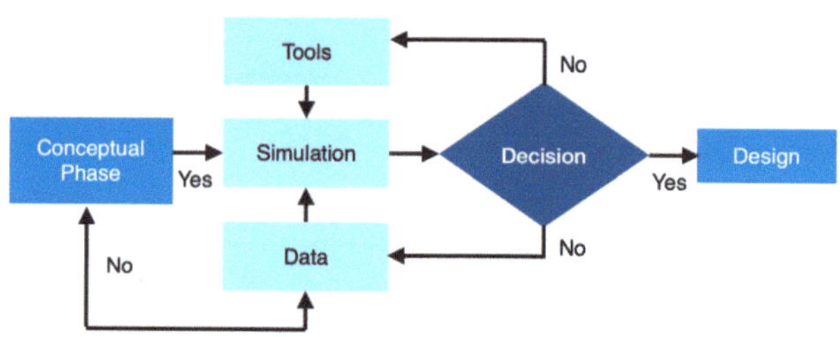

Figure 7. Integrated model, Decision Support System.

3.4.3. Conceptual Phase

In previous studies, during what can be seen as a conceptual phase, the following improvement measures were considered:

- Lighting control in the perimeter zone of exposure.
- Inclusion of heat recovery units in the ventilation system.
- New heat pump and replacement of Air Handling Units (UTAs).
- Replacement of the curtain wall.
- Roof insulation (including the cost of demolishing the slope formation and adding new waterproofing).
- Perimeter shading element in the exterior area to reduce solar incidence on the curtain wall.

In conclusion, all these measures alone have not been sufficient to achieve the objectives in the simulation.

3.5. D Virtual Simulation and Data

An integrated sustainable model concept has been developed from advancements in different fields such as modelling, simulation, computer networking, image processing, and multimedia representation [37]. This audit has explored these areas within the architectural process to address the growing complexity in collecting data, including the architectural

CAD model, prototypical systems, and the modelling of the building according to all gathered information [38–40]. Ultimately, the sustainable integrated environment has become a key framework for achieving sustainable development goals in heritage building conservation [41]. Sustainable design renovations, a relatively new concept, aim to improve the quality of obsolete constructions by minimizing the negative impacts and optimizing the use of natural resources.

3.5.1. The 3D Model

The geometric model of the building has been constructed directly from the TeKton3D TK-CEEP modeler, based on the CAD model created from the original building plans executed in AutoCad software by Autodesk. Once the 3D geometry is defined, the monthly demand data are obtained through the simulation of the building model using specific software [42,43]. For generating the model, the data collected in phases 1 and 2 of the audit (preliminary information and data measured during the building visit) are used, estimating and calculating some of the variables when necessary [44,45]. The following sections list the data required for the construction of the thermal model; see Table 1.

Table 1. Thermal model data.

Data	Main Façade	Basement Wall	Building Roof
Thermal Resistance ($m^2 °K/W$)	0.384	0.179	0.573
Overall Mass (kg)	311.20	611.24	407.50
Total Thickness (m)	0.1856	0.2628	0.3451
Thermal Transmittance ($W/m^2 °C$)	1.804	0.753	1.403
Colour	Medium	–	Medium

The main openings of the building are:

- Doors: Their thermal transmittance has been introduced according to the material they are made of, and their absorption coefficient based on their colour (light, medium, dark, or black). This information has been estimated based on observations during the building visit.
- Glazed openings (windows): The total energy transmittance of the opening, with mobile shading devices (blinds, curtains, etc.) activated, is obtained through the program using values collected in the Building Technical Code (CTE) based on observations and data gathered during the building visit.

Assumed from the information provided by the property and on-site inspections, the HVAC systems are responsible for generating heat transfer fluids for room climate control. The following systems have been defined:

- Direct expansion systems (using refrigerant fluids), such as VRF (Variable Refrigerant Flow) or multi-split systems.
- Air conditioning systems: These are air handlers that discharge into an air duct network, such as UTAs or Rooftops.
- Water condensation systems: Used to temper water from chiller condensers or heat pumps.

3.5.2. Energy Consumption

Once constructed, the model is validated with the actual consumption of the building by comparing the model's results with the measurements taken from historical invoices.

The total annual energy consumption of the building is 162,116 kWh, of which 60,244 kWh is allocated to "other uses". Therefore, the remaining consumption (101,872 kWh) is allocated to heating, lighting, ventilation, and Domestic Hot Water (DHW), in the proportions shown in Figure 8.

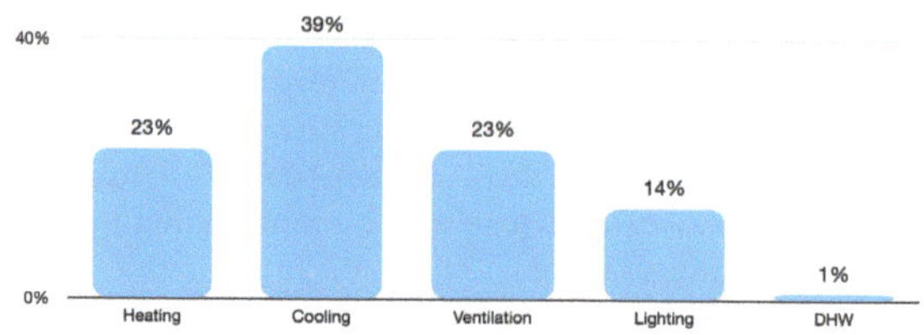

Figure 8. Total energy consumption/specific use.

This breakdown of consumption is an estimate based on the distribution provided by the energy certification software, which does not account for the consumption of "other uses"; see Table 2.

Table 2. Energy consumption model.

Specific Use	Total Energy Consumption (kWh)
Heating	23,805
Cooling	39,787
Ventilation	23,163
Lighting	13,788
DHW	1300

4. Results and Discussion

This research underscores the importance of the continuous monitoring and adaptation of both active and passive measures for this unique pavilion. By addressing current inefficiencies and preparing for future climate scenarios, it aims to maintain the pavilion's functionality and heritage value while minimizing energy consumption.

4.1. Monthly Energy Consumption

The energy consumption data have been represented in Figure 9 to compare the values in winter and summer for the months of January to September. The maximum value occurs in July and the minimum value occurs in August, due to reduced building usage.

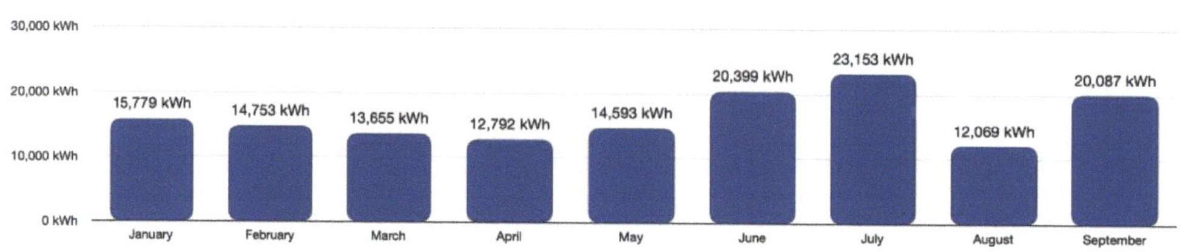

Figure 9. Monthly consumption energy data.

By analysing the load curves, it is observed that in summer, there are significant energy consumptions in the pavilion compared to the winter period, primarily due to high energy usage for interior cooling, as confirmed earlier in Table 2.

4.2. Zones and Data Monitoring

For accurate data monitoring, different thermal zones within the building have been defined. These zones may or may not correspond to physically delineated rooms. A zone can include multiple spaces that share similar thermal conditions (e.g., orientation, climate control). In this case, zoning has been carried out by floor, and each floor is further defined by usage area (e.g., office, restroom, hallways, technical room).

To align the model's consumption with the actual consumption, a setpoint temperature of 20 °C for heating and 22 °C for cooling is established. A sufficient number of temperature and humidity measurements have been taken to define the environmental conditions of the building. Measurements were taken in five thermal zones with different climate control requirements to realistically characterize the building, yielding consistent results across all zones. Thus, a total of five measurement points were used in this model.

The maximum temperature measured was 31.6 °C, and the minimum was 30.4 °C. It was generally found that the comfort conditions regarding indoor temperatures, as established by current regulations and energy-saving recommendations, are not met.

4.3. Materials and Procedures

4.3.1. Building Envelope

- Exterior enclosure: On the ground floor and the first floor, the entire building features a curtain wall with some external latticework elements adorning certain corners, alternating with non-functioning fountains.
  The curtain wall is composed of double glass with an air chamber, covering approximately 1133 m^2. This makes it a significant element for thermal performance and potential intervention. The opaque enclosure consists of a wall made of perforate brick masonry, an air chamber of approximately 40 cm used for installations, and an interior partition. These elements are heavily adorned with detailed Arab craftsmanship, making them a distinctive feature of the building and crucial when analysing possible interventions.
- Roofs: This building features three types of roofs: two are flat and sloped. The two flat roofs are walkable and lack insulation, distinguished only by their flooring finish. Additionally, the building has a wooden dome that crowns the interior courtyard. The exact composition of the dome is unknown, but it is mobile, allowing it to uncover the courtyard for ventilation and cooling.
- Exterior Carpentry (doors and windows): The doors have double glazing with an air cavity. There are only four windows, located on the second floor, which are not significant in terms of the building's envelope.

All technical data required for the construction of the thermal model are listed in Table 1.

4.3.2. Decreasing Energy Consumption Procedures Suggested

According to the study conducted during the measurement period and various virtual evaluations using the TeKton3D TK-CEEP modeler CAD software and local regulations, the following actions are recommended:

- Reducing Heating and Cooling Loads: To reduce the demand for heating and cooling, it is essential to minimize the amount of solar heat that enters the building and obtain natural air currents for cooling indoor spaces [46,47]. This can be effectively achieved by utilizing the central courtyard for passive cooling through natural ventilation.
- Optimizing Lighting Systems: Select an optimal combination of light fixtures and lamps to achieve necessary lighting levels while saving electricity [48].
- Implementing Climate-Responsive Façades: Climate-responsive building façades can be achieved through various strategies [49]. These include shading the building's walls, incorporating insulation or building envelopes where feasible, using reflective glass windows to bounce daylight, and allowing for natural ventilation. These strategies help to regulate the building's temperature and reduce reliance on artificial cooling and heating.
- Using Heat Insulation Envelopes: Properly designed heat insulation envelopes can significantly reduce dependence on mechanical heating and cooling systems [50]. A high-quality envelope is not solely determined by the amount of insulation, but also by its ability to control air infiltration. A low infiltration rate means less air leakage, leading to better temperature regulation and reduced energy usage.
- Employing Occupancy and Light Sensors: Occupancy sensors can be used to automatically switch lights on and off based on room usage, and light sensors can be

incorporated to adjust artificial lighting according to the amount of natural daylight, achieving an efficient balance between sufficient lighting and energy savings.

4.4. Analysis and Results

The conducted research study highlights the critical importance of energy expenditure related to air conditioning, particularly cooling, within the context of building management. This finding underscores the necessity for substantial investment in upgrading outdated building systems, which promises a high return on investment.

4.4.1. Glazing and Curtain Wall Systems

This study identifies that the curtain wall systems are a major source of energy loss. Improving these systems can lead to significant reductions in energy consumption. However, these improvements involve substantial investment in new insulation materials, which are costly and have a long payback period with an unreachable return and a high cost. At this point, a deep reflection is necessary to minimize energy consumption in a sustainable matter.

4.4.2. Proposed Approach: Passive Measures and Façade Redesign

Therefore, bringing back the concept of the envelope and passive measures as the most suitable answer to this problem, this approach involves the redesign and reconfiguration of the main building's façade from an architectural point of view, leading to the consideration of passive design strategies as the more viable answer.

The building's clear and balanced architectural language must not hinder further interventions. Consequently, measures are proposed to reinforce the pavilion's architectural language, making it viable and coherent for its everyday use. By focusing on the envelope, the ventilation and air conditioning systems (as seen in Section 4.3.2) will support the redesign of a new passive façade. A key recommendation is the replacement of the current curtain wall with a new passive façade that offers improved performance. This redesign is analysed in the study with an emphasis on achieving a total transmittance of $U = 1.76\ W/m^2K$ and a solar factor of 0.49. These metrics aim to maximize the energy-saving impact of the façade improvements.

Regarding the types of latticework (mashrabiyas) proposed in this work, they have been equated with the data and improvements obtained from the energy study through a new curtain wall that complies with the requirements set by current regulations. Therefore, we emphasize the replacement of an expensive envelope with one that meets the required qualities, already mentioned in this study. The simulation of this passive façade indicates significant potential for energy savings. By focusing on passive design measures and façade improvements, reliance on mechanical ventilation and air conditioning systems can be reduced. The passive measures become the primary means of enhancing energy efficiency, with mechanical systems serving as additional support.

The research underscores the importance of addressing energy consumption in buildings through a combination of modernizing outdated systems and adopting passive design strategies. While the initial investment in new insulation and passive measures is substantial, the long-term benefits in terms of energy savings and return on investment are clear. The proposed redesign of the building's façade not only aligns with its architectural language but also significantly improves its energy performance, paving the way for more sustainable building practices and coherence with everyday use.

4.5. Implementation of Intervention Strategies

A proposal for a set of passive measures tailored to the building has been made, and this includes shading elements, thermal insulation improvements, and the use of locally sourced materials.

- Insulation in opaque enclosures: This addresses actions from the interior on vertical walls that maintain existing ornamentation and reinforce thermal inertia. On roofs, interventions should focus on the sloping structure with insulation over the waterproofing (inverted roof).

- Analysis of the ensemble of lattices versus glazing, with special emphasis on solar arch façades. Reducing the incidence on curtain walls should help to avoid the need to replace glass and should drastically reduce economic investment (using vinyl where necessary). The architectural design of this new skin is the foundation of this research concept, as shown in Figure 10a,b.

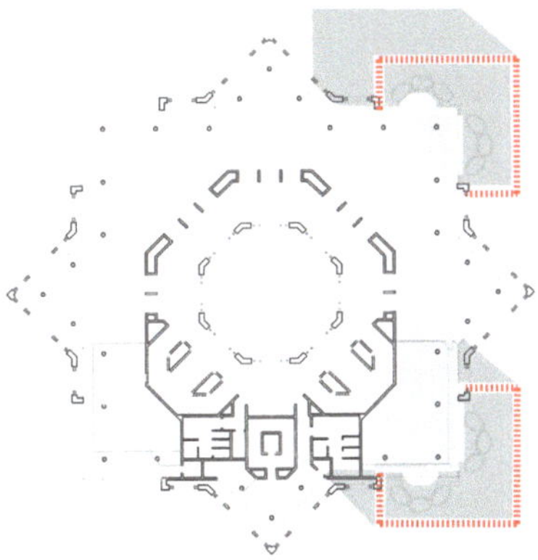
(a)

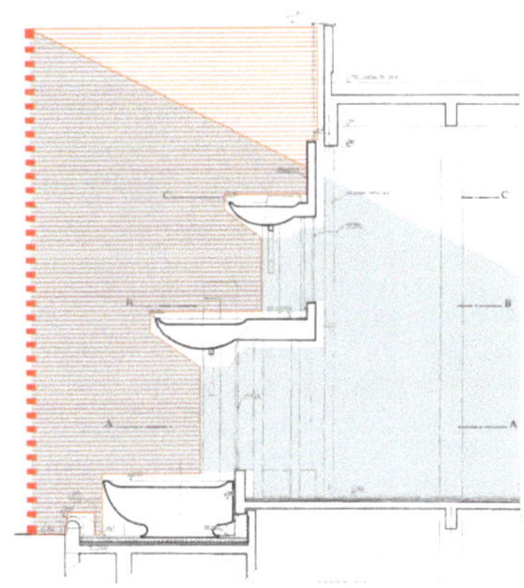
(b)

Figure 10. (**a**) Intervention areas. Low level. Moroccan Pavilion. Isla de la Cartuja (Seville, Spain); (**b**) Section view of curtain wall. Moroccan Pavilion. Isla de la Cartuja (Seville, Spain).

- Improving the ventilation and air conditioning systems: These systems are now sized for the building's new behaviour.
- Other actions: The building will be used responsibly, with the controlled application of lighting, presence detection, and the regulation of light intensities.

The results of this study shed light on various factors that could influence the indoor temperature of this architecture. The proper handling of these values and data is crucial for preserving these structures, which still serve their original purpose today. Additionally, regarding the design process and preservation, industrial heritage buildings are currently and will continue to be affected by climate change. Their ongoing preservation necessitates understanding these impacts on their exceptional universal value and addressing them appropriately.

5. Conclusions

The importance of sustainable architecture practices has increased over time due to the rising environmental concerns. Sustainable design integrates energy-efficient systems, renewable energy, and eco-friendly materials to protect the environment and enhance living spaces by improving the air quality, natural lighting, and thermal comfort. This comprehensive approach makes sustainable architecture essential, not just a passing trend.

This research emphasizes the need for the continuous monitoring and adaptation of active (HVAC systems, smart technologies) and passive (natural ventilation, thermal insulation, shading) measures to maintain the indoor climate of architectural heritage, ensuring energy efficiency and conservation. Establishing a feedback mechanism to refine and adjust interventions based on the observed performance and emerging climatic trends is crucial for the long-term sustainability of these structures.

A case study on the Morocco Pavilion in Seville demonstrates how passive measures, like the use of mashrabiya for solar control and natural ventilation, enhance thermal comfort and energy efficiency while preserving historical value. This research shows how the mashrabiya, a traditional architectural element, can serve as an effective solution for solar control and natural ventilation in buildings during hot summers. The mashrabiya's lattice design reduces heat gain while allowing air circulation, significantly significantly enhancing the building's thermal comfort. This demonstrates that traditional design principles can be relevant in modern architecture. Thus, respecting heritage can be compatible with modern sustainability. By utilizing passive design strategies such as natural ventilation, strategic shading, and thermal mass, and combining them with modern materials and technologies, the renovation can achieve significant energy savings while preserving the building's historical and cultural significance. The Morocco Pavilion's renovation, utilizing passive design strategies and modern materials, achieves energy savings and preserves cultural significance. This approach ensures that heritage buildings meet contemporary sustainability standards and local regulations.

The application of such hybrid techniques ensures that architectural heritage can meet contemporary sustainability standards without compromising its integrity in an area as sensitive as the heritage of a Universal Exhibition, The Expo'92. Comprehensive evaluations of thermal comfort, energy consumption, and existing passive measures are essential. In this research, CAD and specific software for the modelling of a virtual twin in terms of the skin of the Pavilion have been used as advanced simulation tools to represent the building's energy performance under various scenarios, allowing for the development of targeted strategies for improvement. These strategies include enhancing insulation, optimizing natural ventilation, integrating renewable energy sources, and retrofitting the building envelope with high-performance materials.

Equally important is understanding climate variability factors of the city of Seville. Holistic approaches considering immediate and long-term climate impacts are vital for preserving architectural heritage. Continuous monitoring, data collection, and smart building technologies (IoT sensors, automated climate control) enable proactive maintenance and timely interventions. This includes assessing the potential effects of extreme weather events, such as heatwaves, that occur more frequently each year in Seville, with maximum temperatures above 40.0 °C and reaching over 45.0 °C according to the Spanish State Meteorological Agency (AEMET). Implementing adaptive measures, such as responsive shading systems and phase-change materials for thermal regulation, can enhance the building's resilience.

Based on the results obtained and subsequent analysis, it is clear that intervention in the Pavilion's envelope is necessary. This paper highlights and emphasizes this necessity, recognizing and appreciating the importance of improvements that align with the heritage value of the property. This need for intervention serves as a reflection on actions taken regarding historical heritage, underscoring the need to be subtle and delicate in such endeavours and to recognize a set of measures that align with the quality of the construction and historical knowledge from which we must learn. The intervention itself will result from a future study of this catalogue, whose implementation will require further analyses deserving of specific studies.

This research contributes to the growing body of knowledge on sustainable architecture and provides a framework for future projects aiming to balance heritage conservation with modern sustainability goals, ensuring resilience and adaptability to the future.

Author Contributions: M.V.C. and F.L. wrote, reviewed and commented on the manuscript. All authors have read and agreed to the published version of the manuscript.

Funding: This research received no external funding.

Data Availability Statement: All data are available from the corresponding author upon request.

Conflicts of Interest: The author declares no conflicts of interest.

References

1. Betts, R.F. Structuring the Ephemeral: The Cultural Significance of World's Fair Architecture. *Ky. Rev.* **1980**, *2*, 4.
2. Alba-Dorado, M.I. Review: *España 92: Arquitectura y ciudad.* José Ramón Alonso Pereira (comp). Diseño Editorial. *VLC Arquit.* **2020**, *7*, 185–191. [CrossRef]

3. El-Bastawisy, M.M.; Leila, M.M.S.A. Architectural Innovative Technologies for Improving the Built Environment: A Technical Framework. In *Conservation of Architectural Heritage. Advances in Science, Technology & Innovation*; Springer: Cham, Switzerland, 2022; pp. 473–487.
4. Leila, M.M.S.A.; Alaboud, N.S. Reviving the Hejazi Meccan Style to Achieve a Sustainable Saudi Housing - Case Study: Saudi Housing Units in Makkah. *J. Archit. Plan.* **2023**, *35*, 411–437.
5. Bagasi, A.A.; Calautit, J.K.; Karban, A.S. Evaluation of the Integration of the Traditional Architectural Element Mashrabiya into the Ventilation Strategy for Buildings in Hot Climates. *Energies* **2021**, *14*, 530. [CrossRef]
6. Almulhim, A.I.; Cobbinah, P.B. Framing resilience in Saudi Arabian cities: On climate change and urban policy. *Sustain. Cities Soc.* **2024**, *101*, 105172. [CrossRef]
7. Burton, I. Reports on Reports: Our Common Future. *Environ. Sci. Policy Sustain. Dev.* **1987**, *29*, 25–29. [CrossRef]
8. Uzzal-Hossain, D.; Thomas-Ng, S. Critical consideration of building environmental impact assessment towards adoption of circular economy: An analytical review. *J. Clean. Prod.* **2018**, *205*, 763–780. [CrossRef]
9. Huang, Z.; Zhou, H.; Miao, Z.; Tang, H.; Lin, B.; Zhuang, W. Life-Cycle Carbon Emissions (LCCE) of Buildings: Implications, Calculations, and Reductions. *Engineering* **2024**, *35*, 115–139. [CrossRef]
10. United Nations Environment Programme (UNEP). *2021 Global Status Report for Buildings and Construction: Towards a Zero-emission, Efficient and Resilient Buildings and Construction Sector*; Report; United Nations Environment Programme (UNEP): Nairobi, Kenya, 2021.
11. Hyde, R. Refining the principles of passive design. *Archit. Sci. Rev.* **2017**, *60*, 357–359. [CrossRef]
12. Moscoso-García, P.; Quesada-Molina, F. Analysis of Passive Strategies in Traditional Vernacular Architecture. *Buildings* **2023**, *13*, 1984. [CrossRef]
13. Alelwani, R.; Ahmad, M.W.; Rezgui, Y. Public Perception of Vernacular Architecture in the Arabian Peninsula: The Case of Rawshan. *Buildings* **2020**, *10*, 151. [CrossRef]
14. Veeraboina, P.; Yesuratnam, G. Significance of design for energy conservation in buildings: Building envelope components. *Int. J. Energy Technol. Policy* **2013**, *9*, 34–52. [CrossRef]
15. Al-Shareef, F.M. Natural Light Control in Hadjazi Architecture: An Investigation of the Rowshan Performance by Computer Simulation. Ph.D. Thesis, University of Liverpool, Liverpool, UK, 1996.
16. Ahani, F. Natural Light in Traditional Architecture of Iran: Lessons to remember. *WIT Trans. Built Environ.* **2011**, *121*, 25–36.
17. Lomas, K.J. Architectural design of an advanced naturally ventilated building form. *Energy Build.* **2007**, *39*, 166–181. [CrossRef]
18. Eskandari, H.; Saedvandi, M.; Mahdavinejad, M. The Impact of Iwan as a Traditional Shading Device on the Building Energy Consumption. *Buildings* **2018**, *8*, 3. [CrossRef]
19. Aliakbar Heidari, A.; Taghipour, M.; Yarmahmoodi, Z. The Effect of Fixed External Shading Devices on Daylighting and Thermal Comfort in Residential Building. *J. Daylighting* **2021**, *8*, 165–180. [CrossRef]
20. Almerbati, N.; Ford, P.; Taki, A.; Dean, L. From Vernacular to Personalized and Sustainable. In *Living and Learning: Research for a Better Built Environment: 49th International Conference of the Architectural Science Association*; The Architectural Science Association & Genova University Press: Melbourne, Australia, 2015.
21. Özsavaş Akçay, A.; Alotman, H. A Theoretical Framework for the Evaluation from the Traditional Mashrabiya to Modern Mashrabiya. *J. Hist. Cult. Art Res.* **2017**, *6*, 107–121. [CrossRef]
22. Mansourı, A. A Syntactic Approach to the Effect and the Role of Hayat and Riwaq in the Geometric Conception of Traditional Housing Architecture in Iran: Tabriz Houses. *A+Arch Des. Int. J. Archit. Des.* **2018**, *4*, 47–64.
23. Sakkal, M. Geometry of Muqarnas in Islamic Architecture. Ph.D. Thesis, University of Washington, Seattle, WA, USA, 1981.
24. Jasim, H.S. Courtyard is Basic Pattern in Tradition Islamic Architecture: As Prototype. In Proceedings of the 12th International conference "Standardization, Protypes and Quality A Means of Balkan Countries", Izmit, Turkey, 21–24 October 2015.
25. Al-Masrani, S.M.; Al-Obaidi, K. Dynamic shading systems: A review of design parameters, platforms and evaluation strategies. *Autom. Constr.* **2019**, *102*, 195–216. [CrossRef]
26. Hyde, R. *Climate Responsive Design: A Study of Buildings in Moderate and Hot Humid Climates*; Taylor & Francis: London, UK, 2013.
27. Ochoa, C.E.; Capeluto, I.G. Advice tool for early design stages of intelligent facades based on energy and visual comfort approach. *Energy Build.* **2009**, *4*, 480–488. [CrossRef]
28. Attia, S. *Evaluation of adaptive facades: The case study of Al Bahr Towers in the UAE*; In Qscience Proceedings; HBKU Press: Doha, Qatar, 2016.
29. Whybrow, B. KAPSARC-Tensile Building Sking. *Procedia Eng.* **2016**, *155*, 89–102. [CrossRef]
30. Saeid, A.Y.; Al-Kazzaz, D.A. Design algorithms for contemporary pattern-based building envelopes. *Ain Shams Eng. J.* **2024**, *15*, 102656. [CrossRef]
31. Zhang, H.; Arens, E.; Huizenga, C.; Han, T. Thermal sensation and comfort models for non-uniform and transient environments, part III: Whole-body sensation and comfort. *Build. Environ.* **2010**, *45*, 399–410. [CrossRef]
32. Analysis of the impact of sustainability related design parameters in the architectural design process. A case study research. In Proceedings of the 3rd International Conference on Smart and Sustainable Built Environment (SASBE 2009), Rotherdam, The Netherlands, 15 June 2009.
33. Castilla, M.V. Analysis of Performance Strategies for the Preservation of Heritage Buildings: The Case of Puerto de Santa María. *Buildings* **2023**, *13*, 1196. [CrossRef]

34. Barrelas, J.; Silva, A.; de Brito, J.; Tadeu, A. Effects of Climate Change on Rendered Façades: Expected Degradation in a Progressively Warmer and Drier Climate—A Review Based on the Literature. *Buildings* **2023**, *13*, 352. [CrossRef]
35. Stagrum, A.E.; Andenæs, E.; Kvande, T.; Lohne, J. Climate change adaptation measures for buildings—A scoping review. *Sustainability* **2020**, *12*, 1721. [CrossRef]
36. Silva, H.E.; Henriques, F.M.A.; Henriques, T.A.S.; Coelho, G. A sequential process to assess and optimize the indoor climate in museums. *Build. Environ.* **2016**, *104*, 21–34. [CrossRef]
37. Iwaro, J.; Mwasha, A. The impact of sustainable building envelope design on building sustainability using Integrated Performance Model. *Int. J. Sustain. Built Environ.* **2013**, *2*, 153–171. [CrossRef]
38. Thompson, B.P.; Bank, L.C. Use of system dynamics as a decision-making tool in building design and operation. *Build. Environ* **2010**, *45*, 1006–1015. [CrossRef]
39. Rodríguez, F.; Fernández, G. Sustainable Engineering: New Objectives for Construction Projects. *Rev. Ing. Constr.* **2010**, *25*, 147–160.
40. Halog, A.; Manik, Y. Advancing Integrated Systems Modelling Framework for Life Cycle Sustainability Assessment. *Sustainability* **2011**, *3*, 469–499. [CrossRef]
41. Li, D.H.W.; Lau, C.C.S.; Lam, J.C. Predicting daylight illuminance by computer simulation techniques. *Light. Res. Technol.* **2004**, *36*, 113–128. [CrossRef]
42. Wood, T. Climate analysis using autodesk ecotect weather tool and solar tool. In Proceedings of the AIA Montana Annual Fall Conference, Bozeman, MT, USA, 30 September 2010.
43. Castilla, M.V. Architectural Heritage Analysis of the Yuanying Guan Pavilion: Construction and Signification. *Heritage* **2023**, *6*, 2421–2434. [CrossRef]
44. Wang, H.; Zhai, Z. Advances in building simulation and computational techniques: A review between 1987 and 2014. *Energy Build.* **2016**, *128*, 319–335. [CrossRef]
45. Ochoa, C.E.; Aries, M.B.C.; Hensen, J.L.M. State of the art in lighting simulation for building science: A literature review. *J. Build. Perform. Simul.* **2011**, *5*, 209–233. [CrossRef]
46. Cabeza, L.F.; Castell, A.; Medrano, M.; Martorell, I.; Pérez, G.; Fernández, I. Experimental study on the performance of insulation materials in Mediterranean construction. *Energy Build.* **2014**, *42*, 630–636. [CrossRef]
47. Ghosh, A.; Neogi, S. Energy efficient building facade design strategies for composite climate of India: A review. *Front. Archit. Res.* **2017**, *6*, 283–295.
48. Baetens, R.; Jelle, B.P.; Gustavsen, A. Properties, requirements and possibilities of smart windows for dynamic daylight and solar energy control in buildings: A state-of-the-art review. *Sol. Energy Mater. Sol. Cells* **2011**, *94*, 87–105. [CrossRef]
49. Aksamija, A. *Sustainable Facades: Design Methods for High-Performance Building Envelopes*; John Wiley & Sons: Hoboken, NJ, USA, 2015.
50. Granadeiro, V.; Correia, J.R.; Leal, V.M.S.; Duarte, J.P. Envelope-related energy demand: A design indicator of energy performance for residential buildings in early design stages. *Energy Build.* **2013**, *61*, 215–223. [CrossRef]

Disclaimer/Publisher's Note: The statements, opinions and data contained in all publications are solely those of the individual author(s) and contributor(s) and not of MDPI and/or the editor(s). MDPI and/or the editor(s) disclaim responsibility for any injury to people or property resulting from any ideas, methods, instructions or products referred to in the content.

heritage

MDPI

Article

Towards Using Digital Technologies to Balance Conservation and Fire Mitigation in Building Heritage Hosting Vulnerable Occupants: Rapid Evacuation Simulator Verification for the "Omero Museum" (Ancona, Italy)

Marco D'Orazio, Margherita Canafoglia, Gabriele Bernardini * and Enrico Quagliarini

DICEA Department, Università Politecnica delle Marche, Via Brecce Bianche, 60131 Ancona, Italy; m.dorazio@staff.univpm.it (M.D.); e.quagliarini@staff.univpm.it (E.Q.)
* Correspondence: g.bernardini@staff.univpm.it

Abstract: Digital technologies can support designers in balancing architectural heritage conservation and performances using multidisciplinary approaches. Fire safety represents a challenging issue, especially in public historical buildings hosting vulnerable occupants, since heavy modifications are often required to facilitate their evacuation. Digital tools based on evacuation simulation are able to verify the impact of other sustainable, compatible evacuation management and planning approaches, especially considering the use of generic software, which can be used by low-trained technicians according to rapid setups. Nevertheless, simulator reliability should be experimentally verified through case study applications. This work thus offers the experimental verification of a rapid setup-based generic evacuation simulator in the context of a significant case study (the "Omero Museum", Ancona, Italy), placed in a historic building hosting vulnerable occupants (disabled, elderly, children), thanks to a full-scale evacuation drill. The rapid setup described different vulnerable occupants' categories according to literature data. Comparisons between drill and simulation results, using consolidated verification indicators, showed the overall reliability of the proposed approach, and thus encourage additional tests in historical buildings. The proposed setup-based simulator could be combined with other digital tools (virtual reality, BIM-related) to provide full support to fire risk and evacuation assessments when vulnerable occupants are present.

Keywords: building heritage; vulnerable occupants; evacuation simulation verification; case study

Citation: D'Orazio, M.; Canafoglia, M.; Bernardini, G.; Quagliarini, E. Towards Using Digital Technologies to Balance Conservation and Fire Mitigation in Building Heritage Hosting Vulnerable Occupants: Rapid Evacuation Simulator Verification for the "Omero Museum" (Ancona, Italy). *Heritage* **2024**, *7*, 3734–3755. https://doi.org/10.3390/heritage7070177

Academic Editors: Daniela Fico and Daniela Rizzo

Received: 10 May 2024
Revised: 10 July 2024
Accepted: 11 July 2024
Published: 14 July 2024

Copyright: © 2024 by the authors. Licensee MDPI, Basel, Switzerland. This article is an open access article distributed under the terms and conditions of the Creative Commons Attribution (CC BY) license (https://creativecommons.org/licenses/by/4.0/).

1. Introduction

The role of digital technologies in supporting the design of strategies and solutions for the conservation and fruition of architectural heritage is constantly increasing [1–4]. In fact, thanks to multidisciplinary approaches, such technologies are able to properly represent the historic built environment and its hosted artefacts and combine such data by highlighting heritage identity features and vulnerabilities, related possible stressors, occupants and current/future intentions, and then adopt performance-based assessment actions through simulation approaches, thanks to the improvement in the level of knowledge of heritage critical issues in ordinary and extraordinary scenarios [1,3,5–8]. They can thus widely contribute to different challenges, including those related to sustainability issues [9], the preservation of heritage (including materials and surfaces) and its identity features [10,11]), operation and energy efficiency [5,12,13], management and maintenance [14], the continuous use of and balance between adaptation/retrofit/regeneration, and conservation [1,8].

Among these goals, risk reduction and mitigation issues surely relate to one of the most important topics, especially considering possible disasters and emergency conditions, considering events affecting single buildings (e.g., fires) [7,15,16] or the urban area where it is placed (e.g., floods, earthquakes, landslides, climate change) [6,17]. In both cases, digital tools for scenario creation, risk assessment and the evaluation of risk reduction/mitigation/preparedness strategies can actively support designers in evaluating probable conditions and moving towards the optimisation of both physical interventions in architectural heritage and management actions before and during an emergency [15,17–21].

1.1. Fire Safety in Architectural Heritage and Vulnerable Occupants

Fire surely represents a fundamental risk, as also remarked by consolidated safety regulations and research and practice efforts [7,15,16,22] aimed at balancing heritage conservation and occupant safety problems. Considering the building itself, physical vulnerability factors relating to fire risk are widely associated with morphological and construction features, including building technologies and materials, as well as with the historical, artistic and anthropological value of heritage and hosted artefacts [23]. Nevertheless, the complexity of this application topic increases while dealing with buildings open to the public due to the combination of building heritage vulnerabilities and hazards with factors related to the hosted occupants [15,16,21,24–27]. In fact, possible high exposure levels, due to possible significant crowding levels, can be coupled with the presence of vulnerable occupants.

First, occupants in public building heritage are often associated with visitors [13,16]. Typical vulnerabilities are correlated with their possible poor familiarity with the built environment, its layout and risk, and emergency procedures [24]. Moreover, occupant vulnerabilities can be correlated to age (e.g., the elderly, children), motion abilities (e.g., occupants with crutches, walking sticks, or wheelchairs) and sensory abilities (e.g., visibility impairments), which lead to specific effects on evacuation quantities, e.g., motion speed, evacuation choices, path selection and participation in evacuation [28–30]. As a result, besides suffering from limited fruition of the heritage and being treated as ordinary visitors [31], such vulnerable occupants can have different levels of autonomy in evacuation, thus altering fire risk with respect to basic occupancy conditions.

Second, typical protection and mitigation techniques aimed at mitigating damage levels and facilitating the evacuation process (e.g., fire compartmentation, evacuation paths redundancy and dimensions, creating temporary safe spaces where vulnerable occupants can wait for rescuers) may have an heavy impact on the building in view of the complexities of the building heritage and its layout, thus leading to unacceptable modifications [8,16,24]. Non-structural strategies, based on evacuation management and planning, could support fire risk mitigation by limiting the impact of physical interventions in architectural heritage [21,26,32–34]. In this sense, all vulnerable occupants who can move autonomously should be encouraged to properly act by themselves thanks to the support given by members of building emergency staff, thus allowing rescuers to focus on non-autonomous individuals.

1.2. Digital Technologies for Fire Safety Improvement: Simulation Tools

The contribution of digital tools such as simulators, BIM/HBIM and Virtual Reality could be relevant in the definition, application and management of fire safety measures in architectural heritage by also including non-structural strategies such as evacuation management and planning [16,25], particularly when the historical building hosts vulnerable occupants [26].

In particular, evacuation simulation models representing occupant features and emergency behaviours can contribute to this goal [15,21,24,33–35]. In fact, simulators can assess fire risk from a behavioural-based perspective and support the effective investigation of alternatives concerning mitigation strategies in building heritage [35,36]; this can also be performed in combination with tools for structural safety and fire-spreading simulation [18,20]. Moreover, they can be included in digital technologies supporting the whole conservation/restoration process, thus also being coupled with BIM/HBIM and tools for multi-criteria decision-making [8,15,16,37,38]. In particular, generic (e.g., commercial) evacuation simulators can be valuable in supporting the whole process [39]. They are aimed at simulating general-purpose and fire scenarios, allowing for the implementation of standard-based, easy-to-implement setups, which should be specific indeed in respect to the considered emergency scenarios, and thus also to the presence of vulnerable occupants, reducing complexities due to source code modifications or the use of dedicated plugins [40]. In this way, generic evacuation simulators can move towards quick use by non-specifically trained technicians, thus consolidating operational issues [2,17].

Nevertheless, simulators can be properly used when their reliability is demonstrated, thanks to calibration, verification and validation tasks [39]. Most of these activities concern controlled test conditions and basic scenarios in terms of layout (e.g., rooms, corridors, staircases and their combination with a limited extension), according to international guidelines [41,42]. Therefore, evacuation drills [24,43–45] performed in real-world scenarios can provide additional powerful support for the assessment of simulator reliability, by comparing simulation results with experimental data. Nevertheless, the availability of experimental data seems to be still limited essentially because of organisational complexities, especially when referring to full-scale drills [28,30,45]. In particular, those involving building heritage hosting vulnerable occupants are mostly still missing, to the best of the authors' knowledge, and, thus, it can be assumed that a single study can contribute to the whole picture, although partial.

1.3. Work Aims

This work thus offers the experimental verification of a simplified setup for a generic evacuation simulator in the context of a specific case study, i.e., "Omero Museum" (Ancona, Italy). The museum is in a multi-story historical building and usually hosts disabled, elderly and child occupants. A full-scale evacuation drill was performed by involving the presence of such vulnerable occupants, other non-vulnerable adult visitors (without sensory/motion impairments) and emergency staff, who supported occupants according to the evacuation plan actions. Then, a digital model of the museum was implemented in a generic evacuation simulator based on microscopic modelling [46], boosting the capabilities of the simulator application in the context of building heritage for proper evacuation management and risk mitigation strategies design [15,21,23,24,26]. A specific literature-based rapid setup was then defined to consider behaviours and motion quantities depending on occupant vulnerabilities. Finally, consolidated verification indicators [39,41,42] were used to compare the simulation outcomes with the experimental drill results, determining the reliability of the approach for the considered case study.

2. Phases, Materials and Methods

The work is organised according to the framework in Figure 1, which also includes the outlines of the related results sections. In the first phase, a full-scale evacuation drill was organised in the case study (Section 2.1). This full-scale drill was organised according to typical day-use conditions and the established emergency evacuation plan, and the results were analysed (Section 2.2). The second phase concerned the definition of the rapid setup of a commercial evacuation simulator, based on the experimental inputs (Section 2.3). Finally, the third phase concerned a comparison of the drill and simulation results for verification purposes (Section 2.4).

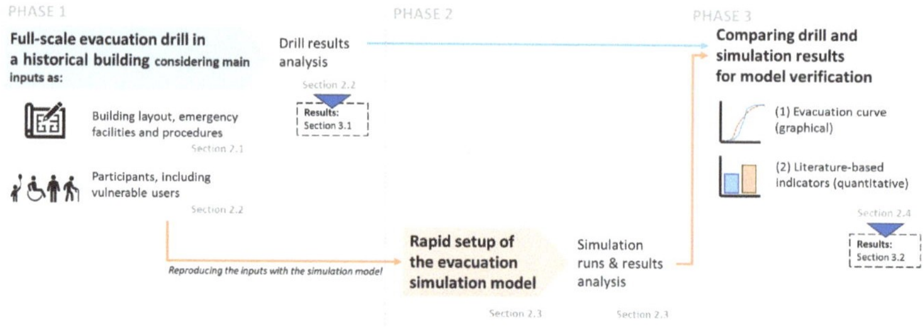

Figure 1. Research framework including references to the methodological and results sections.

2.1. The Case Study

The State Tactile Museum "Omero"[1] of Ancona, Italy, was selected as a relevant case study for this work since it normally hosts vulnerable occupants and is placed in a significant historic building.

The museum was created to promote the cultural growth of all vulnerable people, and, in particular, make art known through touch, and, thus, it is accessible to everyone including people with visual disabilities. Besides such vulnerable people, typical visitors of the museum are children (school groups) and the elderly, who are engaged in different learning and cultural events over the whole year.

The museum is located inside the "Mole Vanvitelliana", located within the Ancona harbour (Figure 2A) [47,48]. Built between 1733 and 1743, and being one of the main landmarks of the city and harbour, the "Mole" is a historic masonry building that is composed of five levels, has a pentagonal shape, and hosts a small temple within the central courtyard (Figure 2B). The "Mole" was designed for multifunctional purposes, i.e., as quarantine places for travellers and goods, port warehouses and fortification. During the XIX century, the "Mole" was adapted into a sugar refinery, and then, in the XX century, it was adapted into a tobacco pressing factory. Complete regeneration actions have been carried out since 1989 by promoting its restoration and reuse for concerts, public events, and exhibitions, including the ones of the "Omero" museum (Figure 2C).

Figure 3 offers an overview of the museum levels, showing that the overall layout relies on a generally axial symmetry, thus being divided into two main sides (left and right). The building entrances are placed on the ground level (ground-level exit on the left side of the building—GEL) and the 1st level (first-level exit on the left side of the building–FEL) and they serve as safety exits. An additional safety exit is placed on the 1st level (first-level exit on the right side of the building—FER). GEL, FEL and FER are then connected to the central courtyard (Figure 2C) by external stairs. The 1st level also hosts the building service (BS) area, which includes the ticket office and the reception, as well as a permanent exhibition area (1st-level exhibition area on the left side of the building—1EL and 1st-level exhibition area on the right side of the building—1ER). The 2nd level hosts additional exhibition areas (in particular, during the test, the right side of the building was assumed to be unavailable due to the presence of fire/smoke, and, thus, only the 2nd-level exhibition area on the left side of the building—2EL was considered) and laboratory areas for school groups (2nd-level laboratory area on the left side of the building—2LL and 2nd-level laboratory area on the right side of the building—2LR). The 3rd level only comprises the administration offices—3OA, with limited access to the public.

Figure 2. The "Mole Vanvitelliana" and the State Tactile Museum "Omero": its location within the City of Ancona (**A**), a general aerial view of the building by marking the areas hosting the museum, (**B**) and an internal view from the central courtyard (**C**). The museum is marked within the light yellow areas (base map from Immagini ©2023 Google, Immagini ©2023 Airbus, Maxar Technologies, and Dati cartografici ©2023).

Figure 3 also traces the following:

- the starting areas (in blue), where occupants were placed when the drill started, including those for free visitors (dashed blue areas), according to Section 2.2, by outlining symbols for their main vulnerability-related typologies, according to Table 1 data;
- the evacuation paths according to the museum emergency plan (green arrows);
- the staircases, by outlining their availability (in green) and unavailability (in light orange), and the related intermediate areas monitored during the drill (i.e., first and second level of staircases on the left side of the building, respectively FSL and SSL), according to Section 2.2;
- the final evacuation exits (by green flag), which are GEL, FEL and FER.

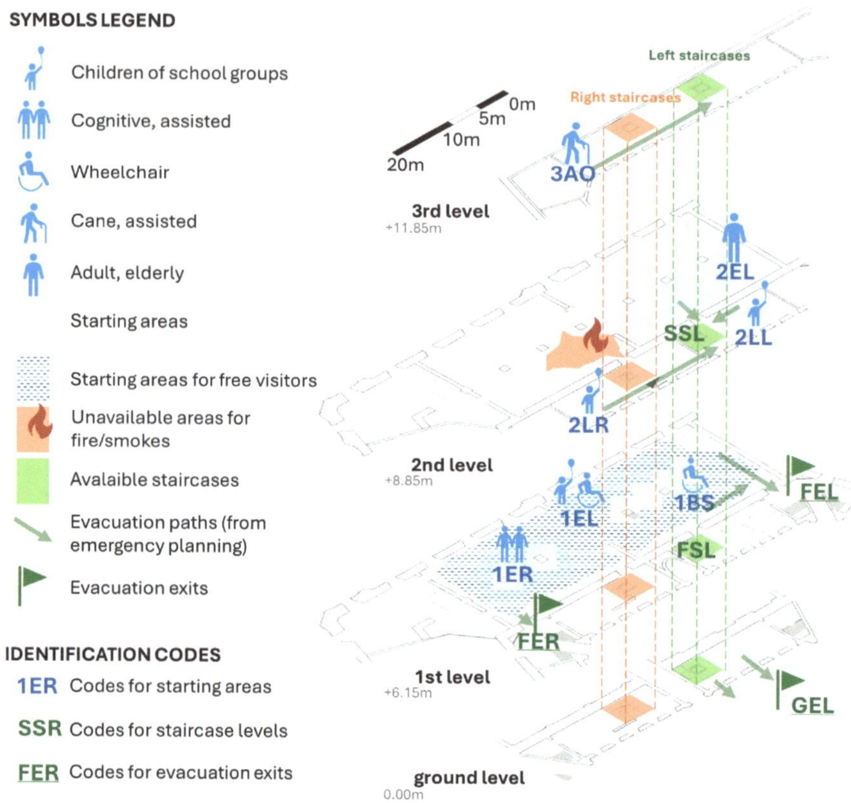

Figure 3. Overview of the "Omero Museum" organisation by level (axonometric view, including metric scale), including details on the experimental drill.

Table 1. Initial position and type of volunteers participating in the evacuation drill. The "notes" column reports general comments, as well as disability types and assistance in motion * Disabled occupant is part of the staff; ^ the occupants are administration office members and evacuate later than the others for safety reasons.

Description (Identification Code in Blue in Figure 3)	Total	Children of School Groups	Adult Visitors (Elderly)	Staff	Disabled Children	Disabled Adults (Elderly)	Notes
1st level, building services (BS)	3	0	0 (0)	2	0	1 (0) *	Wheelchair, unassisted
1st level, exhibition area, left side (1EL)	22	13	7 (0)	1	1	0 (0)	Wheelchair, assisted by an adult visitor
1st level, exhibition area, right side (1ER)	9	0	3 (0)	2	0	4 (0)	Cognitive, assisted
1st level, exhibition area (1EL and 1ER)	7	0	3 (3)	3	0	1 (1)	Cane, assisted; all the occupants are considered free visitors

Table 1. *Cont.*

Description (Identification Code in Blue in Figure 3)	Total	Children of School Groups	Adult Visitors (Elderly)	Staff	Disabled Children	Disabled Adults (Elderly)	Notes
2nd level, exhibition area, left side (2EL)	21	0	19 (10)	2	0	0 (0)	
2nd level, laboratory areas, left side (2LL)	18	13	3 (0)	1	1	0 (0)	
2nd level, laboratory areas, right side (2LR)	21	13	6 (1)	2	0	0 (0)	
3rd level, administration offices (3AO)	6	0	4 (1) ^	1	0	1 (1) ^	Blind, assisted ^
TOTAL	107	39	45 (15)	14	2	7 (2)	

The current museum emergency evacuation layout and plan (including evacuation paths and exits, wayfinding signage, staff rules, and emergency procedures) fully respect Italian regulations regarding fire safety, e.g., DM n. 569, 20 May 1992, [49].

2.2. Experimental Drill

The full-scale evacuation drill was performed in collaboration with the museum administration and emergency management offices, as well as with the firefighters' local command of Ancona. According to Table 1, 107 people voluntarily attended the full-scale drill. For safety issues, the drill was announced, although the starting time was not communicated to them. All the volunteers (except the members of the emergency staff and the administration office) declared being unfamiliar with the museum, and none of them was previously involved in training activities about the emergency procedures, nor informed of the evacuation plan and exit routes.

Table 1 traces the position and typology of drill participants according to the overview of the museum offered in Figure 3. In particular, 61% of participants were vulnerable people, most of them being children from school groups (36% of the total number of participants). For safety reasons, an elderly occupant initially placed on the 3rd level did not participate in the main evacuation process and started moving later, being assisted by a staff member. Furthermore, the emergency staff was composed of adults, and they were homogeneously placed within the building, according to the emergency plan.

The drill was performed during opening times, in typical normal use of the museum. According to the emergency procedures, the building fire alarm rang to point out a fire emergency implying evacuation, and the evacuation time started. The emergency staff assisted the other occupants as defined by the emergency procedures and plan, thus also ensuring the proper selection of evacuation paths. The drill was performed in free-of-smoke conditions, although one of the stairs (see the orange areas in Figure 3) was considered unavailable due to fire and smoke presence. The drill ended when the last building occupant exited from the museum, crossing one of the exits shown in Figure 3.

To obtain a complete overview of the process, fixed video cameras were placed along the evacuation paths, i.e., pointing at the starting areas, the monitored areas along the staircases and the safety exits. Evacuation path choices and timings were retrieved by manually analysing the recorded videotapes and defining evacuation curves for each monitored intermediate area (i.e., SSL, FSL) and exit (i.e., FEL, FER, GEL). In particular, concerning evacuation timing [s], besides pre-movement times by starting areas, the minimum time *Tmin* (first occupant exiting the building), median time *T50*, time at which 95% of occupants arrived at one of the building exits *T95*, and maximum time *Tmax* (last occupant exiting the building, thus equal to the Required Safe Egress Time) [32] were considered to describe the drill conditions from an overall perspective. In addition, the evacuation flows [persons/s] in respect of *T50*, *T95* and *Tmax* were calculated as the ratio between the related number of occupants arriving at the building exits and the time differences between the considered timing and *Tmin*. This kind of indicator traced the rapidity of the process between different moments in the evacuation curves, thus approximating the slope of the curve. Such data were calculated for each exit and the whole process.

Moreover, an occupant evacuation speed assessment was provided by mainly considering data at the monitored intermediate areas and while approaching the evacuation exits to verify the consistency of results with individual speed setup-based values (see Section 2.3 and Appendix A). Speeds (approximated to 0.1 m/s) were organised by occupant vulnerability to fulfil this goal and by local density (approximated to 0.05 persons/m^2) to evaluate if the results were affected or not by relevant crowding conditions with respect to those of typical (quasi-)free-flow motion within the fundamental diagrams of evacuation dynamics [29].

2.3. Rapid Setup Definition for Evacuation Simulation

The evacuation process was then replicated using Oasys MassMotion 11.0 [46]. This simulation tool allows for the easy representation and analysis of the dynamics of crowds in a built environment, in both normal use and emergency conditions, and has been previously verified according to international standards and used by the working group for setup and validation tests in other contexts (e.g., floods) [40]. MassMotion relies on a multi-agent simulation logic, which allows for the representation of evacuation interactions at a microscopic level for each simulated occupant, who is characterised by specific movement (e.g., path choice), size, speed and individual abilities. Evacuation motion rules are based on the Social Forces Model, in which each occupant is subject to repulsion (e.g., to avoid physical contact) and attraction (e.g., to reach an exit) forces with respect to the surrounding built environment elements and occupants [50]. In view of the above, also considering other commercial and generic evacuation simulators [35], it can be considered that the selected tool relies on consolidated and complex logic from the microscopic perspective, that is with respect to the rules of simulated agents, while setup-based logic can be applied by low-trained technicians, who have just to calibrate the basic parameters of the simulator (as discussed in the following sections).

The physical layout of the museum was modelled with Autocad and Revit[2], starting from the floor plans in Figure 3. To avoid changes to the source code, the quick setup of occupants' behaviours in the simulator mainly concerned their features that could be directly defined through existing interfaces, which were calibrated according to the drill data. The setup data are reported in Appendix A and they include the following:

- The general evacuation model parameters, which refer to the starting areas for the evacuation process, where occupants were placed according to a uniform (randomised) distribution in space, and the monitored intermediate areas and exit, defined according to the emergency plan of the building in relation to the starting areas (considering that all the simulated people effectively selected the proper evacuation path). These data are based on Table 1.
- The individual model parameters, which referred to speed and height. In detail, individual speed was randomly assigned depending on the distribution values (i.e., triangular distributions with minimum, mean and maximum values) derived from literature works and standards, depending on the typology of occupants [28,29,46]. Individual height was not used in the simulation since the smoke effects on motion were not considered, according to the free-of-smoke conditions of the drill.

The proposed setup then considered that occupants were randomly generated within the starting areas, according to the overall distribution shown in Table 1 and Appendix A, and, thus, the effective starting evacuation point for each simulated occupant varied in each simulation [46]. Furthermore, a pre-movement time was also associated with the occupants to represent the time necessary to hear the alarm and prepare for the evacuation (i.e., including interactions between occupants and staff to exchange information on the evacuation plan) [28,43,51]. Pursuing a quick setup approach, pre-movement times were defined according to preliminary videotape analysis of the drill (compare Section 2.1).

At least 10 simulation "runs" were performed to take into account behavioural uncertainties due to occupant speeds (see Appendix A) and starting positions. The convergence of the main evacuation indicators for the whole process and each of the building exits and main monitored intermediate areas was then verified [40,41]. In particular, Euclidean Relative Difference *ERD*, Secant Cosine *SC* and Euclidean Projection Coefficient *EPC* (which are described in Table 2) were used for convergence assessment purposes to estimate the impact of the number of "runs" on the specific indicator. In this case, the measure of the convergence of two relative indicators was performed by considering two consecutive average evacuation curves.

Table 2. Adopted literature-based indicators for quantitative comparison between the evacuation curves in simulation and experimental outputs [40,41].

Indicator [Unit of Measure]	Meaning/Interpretation	Formula (x = Simulation Data; y = Experimental Data)
Euclidean Relative Difference *ERD* [-]	Represents the agreement between two curves in terms of angle; the curves can be considered close if *ERD* is close to 0	$ERD = \frac{\lVert x-y \rVert}{\lVert y \rVert}$
Secant Cosine *SC* [-]	Measures the differences in shape between two curves as their first derivative; the shapes are similar if *SC* is close to 1	$SC = \frac{\langle x,y \rangle}{\lVert x \rVert \, \lVert y \rVert}$
Euclidean Projection Coefficient *EPC* [-]	Evaluates the differences between the curves, compared to the translation of the points that compose them, and thus allows one to measure a sort of scale factor; the curves can be considered similar if *EPC* is close to 1	$EPC = \frac{\langle x,y \rangle}{\lVert y \rVert^2}$
Difference between the graphic Areas Under the Curves *DAUC* [%]	Expresses a sort of "rapidity" of the evacuation process given the whole area under the curve, although DAUC should be close to 0% to have similar "rapidity"; acceptable results rely on *DAUC* > 0%, which implies that the simulation curve is "slower" than the experimental one, and, thus, the simulation model predicts values in a conservative approach	$DAUC = \frac{\int x - \int y}{\int y} \cdot 100$

As for drill results, the considered main simulation outputs are based on evacuation paths, curves and timings at the monitored intermediate areas and safety exits.

2.4. Indicators for Comparison and Verification

To verify the reliability of the evacuation model in the case study, two different comparisons were carried out between the experimental drill and the MassMotion simulation results, according to consolidated literature works [40,41]. First, a graphical comparison of the evacuation curves was performed to give a preliminary check of the overall trend.

Then, the literature-based indicator comparison allowed us to quantitatively investigate differences between real-world and setup behaviours. These analyses were carried out for the whole process and each monitored intermediate area/exit of the museum. *T95* [s] was used to evaluate the differences in the overall evacuation process by excluding the effects of behavioural outliers due to specific aspects in crowd motion (extreme values in the distribution of occupant initial position, evacuation path choice and individual speed) [40,41]. In this sense, the percentage difference between *T95* in the simulations and drill was calculated because the simulations were affected by stochastic effects, as discussed in Section 2.3 [46].

The other adopted indicators are summarised in Table 2, providing their meaning and interpretation, as well as the calculation equations. They were used to measure differences between the experimental and simulation curves in terms of shape, distance and scale [40,41] by focusing on the overall effects of the vulnerable occupants' presence rather than just examining their specific performance in evacuation. In particular, *ERD*, *SC* and *EPD* were calculated considering the whole experimental curve and the whole average simulation curve. *DAUC* was calculated at different percentiles of evacuation time, that is, the 5th percentile, to analyse the arrival of the first occupants but excluding behavioural outliers placed too close to an exit or with significantly high evacuation speeds; the 50th percentile, to analyse median and thus recurrent behaviours in cases of non-normal evacuation time distributions; the 95th percentile, in relation to *T95*, as discussed above; and the 100th percentile, that is at *Tmax*, to determine the arrival of the last occupant to the building exits.

To confirm proper verification goals, *ERD*, *SC* and *EPD* were acceptable if they were close to the interpretation values expressed in Table 2, while the percentage acceptability threshold for the percentage difference of *T95* and the *DAUC* values was set up to about 10% [40].

3. Results

3.1. Drill Results

Occupants evacuated the building in an orderly manner, thanks to the direct support of the staff members, who ensured that the emergency procedures of the museum were adopted. A pre-movement phase was noticed [28,51]. Occupants collected information about the emergency and the proper procedure to be adopted, thanks to the direct support of the staff members. Differences among the different areas of the museum (Figure 3) were noticed, leading to pre-movement time values of 40s for 1ER and 1EL, 85s for 2EL, 87s for 2LR, 95s for 2LL, and 110s for 3AO. In general terms, these values seemed to be in line with those of previous works [45]. Nevertheless, in each starting area, occupants started to evacuate almost at the same moment, and, thus, the same pre-movement time was considered for them. These data were then used in simulations, as defined in Section 2.3.

A high level of interaction between staff and occupants (to provide direct help and exchange information) was also noticed in the movement phase, ensuring that the chosen paths were consistent with the ones of the emergency plan. In this sense, as expected, no occupant selected paths towards FER since it was considered unavailable due to fire and smoke. A wider discussion of such interactions is provided in Section 3.2 in comparison with the outcomes of the evacuation simulator.

Table 3 summarises the evacuation process statistics through Section 2.2 indicators. Flow values for the whole building and each exit demonstrated how the evacuation was performed in an orderly manner, being almost constant at the values corresponding to *T50* and *T95*. Nevertheless, the FER exit did not have the same trend due to the limited number of occupants selecting the door and the short time difference between *T50* and *T95*. Differences between the flows for *T95* and *Tmax* were essentially due to the "rapidity" reduction in the evacuation process for the last occupants, e.g., due to their speed and unfavourable initial position or other group phenomena slowing down their movement. In this sense, it is also worth noting that *Tmax* and *T95* for the whole building are essentially affected by data on GEL. Slight differences existed between *T95* and *Tmax*, without considering the evacuation process of the blind, assisted occupant initially placed at 3AO. In fact, this occupant and the assisting adult exited in a much longer time (371 s) for safety reasons during the drill. As a consequence, this value was filtered by further evaluation and the verification process, being "anomalous" in view of the drill participation rules. Moreover, this also points out that the safety of the egress process can be affected by additional critical vulnerability-related conditions. Appendix B also shows *Tmax* values by starting areas.

Table 3. Statistics of the evacuation process for the whole building (total) and each exit (Figure 3). ^ excluding the blind, assisted occupant from 3AO and the assisting adult.

Parameter [Unit of Measure]	Total	GEL	FEL	FER
Occupants [persons]	107	69	29	9
Tmin [s]	50	117	50	58
T50 [s]	149	177	77	60
T95 [s]	231	233	101	68
Tmax [s] ^	237	237	129	70
Flow for T50 [persons/s]	0.54	0.57	0.52	2.00
Flow for T95 [persons/s]	0.56	0.56	0.53	0.80
Flow for Tmax [persons/s]	0.33	0.27	0.37	0.75

Finally, Figure 4 shows a boxplot representation of the individual evacuation speeds of occupants, considering their vulnerability (Figure 4A) and the surrounding density (Figure 4B). Values refer to movement along horizontal paths, considering the monitored intermediate areas and the areas near the evacuation exits. Data of occupants with disabilities (i.e., assisted children in a wheelchair, assisted and unassisted adults with cognitive disability, and assisted elderly with cane) were grouped within the same group (Figure 4A), with the statistical dimension of these samples being lower than those of the other categories (compare with Section 2.1 and Appendix A). Similarly, adults and staff were merged together when there was no specific vulnerability in motion. In general terms, Figure 4A confirms the general speed ranges adopted by the rapid setup, although some maximum values were slightly higher than those in the reference literature works, especially for the elderly and children [28]. Nevertheless, they represented a marginal sample related to extreme values. Minimum speed values of emergency staff and adults seemed to be slightly lower than the setup ones, essentially because they related to occupants who reduced their speed to wait for vulnerable occupants (compare with the behavioural dynamics discussed in Section 3.2). Nevertheless, values seemed to refer to quasi-free-flow conditions for occupant motions, according to the very limited density values shown in Figure 4B. This outcome confirms that the occupant modelling data in the rapid setup could be considered coherent with the experimental data.

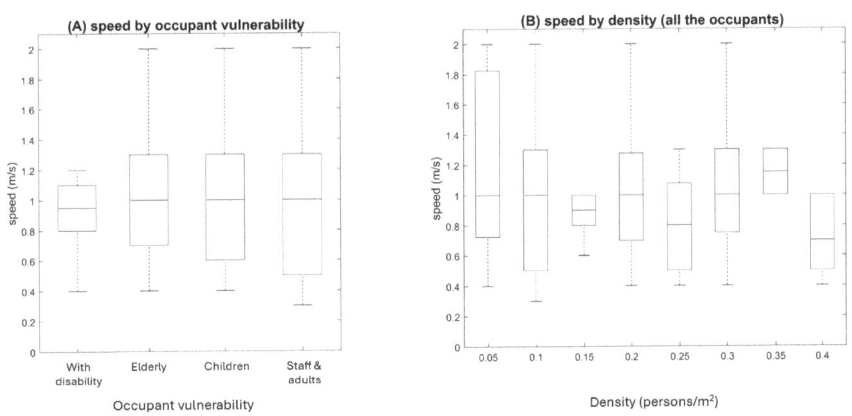

Figure 4. Boxplots of experimental evacuation speeds of occupants by vulnerability (**A**) and density (**B**).

3.2. Simulation Verification Results

Performing 10 simulations allowed us to reach adequate convergence for the *ERD*, *SC* and *EPC* values, as defined in Section 2.3 and shown in Appendix C. Thus, the number of repetitions seemed to be adequate to represent simulation outputs using the proposed setup of MassMotion.

Figure 5 offers a graphic comparison of the whole evacuation curve for the simulation model (grey curve) and experimental drill (blue curve), which shows the same general trend.

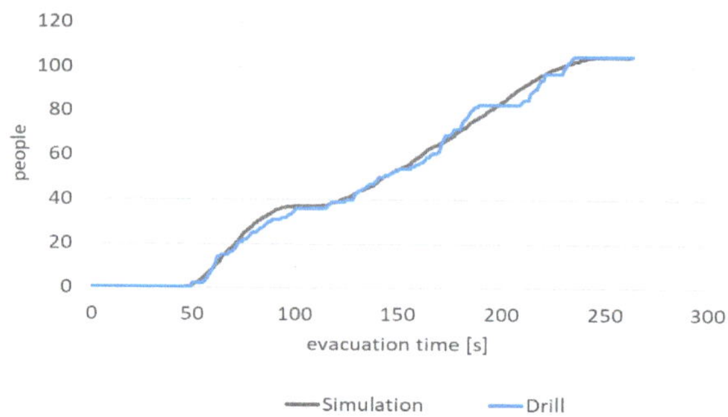

Figure 5. Comparison of the evacuation curves for the whole building in the simulation (grey) and drill (blue).

Figure 6 shows some video frames from the experimental drill, clarifying the main behaviours that provoked differences between the simulation and drill curves. In particular, such behaviours were correlated to the horizontal plateaus in the evacuation process, which were not simulated by the rapid setup simulation results but existed in the drills. In fact, in the drills, plateaus were linked to affiliative behaviours in the evacuation process, which were mainly noticed between vulnerable occupants and staff members. As shown in Figure 6, children tended to increase group cohesion, especially with the staff members, by gathering before moving (Figure 6A, related to the stair access at the second level, near SSL) and looking for staff members' instruction (including directional ones), organising evacuation motion in non-single file (Figure 6B). The group cohesion confirmed the outcomes of previous works [52], although crowd congestion phenomena seemed to be less evident for the case study than in other contexts, in view of the limited density of occupants while moving along the staircases. In this sense, the staff organisation and actions could have played a paramount role in this outcome, and a generally low level of excitement by children seemed to be noticed too.

Nevertheless, the group cohesion phenomenon was not noticed in the simulator, which essentially adopted the quasi-single-file movement of occupants and thus made the curve "more homogeneous" than the drill one.

Figure 7 shows details of the evacuation curves by considering each of the building exits (Figure 7A) and the monitored intermediate areas (Figure 7B) according to simulations (grey) and the drill (blue). All the curves had the same trend and slope, thus confirming their similarity from a graphical perspective. Nevertheless, some differences could be noticed. In particular, the curve related to GEL in the drill (Figure 7A) showed the general presence of plateaus, which affected the whole evacuation curve regarding the aforementioned interaction and grouping behaviours among vulnerable occupants and staff members (see Figure 6B). Similarly, in the simulation, for the FSL gate, 40 s of delay was noticed in comparison with the drill data, while the two curves showed the same trend and slope. This delay could be attributed to two drill-noticed behaviours. First, children and staff members interacted by grouping and waiting before entering the staircase (see Figure 6A). Second, in the drill, the elderly initially placed in 2EL spent additional time interacting with staff members in door selection and use, and the staff members also checked the occupancy status of 2EL before leaving, according to the emergency procedures (see Figure 6D). The second phenomenon implied slight differences in the curves in Figure 7B referring to both SSL and these occupants' arrival to FSL (FSL FROM SSL).

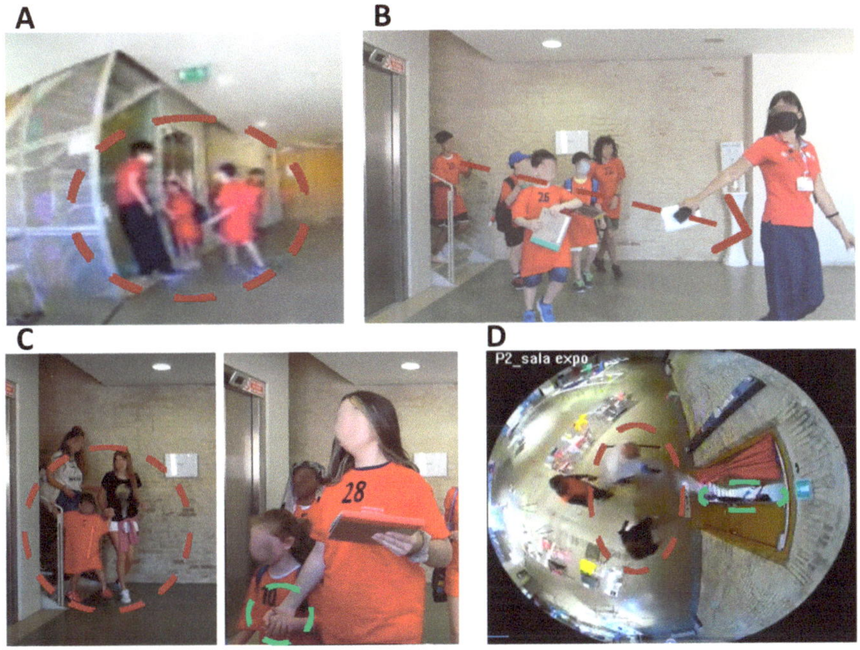

Figure 6. Video frames from the drill concerning vulnerable occupants' behaviours. (**A**) Children interacting with staff members and gathering before using staircases at the 2nd level (in the red circle). (**B**) Interaction between children and members in looking for direction instruction and organising non-single-file movement when placed near GEL (direction by red arrow). (**C**) Younger children hand-assisted by staff members (in the red circle) and by other older children (in the green circle) when placed near GEL. (**D**) Elderly (in the red circle) supported by staff members in exit door selection and use (in green circle) when placed in 2EL.

Considering the Section 2.4 indicators, Table 4 shows that these drill-noticed occupants' organisation in evacuation and related behavioural uncertainties seemed to have quite a limited impact in quantitative terms since they were essentially linked to the aforementioned monitored intermediate areas and exits. In fact, *SC*, *EPC* and *ERD* values seemed to converge to ideal values for the whole evacuation curve and FEL. For FER, a good agreement about both general shape (*SC*) and scale factor (*EPC*) could be noticed. Nevertheless, the experimental curve showed the first occupants passing through the FER exit before what was noticed in simulations. This outcome affected the initial slope of the curve (i.e., between about 50 and 60 s), with the FER experimental values being higher than the simulation values. Therefore, *ERD* did not converge towards 0 and was higher than GEL and FEL. For SSL, FSL and FSL FROM SSL, agreement was confirmed about the shape (*SC*) due to the specific behaviours mentioned above. For GEL, only the angle between the simulation and drill curves (*ERD*) tended to converge due to the similarities in the starting and end points, but differences in *SC* and *EPC* demonstrated the local variability due to the aforementioned behavioural uncertainties.

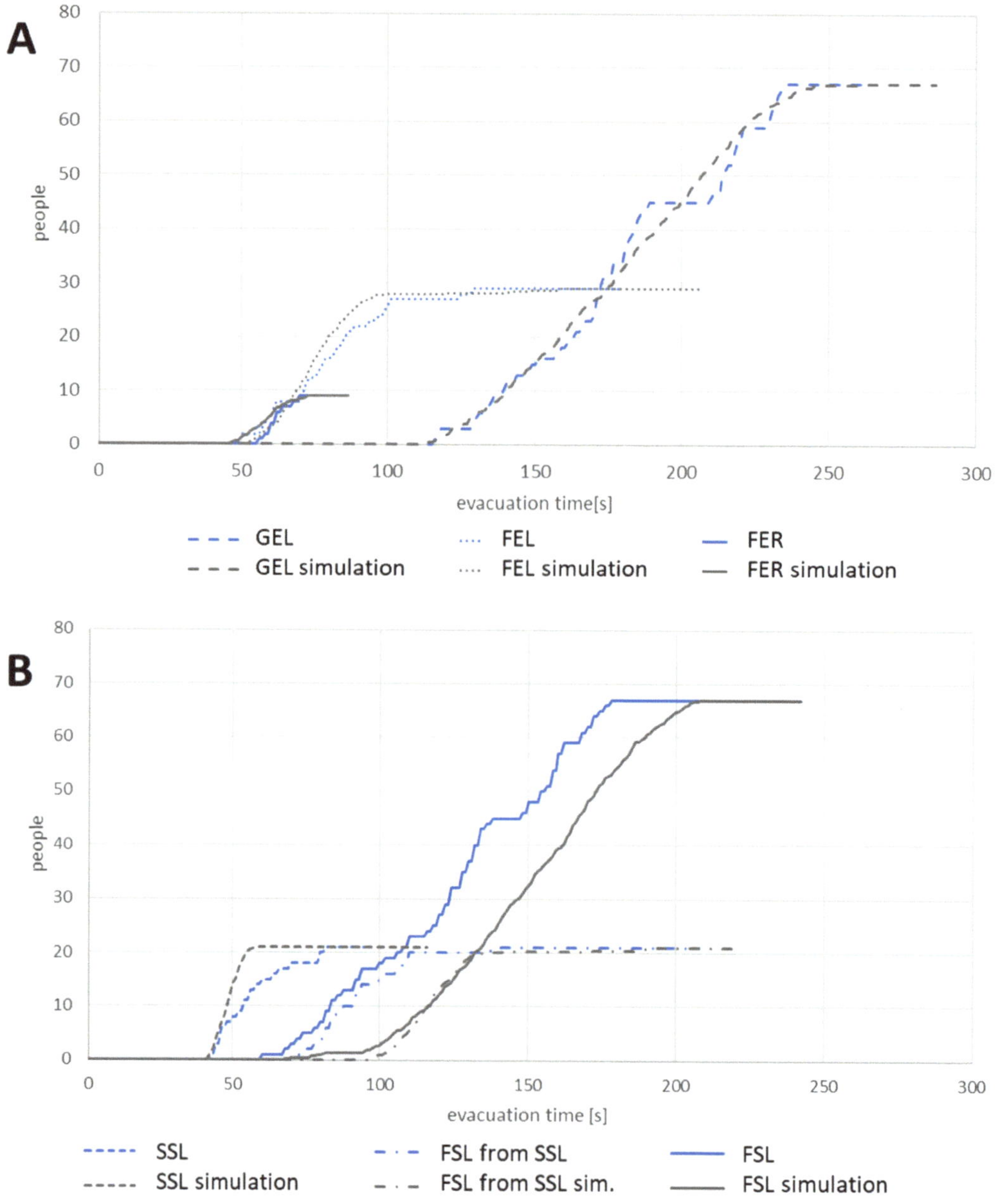

Figure 7. Comparison of the evacuation curves for the single building exits (**A**) and monitored intermediate gates (**B**) in the simulations (grey) and drill (blue). For codes of building exits and monitored intermediate areas, please refer to Figure 3.

According to Table 4, the general modelling reliability was also demonstrated by the percentage differences in *T95*, for the whole curve and the specific building exits, which were all significantly < 10%. Nevertheless, in simulations, *Tmax* = 267 s, thus being >10% with respect to experimental *Tmax* (see Table 3). Although this result confirmed that some behavioural uncertainty and subtitles could affect the simulation results, the rapid setup could be considered valid since the evacuation process was overestimated according to *T95*, *Tmax* and *DAUC*.

Table 4. Literature-based indicators for quantitative comparison between the evacuation curves in the simulation, according to Section 2.4.

Indicator [Unit of Measure]	Total	GEL	FEL	FER	SSL	FSL	FSL FROM SSL
ERD [-]	0.04	0.07	0.09	0.31	0.34	0.29	0.39
SC [-]	1.02	1.18	1	0.96	0.99	0.97	0.93
EPC [-]	1.02	1.37	1.04	1.07	1.27	0.77	0.79
T95 [s]	234	234	127	73	-	-	-
Perc. Diff in T95 [%]	0.4	0.4	1	4	-	-	-
DAUC 5th perc [%]	12	7	20	20	20	17	15
DAUC 50th perc [%]	4.28	8	1	2	6	7	13
DAUC 95th perc [%]	0.13	5	0.16	6	8.22	1.61	8.98
DAUC 100th perc [%]	1.52	17	0.09	12	1.39	8.8	17

4. Discussion

The case study used for the definition and verification of the proposed rapid setup-based generic simulator provides promising insights into the applicability of the proposed approach (Section 4.1) and encourages future research on the matter (Section 4.2) and on model coupling with digital tools for building heritage (Section 4.3)

4.1. Key Findings

The results highlight the general reliability of the setup. In particular, the experimental curve trend for the whole process was close to the simulated curve, as shown by both graphical and literature indicator-based analysis. In simulations, the evacuation time at 95% of the exited occupants overestimated the drill, according to a conservative approach in risk assessment, but the overestimation was modest (around 0.4%) and below the literature threshold of 10%. This confirms the validity of the setup including the occupant behaviour, except for possible behavioural outliers [40,41]. In addition, the analysis of drill evacuation speeds for the occupants seemed to confirm the validity of literature-based assumptions on these selected parameters [27].

4.2. Limitations and Future Works

Considering the drill–simulation results comparisons, the use of a rapid setup seems to be unable to fully represent microscopic interactions between vulnerable occupants and between them and staff members due to grouping and affiliative phenomena during the evacuation process. Deeper setup procedures and source code modifications can be included to improve the assisted evacuation of vulnerable occupants, adapting previous modelling methods for, for instance, healthcare scenarios, where non-autonomous occupants are fully assisted in movement and thus can also move as a "compact" group [33,34]. In particular, staff members should first move towards vulnerable occupants (i.e., those who are placed closer) and organise them into groups, remembering social group concepts [53], e.g., managing their proximity. This can affect the pre-movement time too. Then, close contact among vulnerable occupants should be ensured by, for example, adding social attractive forces or limiting repulsive ones between simulated agents [34,50], modelling proximity-based rules relying on a sort of "maximum" social distance radius criteria [54] and, thus, even creating subgroups in evacuation motion, with occupants moving close one to each other. This can replicate hand-assisted movement and move from quasi-single-file to more "messy" movement of, for example, children. Integrating such approaches can impact the evacuation curve shape, making it more consistent with experimental data.

Considering that this work used a relevant but still single case study application for simulator verifications, it should be noted that specific features of real-world scenarios in terms of layout or emergency plans, other occupant vulnerabilities and their modelling could affect the verification results in other contexts. Thus, additional full-scale drills should be carried out to increase the validity of the comparison results and widely demonstrate the general trends provided by this work.

Nevertheless, as also remarked by previous research involving vulnerable occupants in public buildings [52], this work encourages further research about verification using full- or partial-scale drill results since these data can also be used in the context of other models and other experimental data can be used by replicating the setup methodology of this work. Then, application to real-world scenarios could be achieved by analysing different scenarios in terms of variations in the setup values about pre-movement, path choices and individual speed to verify the impact of these variables on fire evacuation risk. Nevertheless, to this end, actions aimed at experimental data collection and analysis should be coupled with modelling activities to create dependencies among the typology of occupants (among vulnerable ones and among vulnerable and non-vulnerable staff members) for each of these variables.

Therefore, this work represents a first step towards these goals, and additional drills should be performed in future works by involving other typologies of occupants regarding vulnerability and disability [26,28,45,52]. This can promote the development of a common database of related evacuation quantities. At the same time, the capabilities of the quick setup-based simulation tool demonstrate that it can be easily used by fire safety designers and decision-makers for real-world preliminary applications, confirming previous works on the same approach for other kinds of disasters [28].

4.3. Simulation Model Coupling in Digital Tools

As a final output of this research topic, such simulators could then be coupled with other consolidated digital tools [4,10,19,22], such as HBIM, to create a complete design suite exploring the different goals in architectural heritage use, conservation/retrofitting/adaptation/reuse, or Virtual Reality, to include the visualisation of data on common platforms and move towards training designers, stakeholders and occupants (both building staff and end-users, such as visitors) in the context of solutions exploitation (i.e., emergency safety procedures).

5. Conclusions

Digital technologies can support the conservation of architectural heritage, providing representation and simulation data that are useful to understand the impact of different strategies on heritage performances. Among these performances, fire safety surely represents a key issue, especially when considering historical buildings open to the public and hosting vulnerable occupants, such as museums.

Digital tools exploiting evacuation simulation can significantly support designers and stakeholders in identifying the most efficient but sustainable solutions (in terms of invasiveness and compatibility with heritage features) while considering their impact on the typologies of occupants, their mutual interactions within a crowd and their interactions with the built environment and emergency planning. According to this perspective, they could be reliable in understanding the effectiveness of easy-to-apply and low-impact strategies encompassing emergency and evacuation management. Nevertheless, verification tasks should be performed to determine the reliability of evacuation simulators before their application.

In this study, the rapid setup of an existing evacuation simulation model was provided to pursue applicability quickness, and verification was performed thanks to data from a full-scale evacuation drill in a historic museum hosting vulnerable occupants, like children, the elderly, and people with disabilities. The verification performed by drill–simulation results comparisons encourages the future use of such kinds of digital tools in the process of balancing architectural heritage conservations and safety performance assessments. In particular, simulation-based analyses of compatible risk mitigation strategies for historic public buildings would consider management, wayfinding and alter systems, non-invasive layout adaptation, and occupancy areas organisation, and be oriented towards vulnerable occupants and their evacuation behaviours.

Author Contributions: Conceptualisation, M.D., G.B. and E.Q.; methodology, M.D., G.B. and E.Q.; software, M.C. and G.B.; validation, G.B.; formal analysis, M.C. and G.B.; investigation, G.B.; resources, M.D. and E.Q.; data curation, M.C. and G.B.; writing—original draft preparation, M.C. and G.B.; writing—review and editing, M.D. and E.Q.; visualisation, G.B.; supervision, M.D. and E.Q.; project administration, M.D. and E.Q.; funding acquisition, M.D. and E.Q. All authors have read and agreed to the published version of the manuscript.

Funding: This research received no external funding.

Data Availability Statement: Organised data from the experiments are included in this article. Additional raw datasets on the full-scale evacuation drill and simulations will be made available upon reasonable request.

Acknowledgments: The authors thank the Command and Fire Station of Ancona (Italy) of the "Department of Firefighters, Public Rescue and Civil Defence (Corpo Nazionale dei Vigili del Fuoco)", and, in particular, Mariano Tusa for the experiment authorisation and precious support that was provided during the drill and the Museo Omero (AN, Italy) organisation and its staff for the experiment authorisation and proactive participation in the drills. The software license of the MassMotion simulation software was supplied to the researchers by Oaysis Ltd. in the context of the Mutual Non-Disclosure Agreement "Simulating human behaviours during emergencies to improve the safety of buildings: testing and validation of the MassMotion simulation software".

Conflicts of Interest: The authors declare no conflicts of interest.

Ethics Statement: According to other similar studies also performed by our research group at the Università Politecnica delle Marche-DICEA department [24], ethical issues included the need to properly and fully inform volunteers about the experiment procedures and the analysed data, in collaboration with the Command and Fire Station of Ancona and the Museo Omero staff and managers. No sensitive information was collected, and data were anonymised.

Appendix A. Occupant Modelling Details

Table A1. Occupant-related setup within the simulation model, according to Section 2.3 criteria, in terms of egress model parameters (i.e., type, number, starting point and chosen exit, referring to Figure 3). Data are consistent with those of Table 1. * Disabled occupant is part of the staff; ^ the occupants are administration office members.

Type of Occupant	Number of Occupants	Starting Areas	Chosen Exit
Adult in a wheelchair, unassisted *	1	BS	FEL
Staff member	2	BS	FEL
Adult	7	1EL	FEL
Child	13	1EL	FEL
Staff member	1	1EL	FEL
Child in a wheelchair, assisted	1	1EL	FEL
Free visitor	2	BS	FEL
Staff member	2	1ER	FER
Adult	3	1ER	FER
Adult with a cognitive disability	4	1ER	FER
Adult + elderly person	19	2EL	GEL
Staff member	2	2EL	GEL
Free visitor (adult + elderly person)	4	1EL and 1ER	GEL
Elderly person with cane, assisted	1	1EL and 1ER	GEL
Child	13	2LL	GEL
Staff member	1	2LL	GEL
Child with disability	1	2LL	GEL
Adult	3	2LL	GEL
Child	13	2LR	GEL
Adult + elderly person	6	2LR	GEL
Staff member	2	2LR	GEL
Adult ^ + elderly person ^	4	3AO	GEL

Table A2. Literature-based [27] occupant-related setup within the simulation model, in terms of individual model parameters (i.e., speed and height), by distinguishing vulnerable occupants and other occupants. * Same speed for the vulnerable occupant and their assisting member of staff or adult.

Type of Occupant	Min Speed [m/s]	Mean Speed [m/s]	Max Speed [m/s]	Height [m]
Vulnerable occupants:				
Child in a wheelchair, assisted *	0.30	0.60	0.90	1.00
Child	0.60	0.90	1.20	1.00
Adult with cognitive disability, assisted *	0.63	0.93	1.23	1.75
Adult in a wheelchair	0.39	0.69	0.99	1.30
Elderly person	0.80	1.04	1.54	1.75
Elderly person (with cane, assisted) *	0.26	0.81	1.60	1.75
Other occupants:				
Staff member *	1.05	1.35	1.65	1.75
Adult *	0.94	1.24	1.54	1.75

Appendix B. Evacuation Timing Data

Table A3. Maximum evacuation time based on the initial position and chosen exit. ^ Assisted elderly occupant.

Description	Tmax [s]	Chosen Exit
1st level, building services (BS)	128	FEL
1st level, exhibition area, left side (1EL)	99	FEL
1st level, exhibition area, right side (1ER)	70	FER
2nd level, exhibition area, left side (2EL)	188	SSL-GEL
2nd level, laboratory areas, left side (2LL)	231	GEL
2nd level, laboratory areas, right side (2LR)	186	GEL
3rd level, administration offices (3AO)	237; 371 ^	GEL
1st level, exhibition area (1EL and 1ER)	117	FEL

Appendix C. Convergence Measurements

Table A4. Convergence measurements performed on the 10 simulations of the test setup.

Indicator [Unit of Measure]	TOT	GEL	FER	FEL	SSL	FSL	FSL from SSL
ERD [-]	0	0	0.02	0.01	0.06	0.01	0.02
SC [-]	1.02	1	1	1	1	1	1
EPC [-]	1.01	1	1.01	1	1.04	1	0.99

Appendix D. Notations

Table A5. Notations table.

Symbols and Acronyms	Definition	Unit of Measure
FEL, FER	Building exits on the 1st floor, see Section 2.1	-
GEL	Building exits on the ground floor, see Section 2.1	-
SSL, FSL	Monitored intermediate areas, see Section 2.1	-
BS, 1EL, 1ER	Starting areas on the 1st floor, see Section 2.1	-
2EL, 2LL, 2LR	Starting areas on the 2nd floor, see Section 2.1	-
3AO	Starting areas on the 3rd floor, see Section 2.1	-
EPD	Verification indicator concerning "Euclidean Projection Coefficient", see Section 2.4	[-]
ERC	Verification indicator concerning "Euclidean Relative Difference", see Section 2.4	[-]
DAUC	Verification indicator concerning "Difference of the Area Under the Curves", see Section 2.4	[%]
SC	Verification indicator concerning "Secant Cosine", see Section 2.4	[-]
T50	Median evacuation time, that is, the time at which 50% of occupants arrived at one of the building exits, see Section 2.2	[s]

Table A5. *Cont.*

Symbols and Acronyms	Definition	Unit of Measure
T95	Time at which 95% of occupants arrived at one of the building exits, see Sections 2.2 and 2.4	[s]
Tmin	Time needed by the first occupant to exit the building, see Section 2.2	[s]
Tmax	Time needed by the last occupant to exit the building, see Sections 2.2 and 2.4	[s]

Notes

1 Available online: https://www.museoomero.it/en/ (accessed on 15 September 2023).

2 Autocad Version 2024. Available online: https://www.autodesk.it/products/autocad (accessed on 9 May 2024); Revit version 2024, https://www.autodesk.it/products/revit (access on 15 September 2023)–educational license

References

1. Crisan, A.; Pepe, M.; Costantino, D.; Herban, S. From 3D Point Cloud to an Intelligent Model Set for Cultural Heritage Conservation. *Heritage* **2024**, *7*, 1419–1437. [CrossRef]
2. Lucchi, E. Digital Twins for the Automation of the Heritage Construction Sector. *Autom. Constr.* **2023**, *156*, 105073. [CrossRef]
3. Elabd, N.M.; Mansour, Y.M.; Khodier, L.M. Utilizing Innovative Technologies to Achieve Resilience in Heritage Buildings Preservation. *Dev. Built Environ.* **2021**, *8*, 100058. [CrossRef]
4. Lovell, L.J.; Davies, R.J.; Hunt, D.V.L. The Application of Historic Building Information Modelling (HBIM) to Cultural Heritage: A Review. *Heritage* **2023**, *6*, 6691–6717. [CrossRef]
5. Elnagar, E.; Munde, S.; Lemort, V. Energy Efficiency Measures Applied to Heritage Retrofit Buildings: A Simulated Student Housing Case Study in Vienna. *Heritage* **2021**, *4*, 3919–3937. [CrossRef]
6. Iliopoulou, T.; Dimitriadis, P.; Koutsoyiannis, D. Pluvial Flood Risk Assessment in Urban Areas: A Case Study for the Archaeological Site of the Roman Agora, Athens. *Heritage* **2023**, *6*, 7230–7243. [CrossRef]
7. Naziris, I.A.; Mitropoulou, C.C.; Lagaros, N.D. Innovative Computational Techniques for Multi Criteria Decision Making, in the Context of Cultural Heritage Structures' Fire Protection: Case Studies. *Heritage* **2022**, *5*, 1883–1909. [CrossRef]
8. Rebec, K.M.; Deanovič, B.; Oostwegel, L. Old Buildings Need New Ideas: Holistic Integration of Conservation-Restoration Process Data Using Heritage Building Information Modelling. *J. Cult. Herit.* **2022**, *55*, 30–42. [CrossRef]
9. Khalil, A.; Hammouda, N.; El-Deeb, K. Implementing Sustainability in Retrofitting Heritage Buildings. Case Study: Villa Antoniadis, Alexandria, Egypt. *Heritage* **2018**, *1*, 57–87. [CrossRef]
10. Castellazzi, G.; Cardillo, E.; Lo Presti, N.; D'Altri, A.M.; de Miranda, S.; Bertani, G.; Ferretti, F.; Mazzotti, C. Advancing Cultural Heritage Structures Conservation: Integrating BIM and Cloud-Based Solutions for Enhanced Management and Visualization. *Heritage* **2023**, *6*, 7316–7342. [CrossRef]
11. De Fino, M.; Galantucci, R.A.; Fatiguso, F. Condition Assessment of Heritage Buildings via Photogrammetry: A Scoping Review from the Perspective of Decision Makers. *Heritage* **2023**, *6*, 7031–7067. [CrossRef]
12. Lucchi, E. Review of Preventive Conservation in Museum Buildings. *J. Cult. Herit.* **2018**, *29*, 180–193. [CrossRef]
13. Rolim, R.; López-González, C.; Viñals, M.J. Analysis of the Current Status of Sensors and HBIM Integration: A Review Based on Bibliometric Analysis. *Heritage* **2024**, *7*, 2071–2087. [CrossRef]
14. Tahoon, D.; El-Zohairy, A.; Hendawy, H.I. Cost Impact Comparative Analysis via BIM between Heritage Regular Maintenance Projects and Long-Term Restoration Projects—A Case Study. *Heritage* **2023**, *7*, 50–75. [CrossRef]
15. Garcia-Castillo, E.; Paya-Zaforteza, I.; Hospitaler, A. Fire in Heritage and Historic Buildings, a Major Challenge for the 21st Century. *Dev. Built Environ.* **2023**, *13*, 100102. [CrossRef]
16. Naziris, I.A.; Mitropoulou, C.C.; Lagaros, N.D. Innovative Computational Techniques for Multi-Criteria Decision Making, in the Context of Cultural Heritage Structures' Fire Protection: Theory. *Heritage* **2022**, *5*, 1719–1733. [CrossRef]
17. Bonazza, A.; Sardella, A. Climate Change and Cultural Heritage: Methods and Approaches for Damage and Risk Assessment Addressed to a Practical Application. *Heritage* **2023**, *6*, 3578–3589. [CrossRef]
18. Guibaud, A.; Mindeguia, J.-C.; Albuerne, A.; Parent, T.; Torero, J. Notre-Dame de Paris as a Validation Case to Improve Fire Safety Modelling in Historic Buildings. *J. Cult. Herit.* **2023**, *65*, 145–154. [CrossRef]
19. Taileb, A.; Dekkiche, H.; Sherzad, M.F. HBIM: A Tool for Enhancing the Diagnosis of Historical Buildings: The Case of St. George's Memorial Anglican Church, Oshawa. *Heritage* **2023**, *6*, 5848–5866. [CrossRef]
20. Gernay, T. Performance-Based Design for Structures in Fire: Advances, Challenges, and Perspectives. *Fire Saf. J.* **2023**, *142*, 104036. [CrossRef]

21. Caliendo, C.; Ciambelli, P.; Del Regno, R.; Meo, M.G.; Russo, P. Modelling and Numerical Simulation of Pedestrian Flow Evacuation from a Multi-Storey Historical Building in the Event of Fire Applying Safety Engineering Tools. *J. Cult. Herit.* **2020**, *41*, 188–199. [CrossRef]
22. Scorgie, D.; Feng, Z.; Paes, D.; Parisi, F.; Yiu, T.W.; Lovreglio, R. Virtual Reality for Safety Training: A Systematic Literature Review and Meta-Analysis. *Saf. Sci.* **2024**, *171*, 106372. [CrossRef]
23. Marrion, C.E. More Effectively Addressing Fire/Disaster Challenges to Protect Our Cultural Heritage. *J. Cult. Herit.* **2016**, *20*, 746–749. [CrossRef]
24. Bernardini, G. *Fire Safety of Historical Buildings. Traditional versus Innovative "Behavioural Design" Solutions by Using Wayfinding Systems*, 1st ed.; Springer International Publishing: Berlin/Heidelberg, Germany, 2017; ISBN 978-3-319-55744-1.
25. Campinho, M.; Sidani, A.; Couto, A. Tools for Fire Safety in Historic Buildings: Review. In *Occupational and Environmental Safety and Health V*; Studies in Systems, Decision and Control; Arezes, P.M., Melo, R.B., Carneiro, P., Castelo Branco, J., Colim, A., Costa, N., Costa, S., Duarte, J., Guedes, J.C., Perestrelo, G., et al., Eds.; Springer: Cham, Switzerland, 2024; Volume 492, pp. 753–770.
26. Hostetter, H.; Naser, M.Z. Characterizing Disability in Fire: A Progressive Review. *J. Build. Eng.* **2022**, *53*, 104573. [CrossRef]
27. Salazar, L.G.F.; Romão, X.; Paupério, E. Review of Vulnerability Indicators for Fire Risk Assessment in Cultural Heritage. *Int. J. Disaster Risk Reduct.* **2021**, *60*, 102286. [CrossRef]
28. Shi, L.; Xie, Q.; Cheng, X.; Chen, L.; Zhou, Y.; Zhang, R. Developing a Database for Emergency Evacuation Model. *Build. Environ.* **2009**, *44*, 1724–1729. [CrossRef]
29. Bosina, E.; Weidmann, U. Estimating Pedestrian Speed Using Aggregated Literature Data. *Phys. A Stat. Mech. Its Appl.* **2017**, *468*, 1–29. [CrossRef]
30. Tong, Y.; Bode, N.W.F. Simulation Investigation on Crowd Evacuation Strategies for Helping Vulnerable Pedestrians at Different Stages of Egress. *Int. J. Disaster Risk Reduct.* **2023**, *84*, 103479. [CrossRef]
31. Kasemsarn, K.; Sawadsri, A.; Harrison, D.; Nickpour, F. Museums for Older Adults and Mobility-Impaired People: Applying Inclusive Design Principles and Digital Storytelling Guidelines—A Review. *Heritage* **2024**, *7*, 1893–1916. [CrossRef]
32. Tinaburri, A. Principles for Monte Carlo Agent-Based Evacuation Simulations Including Occupants Who Need Assistance. From RSET to RiSET. *Fire Saf. J.* **2022**, *127*, 103510. [CrossRef]
33. Abir, I.M.; Ibrahim, A.M.; Toha, S.F.; Shafie, A.A. A Review on the Hospital Evacuation Simulation Models. *Int. J. Disaster Risk Reduct.* **2022**, *77*, 103083. [CrossRef]
34. Fu, L.; Qin, H.; He, Y.; Shi, Y. Application of the Social Force Modelling Method to Evacuation Dynamics Involving Pedestrians with Disabilities. *Appl. Math. Comput.* **2024**, *460*, 128297. [CrossRef]
35. Kuligowski, E.D. Computer Evacuation Models for Buildings. In *SFPE Handbook of Fire Protection Engineering*; Springer: New York, NY, USA, 2016; pp. 2152–2180.
36. Ministero dell'Interno. *DM 03/08/2015: Codice Di Prevenzione Incendi (Testo Coordinato Dell'allegato I Del DM 3 Agosto 2015 e Ss.Mm.Ii)*; Ministero dell'Interno: Rome, Italy, 2015.
37. Tang, S.; Shelden, D.R.; Eastman, C.M.; Pishdad-Bozorgi, P.; Gao, X. A Review of Building Information Modeling (BIM) and the Internet of Things (IoT) Devices Integration: Present Status and Future Trends. *Autom. Constr.* **2019**, *101*, 127–139. [CrossRef]
38. Huang, Y.; Guo, Z.; Chu, H.; Sengupta, R. Evacuation Simulation Implemented by ABM-BIM of Unity in Students' Dormitory Based on Delay Time. *ISPRS Int. J. Geo-Inf.* **2023**, *12*, 160. [CrossRef]
39. Ronchi, E. Developing and Validating Evacuation Models for Fire Safety Engineering. *Fire Saf. J.* **2021**, *120*, 103020. [CrossRef]
40. Quagliarini, E.; Bernardini, G.; Romano, G.; D'Orazio, M. Simplified Flood Evacuation Simulation in Outdoor Built Environments. Preliminary Comparison between Setup-Based Generic Software and Custom Simulator. *Sustain. Cities Soc.* **2022**, *81*, 103848. [CrossRef]
41. Ronchi, E.; Kuligowski, E.D.; Reneke, P.A.; Peacock, R.D.; Nilsson, D. *The Process of Verification and Validation of Building Fire Evacuation Models*; NIST Technical Note; National Institute of Standards and Technology: Gaithersburg, MD, USA, 2013; Volume 1822.
42. Grandison, A.; Deere, S.; Lawrence, P.; Galea, E.R. The Use of Confidence Intervals to Determine Convergence of the Total Evacuation Time for Stochastic Evacuation Models. *Ocean. Eng.* **2017**, *146*, 234–245. [CrossRef]
43. Haghani, M.; Sarvi, M. Crowd Behaviour and Motion: Empirical Methods. *Transp. Res. Part B Methodol.* **2018**, *107*, 253–294. [CrossRef]
44. Galea, E.R.; Xie, H.; Deere, S.; Cooney, D.; Filippidis, L. Evaluating the Effectiveness of an Improved Active Dynamic Signage System Using Full Scale Evacuation Trials. *Fire Saf. J.* **2017**, *91*, 908–917. [CrossRef]
45. Hostetter, H.; Naser, M.Z.; Randall, K.; Murray-Tuite, P. Evacuation Preparedness and Intellectual Disability: Insights from a University Fire Drill. *J. Build. Eng.* **2024**, *84*, 108578. [CrossRef]
46. MassMotion. MassMotion Guide. Available online: https://www.oasys-software.com/wp-content/uploads/2017/12/MassMotion.pdf (accessed on 21 April 2024).
47. Comune di Ancona; Pinacoteca, F.P.; Galleria d'Arte Moderna. *Il Lazzaretto di Luigi Vanvitelli: Indagine su Un'opera*; Galleria comunale d'Arte Moderna: Rome, Italy, 1980; ISBN ANA0001260.
48. VV.AA. *Il Lazzaretto Tra Mare e Città: Dalla Conoscenza al Restauro Verso Il Riuso della Mole di Luigi Vanvitelli. Atti del Convegno: Ancona, 9–10 Marzo 1990*; Cassa di risparmio di Verona, Vicenza, Belluno e Ancona: Ancona, Italy, 1990.

49. Ministero dell'Interno. *Testo Coordinato del DM 10 Marzo 1998 Criteri Generali di Sicurezza Antincendio e per la Gestione Dell'emergenza nei Luoghi di Lavoro*; Ministero dell'Interno: Rome, Italy, 2021.
50. Helbing, D.; Farkas, I.; Vicsek, T. Simulating Dynamical Features of Escape Panic. *Nature* **2000**, *407*, 487–490. [CrossRef] [PubMed]
51. Forssberg, M.; Kjellström, J.; Frantzich, H.; Mossberg, A.; Nilsson, D. The Variation of Pre-Movement Time in Building Evacuation. *Fire Technol.* **2019**, *55*, 2491–2513. [CrossRef]
52. Rostami, R.; Alaghmandan, M. Performance-Based Design in Emergency Evacuation: From Maneuver to Simulation in School Design. *J. Build. Eng.* **2021**, *33*, 101598. [CrossRef]
53. Huang, L.; Li, W.; Gong, J. Simulation of the Emergency Evacuation about Social Groups in a Complex Subway Station. *Phys. A Stat. Mech. Its Appl.* **2024**, *637*, 129535. [CrossRef]
54. Ronchi, E.; Lovreglio, R. EXPOSED: An Occupant Exposure Model for Confined Spaces to Retrofit Crowd Models during a Pandemic. *Saf. Sci.* **2020**, *130*, 104834. [CrossRef]

Disclaimer/Publisher's Note: The statements, opinions and data contained in all publications are solely those of the individual author(s) and contributor(s) and not of MDPI and/or the editor(s). MDPI and/or the editor(s) disclaim responsibility for any injury to people or property resulting from any ideas, methods, instructions or products referred to in the content.

heritage

MDPI

Article

Deep Learning Enhanced Multisensor Data Fusion for Building Assessment Using Multispectral Voxels and Self-Organizing Maps

Javier Raimundo [1,*], Serafin Lopez-Cuervo Medina [1], Julian Aguirre de Mata [1], Tomás Ramón Herrero-Tejedor [2] and Enrique Priego-de-los-Santos [3]

1 Departamento de Ingeniería Topográfica y Cartográfica, Escuela Técnica Superior de Ingenieros en Topografía, Geodesia y Cartografía, Universidad Politécnica de Madrid, Campus Sur, A-3, Km 7, 28031 Madrid, Spain; s.lopezc@upm.es (S.L.-C.M.); julian.aguirre@upm.es (J.A.d.M.)
2 Departamento de Ingeniería Agroforestal, Universidad Politécnica de Madrid, Campus Ciudad Universitaria, Av. Puerta de Hierro, nº 2–4, 28040 Madrid, Spain; tomas.herrero.tejedor@upm.es
3 Department of Cartographic Engineering, Geodesy and Photogrammetry, Universitat Politècnica de València, Camí de Vera s/n, 46022 Valencia, Spain; epriego@cgf.upv.es
* Correspondence: ajraimundo@alumnos.upm.es

Abstract: Efforts in the domain of building studies involve the use of a diverse array of geomatic sensors, some providing invaluable information in the form of three-dimensional point clouds and associated registered properties. However, managing the vast amounts of data generated by these sensors presents significant challenges. To ensure the effective use of multisensor data in the context of cultural heritage preservation, it is imperative that multisensor data fusion methods be designed in such a way as to facilitate informed decision-making by curators and stakeholders. We propose a novel approach to multisensor data fusion using multispectral voxels, which enable the application of deep learning algorithms as the self-organizing maps to identify and exploit the relationships between the different sensor data. Our results indicate that this approach provides a comprehensive view of the building structure and its potential pathologies, and holds great promise for revolutionizing the study of historical buildings and their potential applications in the field of cultural heritage preservation.

Keywords: multisensor; data fusion; voxel; multispectral; building; point cloud; cultural heritage

Citation: Raimundo, J.; Medina, S.L.-C.; Mata, J.A.d.; Herrero-Tejedor, T.R.; Priego-de-los-Santos, E. Deep Learning Enhanced Multisensor Data Fusion for Building Assessment Using Multispectral Voxels and Self-Organizing Maps. *Heritage* **2024**, 7, 1043–1073. https://doi.org/10.3390/heritage7020051

Academic Editors: Daniela Fico and Daniela Rizzo

Received: 2 January 2024
Revised: 1 February 2024
Accepted: 2 February 2024
Published: 17 February 2024

Copyright: © 2024 by the authors. Licensee MDPI, Basel, Switzerland. This article is an open access article distributed under the terms and conditions of the Creative Commons Attribution (CC BY) license (https://creativecommons.org/licenses/by/4.0/).

1. Introduction

Picture a historical building as an enigmatic puzzle, each crack, deformation, and hidden corrosion forming the cryptic pieces waiting to be deciphered. As we stand at the intersection of technology and architectural history, our pursuit is to decode this intricate puzzle. Multisensor data fusion emerges as our toolkit, seamlessly blending the precision of laser scanners with the spectral insight of photographic cameras. Here, we introduce a revolutionary approach, not just to study buildings, but to unravel the mysteries concealed within their weathered walls.

The state-of-the-art in multisensor data fusion in the study of building pathologies has undergone significant advances in the last decades. The convergence of various technologies and sensors has enabled researchers to assess structural conditions and pathologies in buildings more fully and accurately [1,2].

Studies on potential issues in historical buildings employing geomatic sensor technology have garnered diverse attention in late years [3,4]. First approaches were dedicated only to generate three-dimensional models with different levels of detail. These three-dimensional models were complemented by spectral information so that visual representations could be made to highlight areas that could present some pathology [5,6].

Various studies have focused on different architectural elements and their specific pathologies. For example, in the study of façades [7–9] or concrete structural elements [10–12].

The data obtained were sometimes integrated into Building Information Models (BIM) for further study [13,14].

In turn, different research works have concentrated on the use of singular sensors. There are research works focused on the use of infrared thermal cameras [15–19], multispectral photographic cameras [20,21] and laser scanners [22–24].

The combined use of active and passive sensors such as terrestrial laser scanners and photographic cameras with different spectral sensitivities has provided a wealth of information. This allows the generation of detailed three-dimensional models, point clouds with spectral information, and visual representations highlighting the affected areas. This combination of different sensor types provides a complete view of the structure and its potential pathologies. Terrestrial laser scanners capture highly detailed three-dimensional point clouds, thereby facilitating the detection of structural deformations and cracks [25,26]. Photographic cameras provide valuable spectral information that can indicate signs of corrosion, moisture, or other problems that can not be visible to the naked eye.

The use of 3D point clouds requires the management of an immense volume of data, which requires substantial data-processing infrastructure for handling and visualization. Additionally, 3D point clouds are inherently unstructured, making the task of locating specific points within these clouds nontrivial [27]. One consequence of this lack of structure is that point clouds do not contain topological information, thereby precluding an understanding of neighborhood relationships among points [28].

Point clouds also exhibit non-uniform densities and distributions. The point density achieved is not always optimal for the intended applications. Some areas will display redundant information, particularly in the overlapping zones between scans. Conversely, other areas will present lower point densities. This results in heterogeneity in both the density and accuracy of points within the cloud [29].

In terms of analysis, multisensor data fusion has enabled the early detection and monitoring of structural pathologies, such as cracks, deformations, and corrosion. In addition, the integration of information from airborne and ground sensors has enabled more complete coverage of structures, both in terms of physical access and data diversity. Traditionally, the assessment of structural problems involves visual inspections, manual measurements, and, at best, the use of a single sensor to capture limited data [30]. Modern data fusion systems allow the creation of accurate three-dimensional digital models that can be compared to the original designs to identify discrepancies. This is especially valuable for assessing the performance of historic structures and monuments where preservation is critical [31].

The integration of various sensors and techniques also leads to the creation of diverse point cloud datasets, which in turn contributes to the formation of large data repositories [29]. This not only affects processing speed but also requires the conversion of substantial volumes of point data into dependable and actionable insights. Traditional methods for customizing point clouds for specific applications are becoming increasingly time consuming and require manual intervention. The growing complexity and volume of data, often spread across multiple stakeholders or platforms, pose a challenge to human expertise in effective data management [32]. To facilitate more effective decision making, it is crucial to efficiently convert voluminous point cloud data into streamlined processes, thereby heralding a new age of decision-support services. Strategies must be developed for extensive automation and structuring to eliminate the need for task-specific manual processing and to promote sustainable collaboration.

Voxels have been used in a variety of distinct fields, including geology [33], forest inventory [34], and medical research [35]. In the context of their application to point cloud management, multiple studies [36–40] have substantiated the efficacy of voxels as an appropriate tool for handling point cloud data.

Voxelization is more beneficial for managing raw 3D point clouds. With the points being placed in a regular grid pattern, it is now possible to structure the point cloud in a tree format that allows for significantly reduced computing time. Logical inference is also possible from a voxel structure because the known relationship with neighboring points allows for semantic reasoning [41].

In voxelization of point clouds, one defining parameter of the voxel structure is the elemental voxel size. This parameter governs the resolution of the phenomena under study through the data structure, impacting various applications, such as finite elements [42], structural analyses [43,44], and dynamic phenomena [45]. For instance, if a study or simulation requires a resolution of 5 cm, this would dictate the voxel elemental size established during the voxelization process. In addition, voxel size determines the extent of element reduction relative to the number of points present in the original clouds [29].

The application of deep learning algorithms to voxel structures is an active research area across various fields, such as computer vision, robotics, geography, and medicine. Within the spectrum of deep learning approaches applied to voxels, some notable methodologies as 3D Convolutional Neural Networks (3D CNNs) [46,47], Voxel-based Autoencoders [48,49], Generative Networks for Voxels [50–52], and Semantic Segmentation [28,53] have been used.

However, despite these advances, challenges still exist in terms of accurate calibration, data correction, and effective integration of the collected information. Additionally, the interpretation of merged data requires expertise in remote sensing and structural analysis to avoid false diagnoses. Accurate integration of data from different sources requires careful calibration and correction [54]. In addition, proper interpretation of results remains crucial, as the presence of detailed data does not always guarantee a complete understanding of the underlying causes of pathologies.

In this paper, we present a novel approach to multisensor data fusion using multispectral voxels. This data fusion allows for optimal and efficient management of the sensors information, allowing later applications of deep learning algorithms. With this, a workflow is determined that allows the study of buildings as well as the possible pathologies that could be present on them.

This study is organized as follows. Initially, the sensors that were used in a data acquisition campaign on a prominent Spanish Cultural Heritage building are described. Subsequently, the methodology that was designed for handling, processing, and fusing the data using our concept of multispectral voxels is outlined. The deep learning Self-organizing map algorithm will be applied to our multispectral voxel structure. Finally, we discuss the results.

2. Materials and Methods

2.1. Sensors

Within the context of multisensor fusion, a data-capture campaign with a range of different sensors featuring distinct spectral sensitivities and types were chosen. To this end, active and passive sensors were used. The set of active sensors comprises terrestrial laser scanners, whereas the passive sensor group encompasses a variety of photographic cameras deployed in both terrestrial and aerial arrangements using Unmanned Aerial Vehicles (UAVs). Each camera had distinct spectral sensitivities.

This meticulous selection of sensors encompasses various conventional modalities of geospatial data capture, which are commonly employed in the analysis of buildings and architectural structures. The acquired data were subjected to a series of processing techniques, with photogrammetry as one of the principal methodologies. Consequently, the final output consisted of multiple point cloud datasets, each characterized by unique and significant spectral properties.

While developing the data capture strategy, it was acknowledged that not all sensors were capable of capturing data encompassing the entire built structure. Sensors located on the ground can only provide information about the lower sections of the building, while sensors attached to aerial vehicles can capture data covering the entire architectural unit. This distinction is essential for accurate interpretation and cohesive integration of the collected data during the analytical process.

The employed sensors are described herein, including photographic cameras, unmanned aerial sensors (UAVs), and laser scanners:

- The terrestrial digital imaging device deployed was a Sony NEX-7 photographic camera (Sony Corporation, Tokyo, Japan), a model equipped with a 19 mm optical lens of fixed length. This particular sensor captures spectral components within the red (590 nm), green (520 nm), and blue (460 nm) bands of the visible spectrum. The spectral response of this camera is illustrated in Figure 1a, and its specific parameters are outlined in Table 1.

Table 1. Sony NEX 7 camera sensor parameters.

Parameter	Value
Sensor	APS-C type CMOS sensor
Focal lenght (mm)	19
Sensor width (mm)	23.5
Sensor lenght (mm)	15.6
Effective pixels (megapixels)	24.3
Pixel size (micrometers)	3.92
ISO sensitivity range	100–1600
Image format	RAW (Sony ARW 2.3 format)
Weight (g)	350

- A modified version of the digital image camera, the Sony NEX-5N model, was also employed in terrestrial positions. It was outfitted with a 16 mm fixed-length lens, accompanied by a near-infrared (NIR) filter and an ultraviolet (UV) filter in different shot sessions. Detailed information on this sensor is provided in Table 2.

Table 2. Sony NEX 5N camera sensor parameters.

Parameter	Value
Sensor	APS-C type CMOS sensor
Focal lenght (mm)	16
Sensor width (mm)	23.5
Sensor lenght (mm)	15.6
Effective pixels (megapixels)	16.7
Pixel size (micrometers)	4.82
ISO sensitivity range	100–3200
Image format	RAW (Sony ARW 2.2 format)
Weight (g)	269

By removing the internal infrared filter from the Sony NEX-5N camera, the sensitivity of the sensor was enhanced to encompass specific regions of the electromagnetic spectrum, including the near-infrared (820 nm) and ultraviolet (390 nm) wavelengths. The spectral responsiveness of this modified camera is shown in Figure 1b. Notably, the modified sensor exhibits sensitivity to distinct segments of the electromagnetic spectrum in contrast to its unmodified counterpart, the Sony NEX-7, which retains the internal infrared filter (as illustrated in Figure 1a) [55]. This modified camera was equipped with a collection of filters to obtain data corresponding to ultraviolet (UV) and near-infrared (NIR) spectral bands. The transmission curves of the filters employed during the data acquisition process are shown in Figures 2 and 3.

- The terrestrial laser scanner employed in this study was the Faro Focus S350 laser scanner, developed by Faro Technologies (Lake Mary, FL, USA). It has proven to be an invaluable tool for archaeological and building studies applications. Its cutting-edge phase-based laser scanning technology enables it to accurately measure distances and efficiently gather an extensive dataset of millions of data points in a short span. Particularly relevant in archaeological studies, this scanner aids in documenting historical sites and structures with high precision. The Faro Focus S350 laser beam wavelength is 1550 nm (Table 3). This allowed us to obtain building spectral information in the short-wavelength infrared spectral (SWIR) band [56]. Concerning the use of data derived from the laser scanner, the registered return signal intensity value has been used as its application has been demonstrated to be effective in buildings made of stone [22].

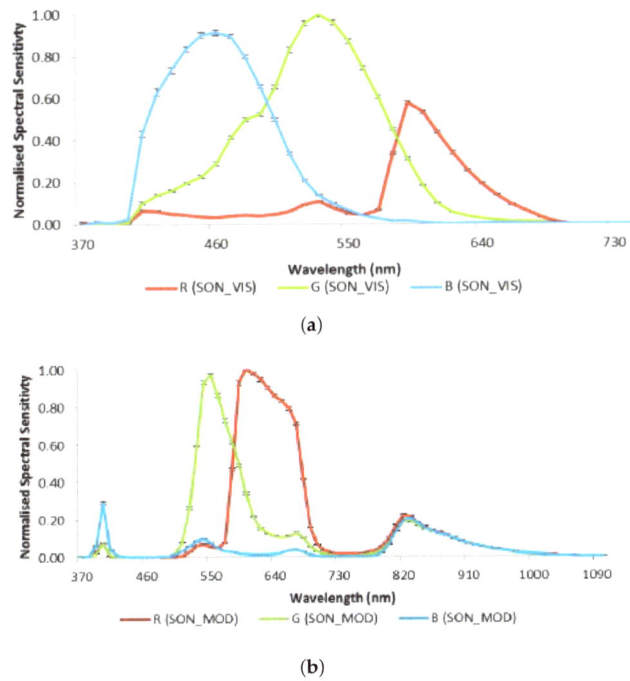

Figure 1. Spectral responses for Sony NEX cameras, from [55]: (**a**) Unmodified camera, normalized to the peak of the green channel. (**b**) Modified camera, normalized to the peak of the red channel.

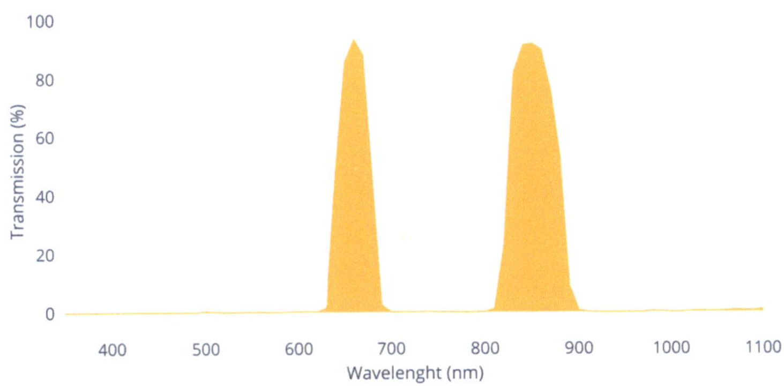

Figure 2. Midopt DB 660/850 Dual Bandpass filter light transmission curve (Midwest Optical Systems, Inc., Palatine, IL, USA).

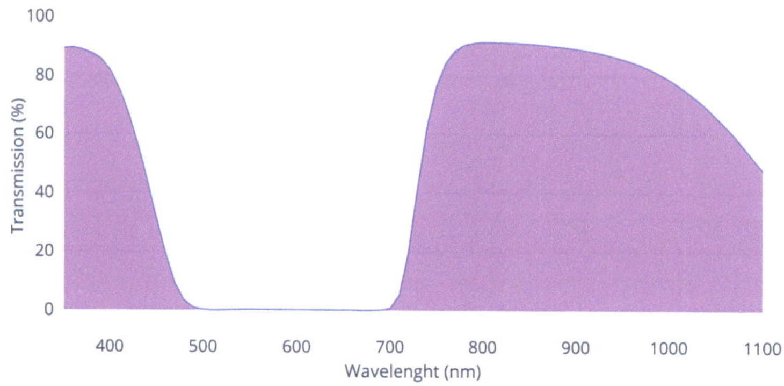

Figure 3. ZB2 filter light transmission curve, from Shijiazhuang Tangsinuo Optoelectronic Technology Co., Ltd., Shijiazhuang, Hebei, China.

- For data collection in the upper regions of the building, where ground-based sensors lack access, an Unmanned Aerial Vehicle (UAV) was deployed. The chosen instrument was the Parrot Anafi Thermal (Table 4), which mounts a dual-camera system comprising an RGB sensor (for the visible spectrum) and an infrared thermal sensor. This configuration facilitates the simultaneous capture of conventional RGB and infrared thermal images within a single flight mission. Referring to the thermal infrared data, the Parrot Anafi Thermal incorporates a factory-calibrated uncooled microbolometer-type FLIR Lepton 3.5 thermal module [57], enabling the establishment of an absolute temperature for each image pixel. Consequently, precise surface temperatures of the building points were determined. This aids in discerning the discontinuities and potential pathologies resulting from the heterogeneity of construction materials [19].

Table 3. Faro Focus S 350 Laser Scanner sensor parameters.

Parameter	Value
Range	0.6–350 m
Angular FOV (vertical)	300°
Angular FOV (horizontal)	360°
Beam Divergence	0.3 mrad
Beam Diameter at Exit	2.12 mm
Laser Wavelength	1550 nm

Table 4. Parrot Anafi Thermal sensor parameters.

Parameter	Value
Thermal sensor spectral range	8 µm–14 µm
Thermal sensor size	160 × 120 pixels
Pixel pitch	12 µm
Thermal Sensitivity	0.050 °C

2.2. *Data Capture Campaign*

For data acquisition, an emblematic building of Spanish historical heritage was selected. The chosen structure was the Visigothic Church of the Santa Maria de Melque. Data from all the described sensors were directly collected on this 7th-century A.D. building [58], located in the province of Toledo, Spain. The archaeological complex of Santa Maria de Melque (N 39.750878°, W 4.372965°) is located approximately 30 km southwest of the city of Toledo, in close proximity to the Tagus River [59] (Figure 4).

The structure of the Visigothic church of Santa Maria de Melque was constructed using masonry of immense granite blocks assembled without mortar, distinguished by its barrel vault covering the central nave. The layout of the aisles in the form of a Greek cross, the straight apse, and the arrangement of architectural elements reveal both Roman and Byzantine influences, reflecting the rich cultural diversity of the period [60].

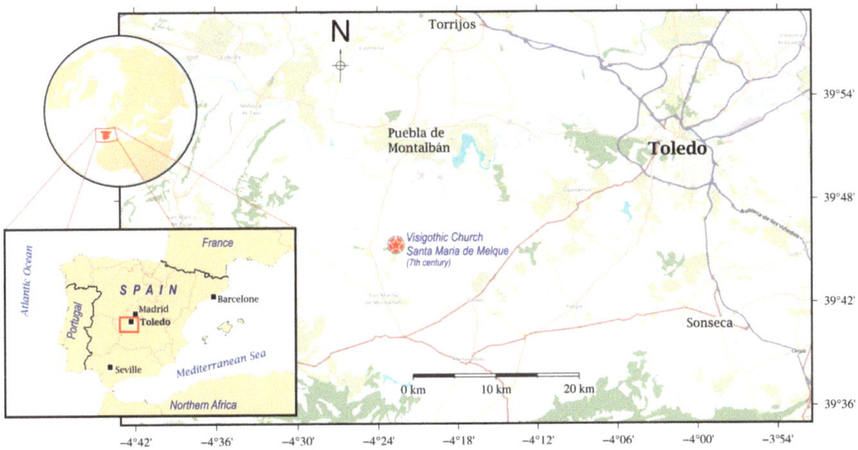

Figure 4. Location of the building surveyed.

The data collection campaign was carried out only on the exterior of this building in February 2022. The campaign covered the entire architectural structure and its ornamental elements. This location was chosen for its historical and architectural relevance, which allowed us to obtain precise and detailed information on the current state of the building and its possible pathologies. Figure 5 shows the studied building.

Figure 5. Visigothic Church of Santa Maria de Melque, seen from the southeast.

Prior to capturing the dataset, a meticulous preparatory phase was undertaken, involving the careful placement of a series of precision targets across the walls surface. These identified markers collectively formed a robust set of control points that served as pivotal anchors for the subsequent data collection process. By ensuring uniformity and accuracy at these measurement points, a reliable common geometric reference system was established, effectively standardizing the subsequent data acquisition process across all data collection sessions. This rigorous methodology not only underscored the accuracy of the data collected but also facilitated seamless comparability and analysis of the collected visual information (Figure 6).

Figure 6. Target signals attached to walls and on terrain.

The Terrain Target Signals (TTS) were measured using GNSS techniques. These marks were materialized by means of a metal nail in the ground, in order to be revisited on future occasions. Subsequently, a flat black-and-white square sign measuring 30 cm on each side was adopted (Figure 6).

For the observation, dual-frequency Topcon Hiper GNSS receivers were used, with calibrated antennas (GPS, GLONASS), tripod and centering system on the point mark, measuring a total of 12 TTS, forming a network of 30 vectors. The observation time at each point was 15–20 min, with at least a double observation session at each point with different receivers, thus configured in relative static surveying with high repeatability.

In the subsequent processing of GNSS data, to obtain greater precision in the results, the precise geodetic correction models of the ionosphere from the CODE (Center for Orbit Determination in Europe) and precise ephemeris from the IGS (International GNSS Service) for both constellations were downloaded. Along with these Topcon field GNSS data, data from continuous CORS (Continuously Operating Reference Stations) of the Spanish National Positioning System ERGNSS-IGN (Instituto Geográfico Nacional) were also processed. This was done in order to link the local GNSS measurements to a geodetic reference frame that guaranteed high precision stability and temporal permanence for future actions in the environment.

To compute the GNSS vectors, the Leica Infinity version 3.0.1 software was used, using the absolute antenna calibration models and using the VMF (Vienna Mapping Functions) [61]. Both for the observation and for the adjustment and compensation of the coordinates of all the TTS points that make up the network, the same methodology used in the implementation of geodetic precision control networks in engineering was followed [62]. The final adjustment of the network, the calculation of the coordinates and the estimation of its accuracy (Table 5) was performed with Geolab version px5 software, with a complete constraint in the ETRS89 (ETRF2000) frame.

Wall Surface Signals (WSS) were placed around the entire building and attached to the wall. They are intended to tie points supporting the georeferencing process in all point clouds. WSS were 12 bit coded precision targets (Figure 7). Thus, each WSS can be clearly identified by photogrammetry software, almost in automatic mode, and operators.

In Figure 8, we show the distribution of targets on terrain (TTS) and on the walls (WSS). The coordinates of the precision targets attached to the walls (Wall Surface Signals) are expressed in Table A1.

Figure 7. Example of four 12 bit coded precision targets used as Wall Surface Signal (WSS).

Figure 8. Distribution of control points.

Table 5. Terrain Target Signals (TTS) coordinates and their standard deviations. UTM 30N coordinate system, ETRS89 (ETRF2000) geodetic frame. All values in meters.

Station	UTM 30 Coordinates		Height	Standard Deviations		
	Northing	Easting		North	East	Height
D001	4401021.383	382387.421	567.625	0.004	0.004	0.009
D002	4401007.299	382397.872	566.853	0.002	0.002	0.004
D003	4401011.239	382391.663	566.584	0.021	0.015	0.054

Table 5. *Cont.*

Station	UTM 30 Coordinates		Height	Standard Deviations		
	Northing	Easting		North	East	Height
D004	4400996.864	382365.065	568.053	0.003	0.002	0.006
D005	4401011.720	382359.282	568.848	0.003	0.003	0.007
D006	4401021.414	382377.979	565.803	0.014	0.005	0.017
D007	4401014.232	382362.130	567.291	0.007	0.005	0.012
D008	4401021.499	382383.108	565.844	0.010	0.005	0.014
D009	4400998.691	382375.501	567.076	0.007	0.005	0.013
D010	4400990.703	382374.625	568.205	0.002	0.002	0.005
D011	4401024.085	382363.273	567.641	0.005	0.004	0.009
D012	4401028.664	382375.382	566.752	0.006	0.004	0.011

2.3. Point Clouds

In the course of this research, a range of sensors were used for the collection of geomatic data across multiple spectral bands. The processing of each dataset has been carried out with great care, resulting in the generation of several georreferenced three-dimensional point clouds. These point clouds offer a comprehensive and diverse view of the building under examination from various angles (as presented in Table 6). Each point cloud, except the laser scanner point cloud, comes from a photogrammetric process. The photogrammetric software used was Agisoft Metashape version 1.8.5. All derived photogrammetry point clouds are the result of processing a single sensor in a single process. Only in RGB point cloud, unmodified terrestrial RGB camera and UAV RGB image datasets were combined in the same process to obtain the RGB point cloud.

Figure 9 illustrates a subset of the point clouds views resulting from the use of the aforementioned sensors. First, Figure 9a presents a visualization of the point cloud obtained via RGB sensors using both conventional terrestrial and aerial (UAV) cameras. This representation unveils architectural details as well as visible features perceptible to the naked eye, furnishing a meticulous view of the surface of the structure. This point cloud has been selected as the starting point for the development of the voxelized data structure of the building in question, which will be subjected to analysis in subsequent stages.

Table 6. Point Clouds.

Band	Number of Points
Red, Green and Blue (visible spectrum)	46,512,981
Near infrared	39,472,866
Ultraviolet	352,900,707
Short-Wavelength Infrared (Laser Scanner)	54,735,907
Thermal infrared	6,412,099

Figure 9b shows the point cloud corresponding to the UV spectral band. This view allowed us to identify elements that are typically imperceptible within the visible range. In this Figure 9b we also can see that not the entire building has point cloud information in this band.

Finally, in Figure 9c,d, we depict the point cloud generated by our infrared thermal sensor. This visual representations highlight the heat distribution on the building surface, endowing us with an exceptional perspective concerning potential issues related to heat retention and the variability of construction materials.

Beyond the previously outlined difficulties concerning data structure, heterogeneity, and inconsistent density, point clouds generated by multiple sensors emanate from disparate data origins and convey unique informational facets. Consequently, it becomes essential to formulate strategies for the fusion of both geometric and spectral elements with the aim of optimizing data processing and extracting valuable conclusions from the varied information contained within.

(**a**) RGB point cloud (south façade).

(**b**) Ultraviolet band point cloud (south façade).

(**c**) Infrared thermal point cloud (south façade).

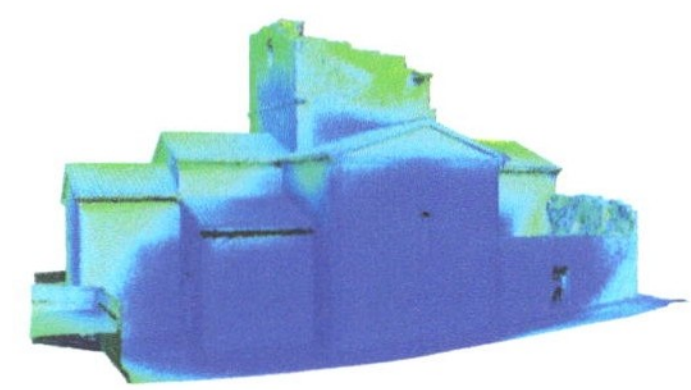

(**d**) Infrared thermal point cloud (north façade).

(**e**) Laser Scanner point cloud.

Figure 9. Different point clouds obtained from data adquisition campaing.

2.4. Voxelization

To facilitate an optimal analysis, it is necessary to fuse the numerous point clouds that have been generated. To accomplish this, we have employed voxels as the data structure, specifically the multispectral voxel variant developed in our previous research [29].

Voxels, abbreviated from "volumetric elements", serve as the fundamental abstract units in three-dimensional space, each having predetermined volume, positional coordinates, and attributes [63]. They offer significant utility in providing topologically explicit representations of point clouds, thereby augmenting the informational content.

In this study, we examined how the selected elemental voxel size affects the point distribution within the point clouds. Voxel structures with various elemental sizes (50, 25, 10, 5, and 3 cm) were created. To manage the point clouds and their voxelization, we used the Open3d open-source library [64]. We used the RGB point cloud as the initial dataset to create voxel structures, given its comprehensive coverage of the entire building. After establishing voxel structures based on these sizes, we located the points from each respective cloud enclosed within individual voxels. In instances where one specific voxel enclosed multiple points from a particular spectral band, the voxel's spectral property was set to the mean value of those enclosed points. Additional statistical metrics, such as the maximum, minimum, and variance values, were also calculated. Given that the voxel adopts the spectral attributes of enveloped points, it effectively becomes a multispectral voxel [29]. The concept of a multispectral voxel used in this work is illustrated in Figure 10.

Figure 11 illustrates how point clouds are distributed along the 5 cm elemental voxel structure. For other chosen voxel elemental sizes, please refer to Appendix A, where we provide the respective distributions. The histograms in Figure 11 show the number of points per voxel in each spectral band.

We have also carried out an analysis of the influence of the size of the elementary voxel with respect to the number of points contained. We can observe their distribution in the form of histograms in Figure 12. We note that the distribution of the spectral bands contained in the multispectral voxels does not depend specifically on the elemental voxel size, as the distributions outlines are similar.

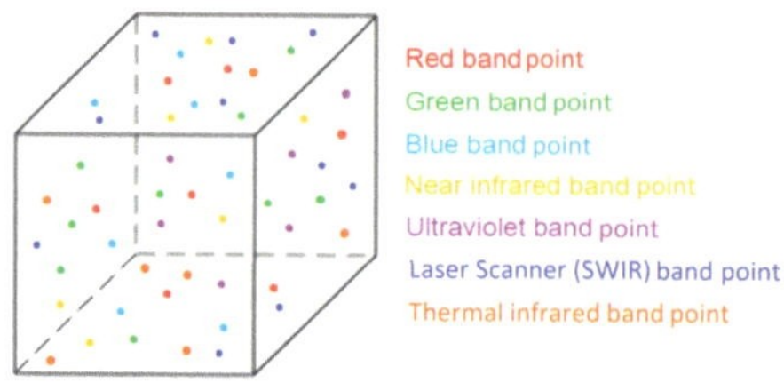

Figure 10. Multispectral voxel concept.

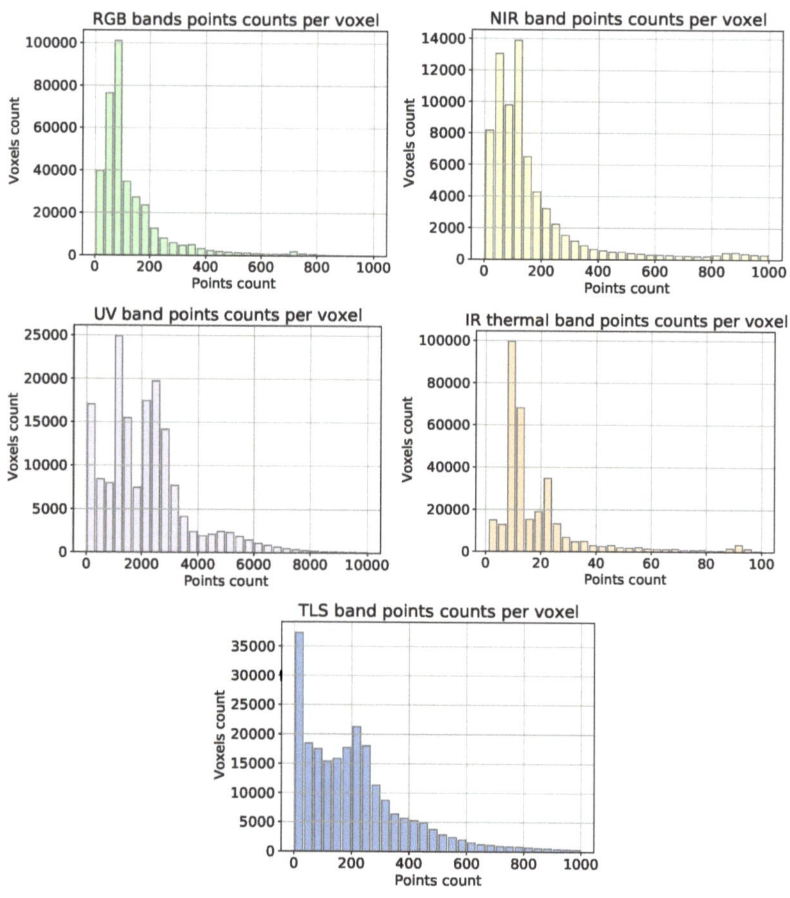

Figure 11. Number of Points per 5 cm Voxel.

Once the structure of multispectral voxels is established, it is crucial to process this data to gain insights and draw conclusions about the target building under study. In this research work, we have opted for the implementation of Self-Organizing Maps (SOMs) algorithms, due to their multiple advantages in handling voxel data. SOMs provide an efficient summarization of data, facilitating understanding, especially when dealing with large datasets, thus excelling in dimensionality reduction tasks. They do not require a separate training phase and are adept at exploiting spatial attributes while maintaining local neighbourhood relationships. With a design that is more intuitive than other neural network types, SOMs ease the interpretation of generated outcomes, making them particularly suitable for tackling categorization problems.

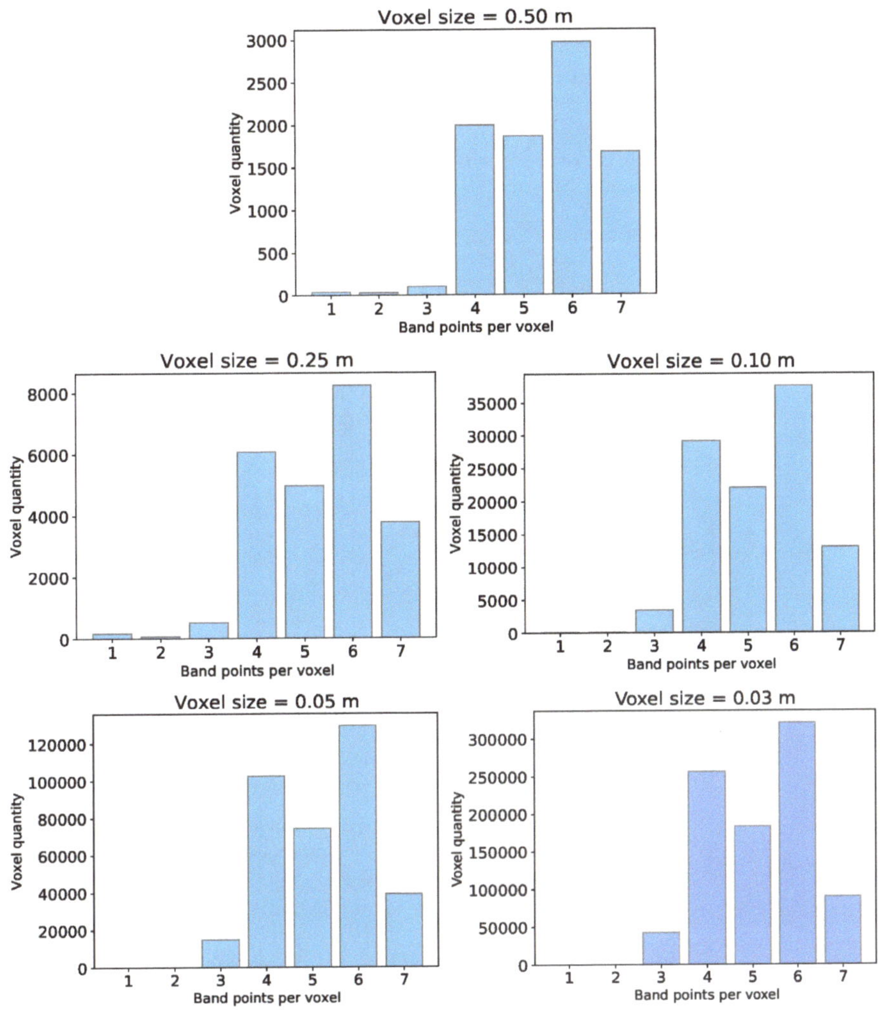

Figure 12. Distribution of various spectral bands according to different voxel elemental sizes.

2.5. *Self-Organizing Maps*

The Self-Organizing Map (SOM) executes a transformation from an input space of higher dimensionality to a map space of lower dimensionality, using a two-layered, fully interconnected neural architecture. The input layer comprises a linear array of neurons and the elementary units of an Artificial Neural Network (ANN), and the number of these neurons corresponds to the dimensionality of the input data vector (n). The output layer, also known as the Kohonen layer, is composed of neurons, each possessing a weight vector that matches the dimensionality of the input data (n). These neurons are organized within a rectangular grid of arbitrary dimensions (k). The weight vectors are collectively represented in a weight matrix configured as an $n \times k \times k$ array [65]. Figure 13 illustrates the typical Kohonen map architecture.

Among the various features provided by the use of SOMs are the following:

- Group together similar items and separate dissimilar items.
- Classify new data items using the known classes and groups.
- Find unusual co-ocurring associations of attribute values among items.
- Predict a numeric attribute value.
- Identify linkages between data items based on features shared in common.
- Organize information based on relationships among key data descriptors.

It has been proven that SOMs need complete data records [66]. SOMs are highly sensitive to Nan values and empty fields in the input layer. For this purpose, only full voxels, that is, those with at least one point contained in each of the point clouds to be fused, were classified in our training.

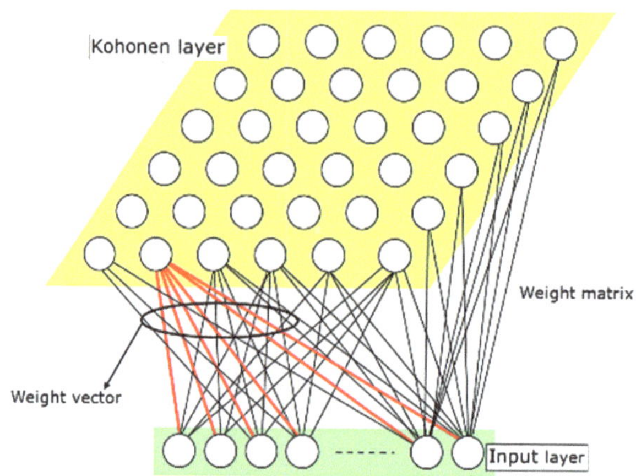

Figure 13. Typical Kohonen map architecture (source: [65]).

SOM Quality Indices

The various parameters that define a SOM can lead to different neural spaces. SOM quality measures are required to optimize training, such that meaningful conclusions can be drawn from them. Among these, the most significant quality indices are:

- Quantization error is the average error made by projecting data on the SOM, as measured by euclidean distance, i.e., the mean euclidean distance between a data sample and its best-matching unit [67]. Best value of quantization error is zero.
- Topographic product (TP) [68] measures the preservation of neighborhood relations between input space and the map. It depends only on the prototype vectors and map topology, and can indicate whether the dimension of the map is appropriate for fitting the dataset, or if it introduces neighborhood violations, induced by foldings of the map [67]. Topographic product will be <0 if map size is small and >0 if map size is big. Best value will be the one with the lower absolute value.

In this work, SOM quality indices have been calculated using an open-source library, SOMperf [67].

To determine the optimal neuron map size (k), Topographic Product (TP) [68] has been used. An approximation corresponding to a rule of thumb, according to Equation (1):

$$k = 5\sqrt{N}, \tag{1}$$

being N the number of full voxels in the voxelized structure.

In Figure 14 we summarize in a flowchart the handling of the processed data during our multisensor data fusion strategy.

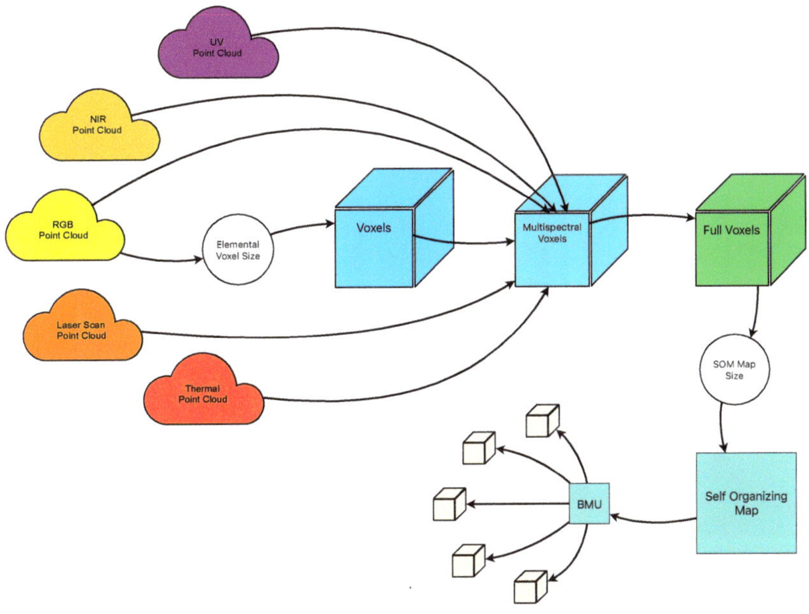

Figure 14. Multisensor data fusion with multispectral voxels and SOM flowchart.

3. Results

SOM trainings was performed for each voxel size. For this purpose, the optimal map size was determined such that the topographic product was minimum (in absolute value). Table 7 lists the different parameters according to the voxel size, as well as the obtained topographic product and quantization error.

Table 7. SOM map sizes.

Voxel Size (m)	Num. Voxels	Num. Full Voxels	% Full Voxels	Map Size by Equation (1)	Optimal Map Size	TP (abs) $\times 10^{-4}$	Quantization Error
0.50	8743	1681	19.22	8	17	0.49	0.161
0.25	23,962	3775	15.75	17	20	0.37	0.148
0.10	105,552	13,038	12.35	24	30	21.33	0.127
0.05	360,371	39,052	10.83	32	34	8.17	0.118
0.03	893,448	89,954	10.06	38	32	6.51	0.123

One of the first steps prior to training is the normalization of the data. Normalization helps the features to have a similar scale, facilitating the training process and faster convergence. In addition, normalization helps to avoid problems associated with large gradients during training, which negatively affect model convergence. In our work, we conducted a Min-max rescale normalization.

To illustrate, we will concentrate on the presentation of the SOM outcomes for a voxel size of 5 cm (0.05 m). The varied results of the maps according to the voxel size are organized in Appendix A.

As shown in Figure 13, the results of the training and classification of a self-organized map are provided by a matrix of neurons, which are assigned a series of weights given by the input layer. One way to show the relationships between neurons is to express the distances between them (Figure 15). The lighter parts in that matrix represent the parts of the map where the nodes are far away from each other, and the dark parts represent less distance between nodes.

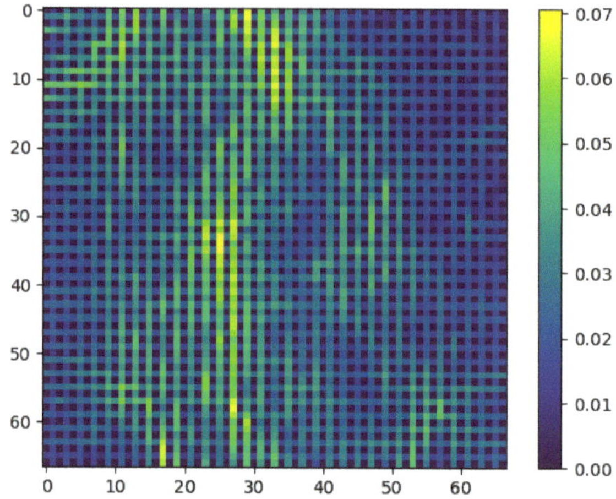

Figure 15. U-Matrix SOM voxel 0.05 m.

Each activated neuron represents a set of voxels. There are also neurons that have not been activated by any of the vectors given by the multispectral voxels within the input layer. Figure 16 shows how many patterns (x-axis) have been recognized by how many neurons (y-axis).

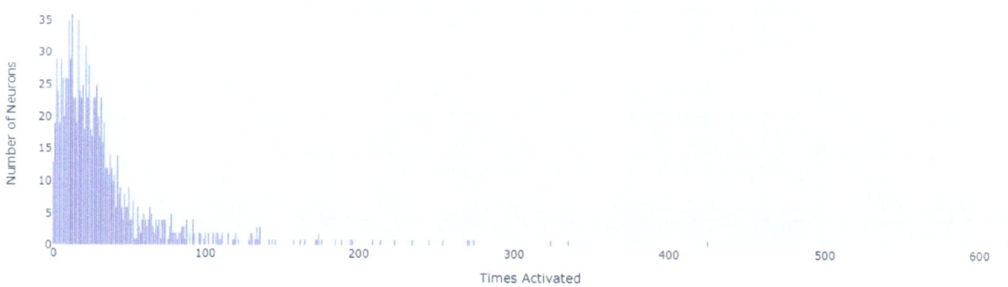

Figure 16. Activated neurons from SOM voxel 0.05 m.

The distribution of activations can be seen in the activation map depicted in Figure 17.

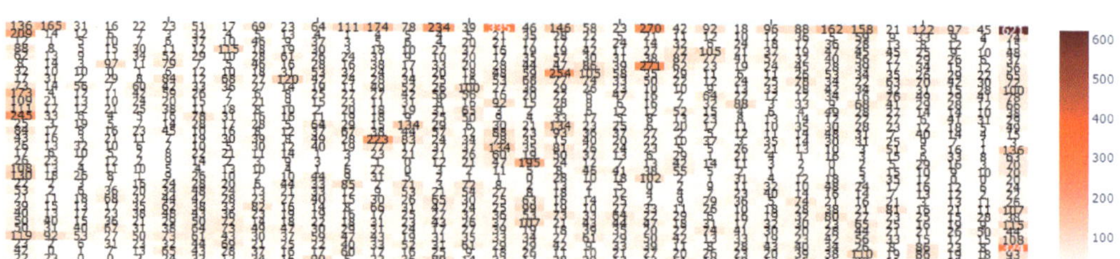

Figure 17. Heatmap of activated neurons from SOM voxel 0.05 m.

In the same way, we show how the activations of the neurons are distributed within each codebook vector (Figure 18).

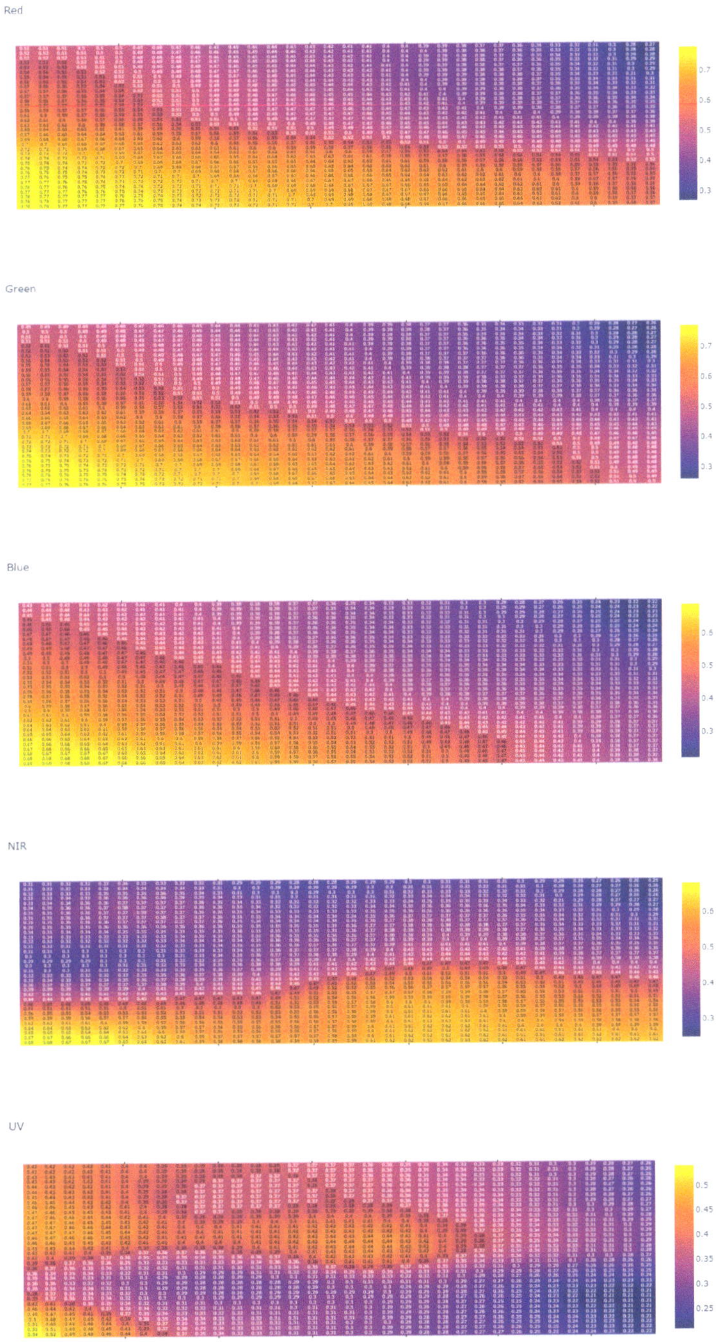

Figure 18. *Cont.*

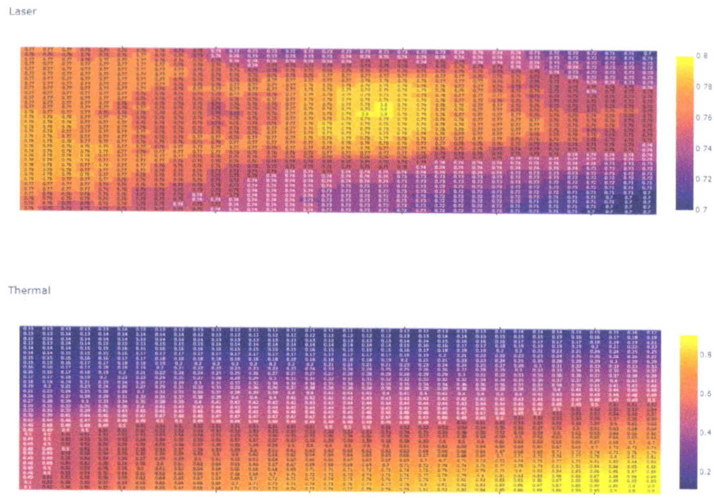

Figure 18. SOM codebooks for voxel 5 cm.

4. Discussion

The use of Self-Organizing Maps (SOM) for the training and classification of multispectral voxel structures results in the generation of a neural map. In this map, each neuron points towards a group of voxels that exhibit similar characteristics, as indicated by a vector of weights assigned to each property (spectral band) of the input layer. Every voxel in the input layer has a Best Machine Unit (BMU). The BMU is the neuron whose weight vector is closest in the space to the pattern.

In the SOM heat map voxel 5 cm (Figure 17), we can identify which neurons are those that determine clearer patterns because they have been activated a greater number of times. We highlight the neurons in the corners of the heat map.

- Neuron (0, 0), activated by 274 voxels. The red band is the predominant band, together with the laser. The UV and thermal bands have less weight for these voxels (Figure 19a). These activated voxels are mostly located on the south building façade.
- Neuron (0, 33), activated by 425 voxels. Its most important band is the laser band. The rest of the bands (RGB and UV) show a medium weight, whereas the thermal information has little influence (Figure 19b). Its voxels are located on the western planes not illuminated by the Sun during data acquisition.
- Neuron (33, 0), activated by 621 voxels. Its characteristic band is the thermal band, with a minimum in the UV band. The red and green bands, along with the NIR have medium weights. The laser band also has considerable weight (Figure 19c). The voxels targeted by this neuron are located on the northern façade.
- Neuron (33, 33), activated by 136 voxels. Larger weight of the laser band. The rest of the bands (RGB, NIR, UV, and thermal) have lower weights (Figure 19d).

However, each neuron does not act in isolation, but interconnects with nearby neurons (Figure 15). Nearby neurons interrelate similar multispectral voxels. As shown in Figure 18, neurons can be zoned using cross-band interrelated weights. In our analysis, we have observed a striking similarity in the distribution of weights of the neurons in the heatmap, corresponding to the red, green, and blue bands, located in the lower left corner. This finding can be attributed to the fact that the built-up area of the building, as depicted in Figure 5, primarily exhibits a red-grayish tone, derived from the granite construction.

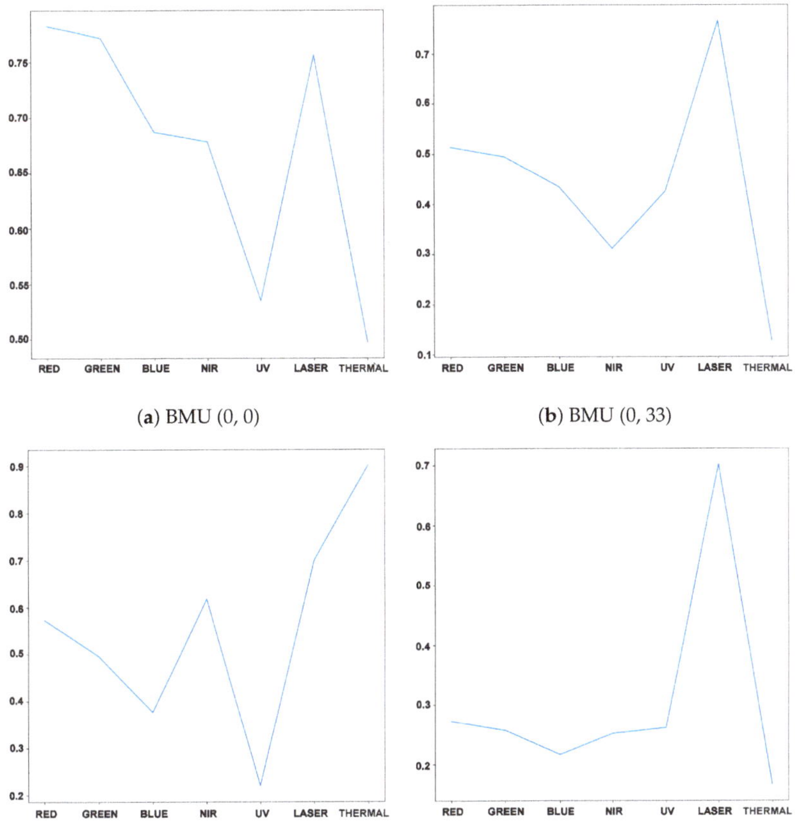

(**a**) BMU (0, 0) (**b**) BMU (0, 33)

(**c**) BMU (33, 0) (**d**) BMU (33, 33)

Figure 19. Characteristic Graph on some neurons of SOM voxel 5 cm.

This distribution of weights presents an opportunity for further analysis. Employing a clustering algorithm, as the K-means method, we have successfully segmented the BMUs. Using, for example, a 12-zone clustering, we are able to identify similar BMUs in a manner that considerably streamlines analysis and enhances decision-making (Figure 20).

The primary focus of our methodology is to identify and study building pathologies through the fusion of geomatic sensor data. We identified an area with dampness in the building that was examined. This area is located in the lower part of one of the northern façades (Figure 21). Dampness in buildings can lead to various issues such as the formation of salt efflorescence. These are white deposits on the surfaces resulting from the migration of soluble salts in water through the pores of the construction materials. This phenomenon often occurs when water carries salts from the ground to walls. In addition to efflorescence, moisture can trigger mold growth, material deterioration, and structural problems [69].

In order to evaluate our hypothesis regarding the potential application of multispectral voxels in multisensor data fusion for the purpose of studying building pathologies through the use of self-organizing maps (SOMs), we specifically focused on the voxels that represented areas of the building with pathology present. We then identified the best matching units (BMUs) for these voxels and discovered that they were located in close proximity to the neuron (33, 20). To confirm the validity of their weight vector, we obtained the corresponding characteristic graph, shown in Figure 22. The graph indicates a strong correlation between the visible spectrum bands and the laser scanner band.

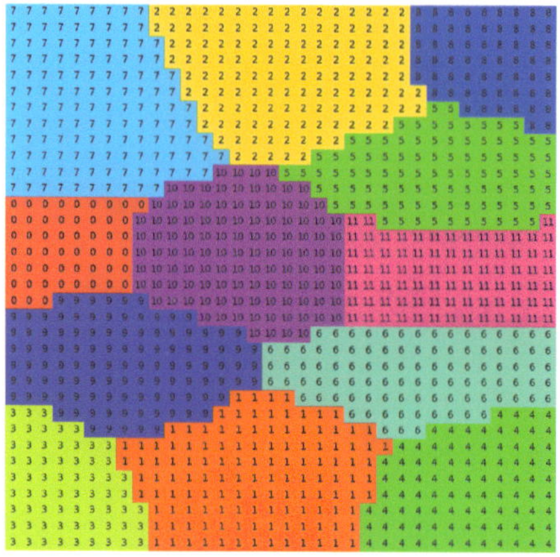

Figure 20. Clustering of BMU Hits Matrix by K-means algorithm (12 clusters).

Figure 21. Pathology in north façade of the building.

Earlier research in remote sensing building pathologies studies discovered that the infrared spectral range, particularly the 778, 905, and 1550 nm wavelength bands, is ideal for detecting dampness. The study [1] found that the best results were obtained using short-wave infrared data (1550 nm). Visible wave bands can detect efflorescence, as surfaces with this phenomenon have higher reflectance in these spectral ranges. For chlorides or sulphates being the cause, the short-wave infrared range is more appropriate. In accordance with the research conducted by [1], our methodology demonstrates a higher level of effectiveness as it does not rely on the derivation of information from point clouds to images to localize and study pathologies. Instead, we work directly with the original 3D point clouds, integrating geospatial 3D geometry with multispectral data information.

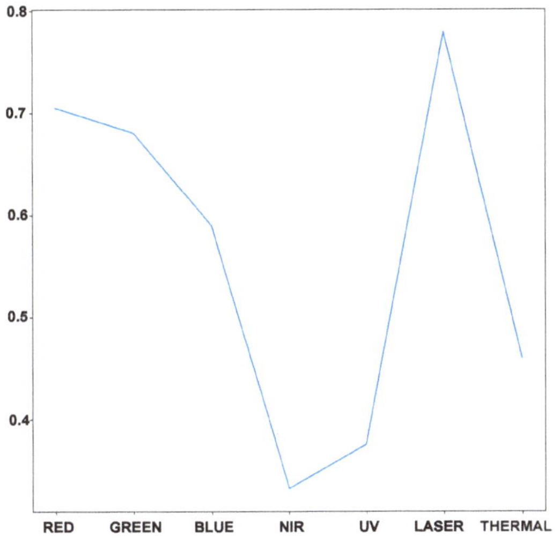

Figure 22. Characteristic graph of BMU (33, 20).

5. Conclusions

Given the vast amount of data provided by different geomatic sensors and the diverse nature of the information they offer, it is challenging to manage multisensor data and generate knowledge from it. Therefore, new approaches are required to effectively and efficiently express this information from these sensors. Our methodology combines the power of voxels in their multispectral character, which is necessary for data fusion, with deep-learning algorithms. In this case, the methodology was tested in a study of a cultural heritage building. Using this methodology, we were able to geospatially locate areas (through their respective voxels) that exhibit certain characteristics, showing pathologies. This allows to identify, study, and derives information from them about any building that has been studied with different sensors, regardless of their nature. Our methodology provides a powerful tool for analyzing architectural structures using various sensors fusing their data. The use of multisensor multispectral voxels locates possible pathologies such as damp and efflorescence in buildings. Further studies can verify correlations between the SOM and different fused sensors, defining their characteristic curves in such a way as to identify the whole catalogue of the most common pathologies in Historical Heritage buildings.

Author Contributions: All authors have made significant contributions to this manuscript. Conceptualization, J.R., S.L.-C.M., T.R.H.-T. and E.P.-d.-l.-S.; methodology, J.R.; software, J.R.; validation, J.R.; formal analysis, J.R., S.L.-C.M. and T.R.H.-T.; investigation, J.R.; field observations: J.R., S.L.-C.M., J.A.d.M. and E.P.-d.-l.-S.; resources, J.R., S.L.-C.M., T.R.H.-T. and E.P.-d.-l.-S.; data curation, J.R. and S.L.-C.M.; writing–original draft preparation, J.R.; writing–review and editing, J.R., S.L.-C.M., J.A.d.M., T.R.H.-T. and E.P.-d.-l.-S.; supervision, S.L.-C.M., J.A.d.M., T.R.H.-T. and E.P.-d.-l.-S. All authors have read and agreed to the published version of the manuscript.

Funding: This research was founded by Grupo Acre at Fundación Premio Arce reference FPA1900001501.

Data Availability Statement: The data presented in this study are available on request from the corresponding author.

Acknowledgments: We extend our gratitude to the dedicated staff of the Geovisualización, Espacios Singulares y Patrimonio (GESyP) research group. Additionally, the first author, Javier Raimundo, expresses his appreciation to the Consejo General de la Arquitectura Técnica de España (CGATE) for their valuable support. Special thanks to Noemi Martinez for her assistance in the field data capture campaign.

Conflicts of Interest: The authors declare no conflict of interest.

Abbreviations

The following abbreviations are used in this manuscript:

TLS	Terrestrial Laser Scanner
NIR	Near Infrared
RGB	Red, green and blue
IR	Infrared
UV	Ultraviolet
TTS	Terrain Target Signals
WSS	Wall Surface Signals
GNSS	Global Navigation Satellite Systems
CODE	Center for Orbit Determination
IGS	International GNSS Service
CORS	Continuously Operating Reference Stations
IGN	Instituto Geografico Nacional
VMF	Vienna Mapping Functions
SOM	Self Organizing Map
ANN	Artificial Neural Network
SWIR	Short-Wavelength Infrared
TP	Topographic Product
BIM	Building Information Modelling
BMU	Best Matching Unit

Appendix A

Appendix A.1. Surface Targets Coordinates

Here, we show the coordinates for the signals attached on building surface (Table A1).

Table A1. Wall Surface Signals (WSS) coordinates. UTM 30N coordinate system, ETRS89 (ETRF2000) geodetic frame. All values in meters.

Target Signal	UTM 30 Coordinates		Height
	Northing	Easting	
target 1	382360.015	440005.730	567.466
target 2	382368.957	4401021.247	566.398
target 3	382387.272	4401026.984	567.755
target 4	382396.179	4400998.666	567.397
target 5	382360.322	4400996.444	568.799
target 6	382388.457	4401003.889	566.041
target 7	382379.653	4400999.527	567.669
target 10	382382.008	4401000.464	568.380
target 12	382385.244	4401004.683	568.189
target 13	382383.552	4401018.785	567.828
target 15	382385.290	4401017.194	566.884
target 16	382367.348	4401015.839	569.020
target 18	382385.407	4401002.420	566.875
target 20	382380.706	4401019.521	567.237
target 21	382378.862	4401018.659	566.457
target 25	382374.552	4401019.167	568.003
target 26	382363.300	4401013.857	569.381
target 27	382391.098	4401011.804	568.204
target 28	382366.036	4401007.650	569.078
target 29	382378.289	4400999.028	568.894
target 30	382363.924	4401014.202	568.630
target 34	382372.581	4401003.133	569.414
target 35	382382.719	4401020.449	567.849
target 36	382370.265	4401017.214	567.388
target 37	382370.390	4401002.189	568.604
target 43	382368.778	4401001.428	569.551
target 45	382388.164	4401010.347	568.259
target 47	382389.463	4401015.952	568.259
target 48	382390.180	4401014.380	568.103
target 50	382385.086	4401001.835	568.112

Appendix A.2. Points Histograms

In Appendix A.2, we present the distribution of points per voxel through histograms, based on the voxel's elemental size.

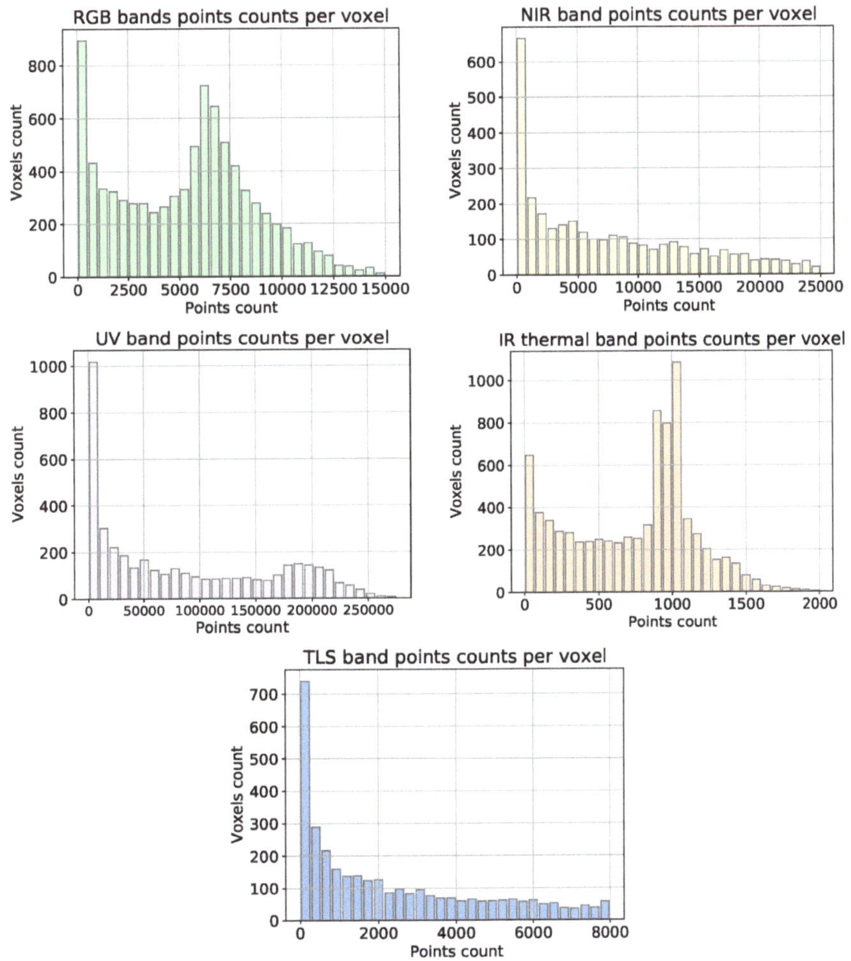

Figure A1. Points per voxel. Voxel size: 50 cm.

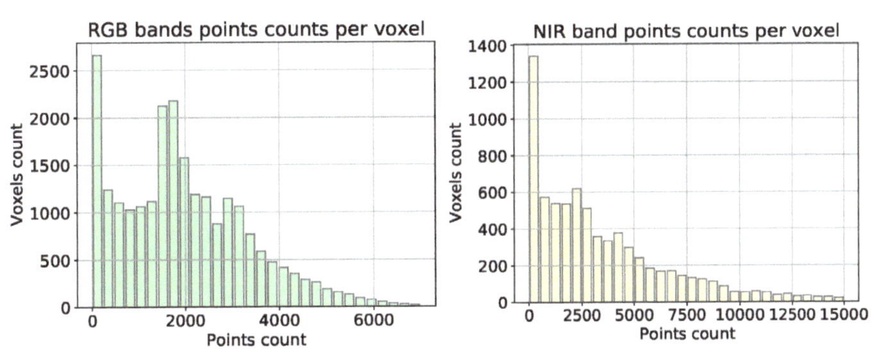

Figure A2. *Cont.*

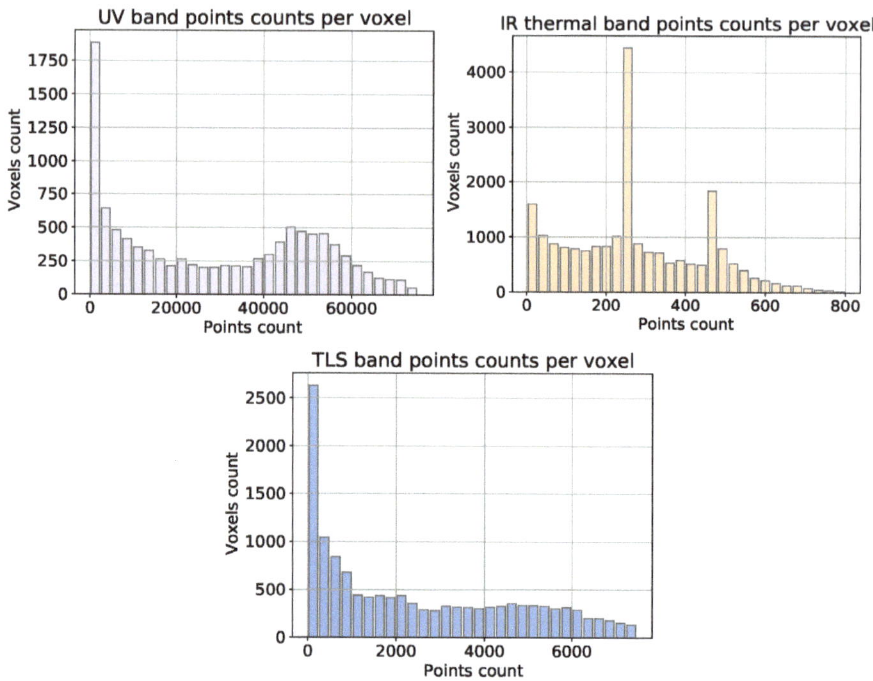

Figure A2. Points per voxel. Voxel size: 25 cm.

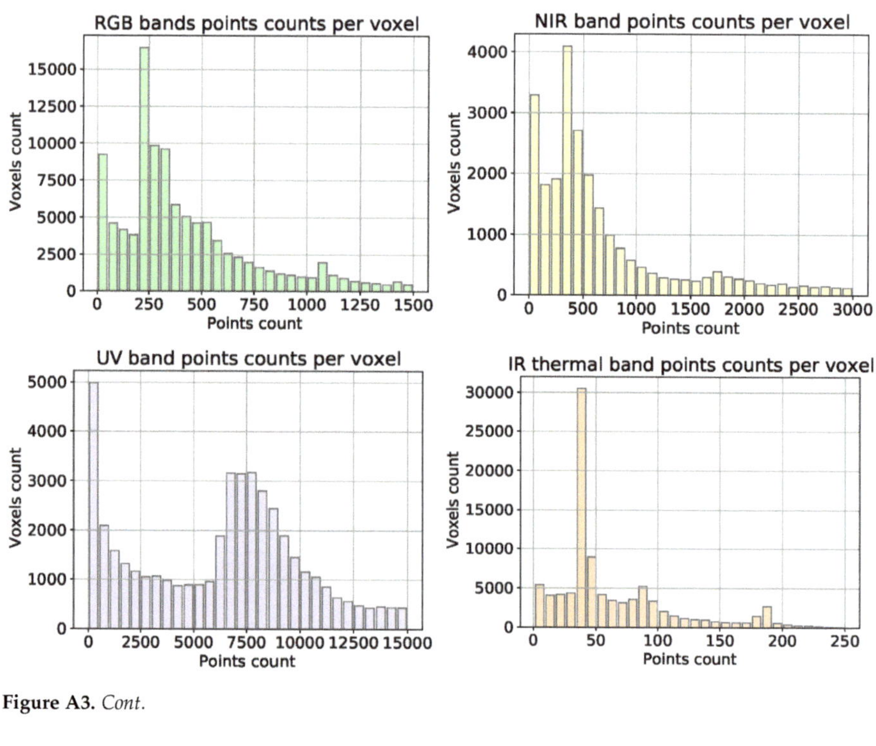

Figure A3. *Cont.*

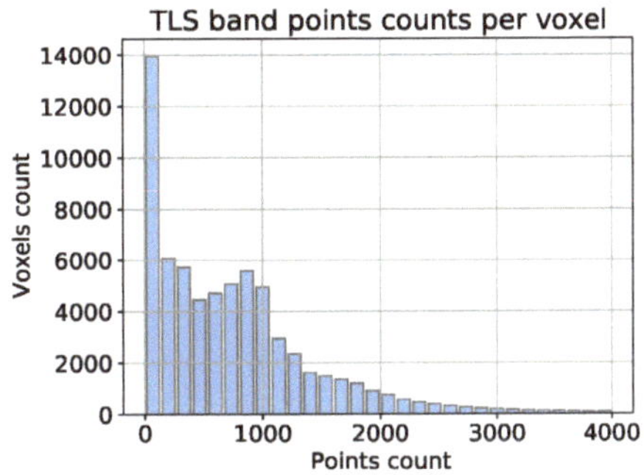

Figure A3. Points per voxel. Voxel size: 10 cm.

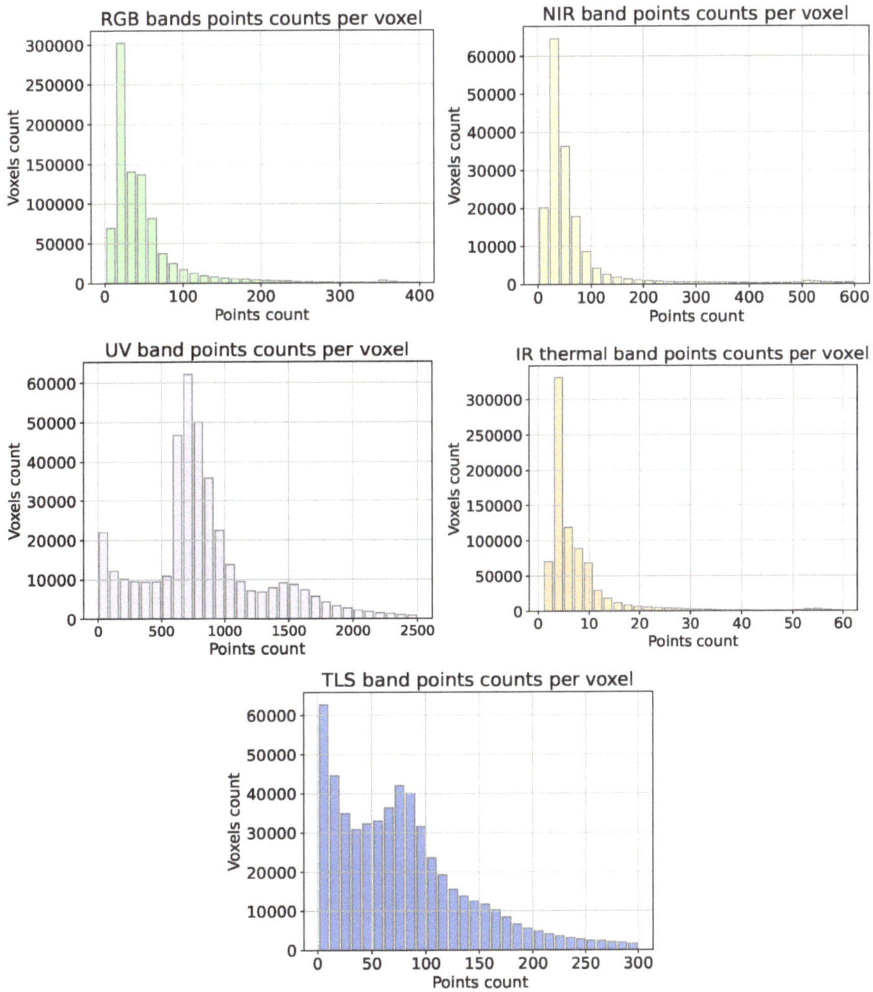

Figure A4. Points per voxel. Voxel size: 3 cm.

Appendix A.3. SOM Codebooks

Here, in Appendix A.3, we express the different heat map codebooks , which allows to see map distribution of all of its characteristics. Different voxel elemental sizes have been trained, expressing coherent results within the Results Section 3 and Discussion Section 4.

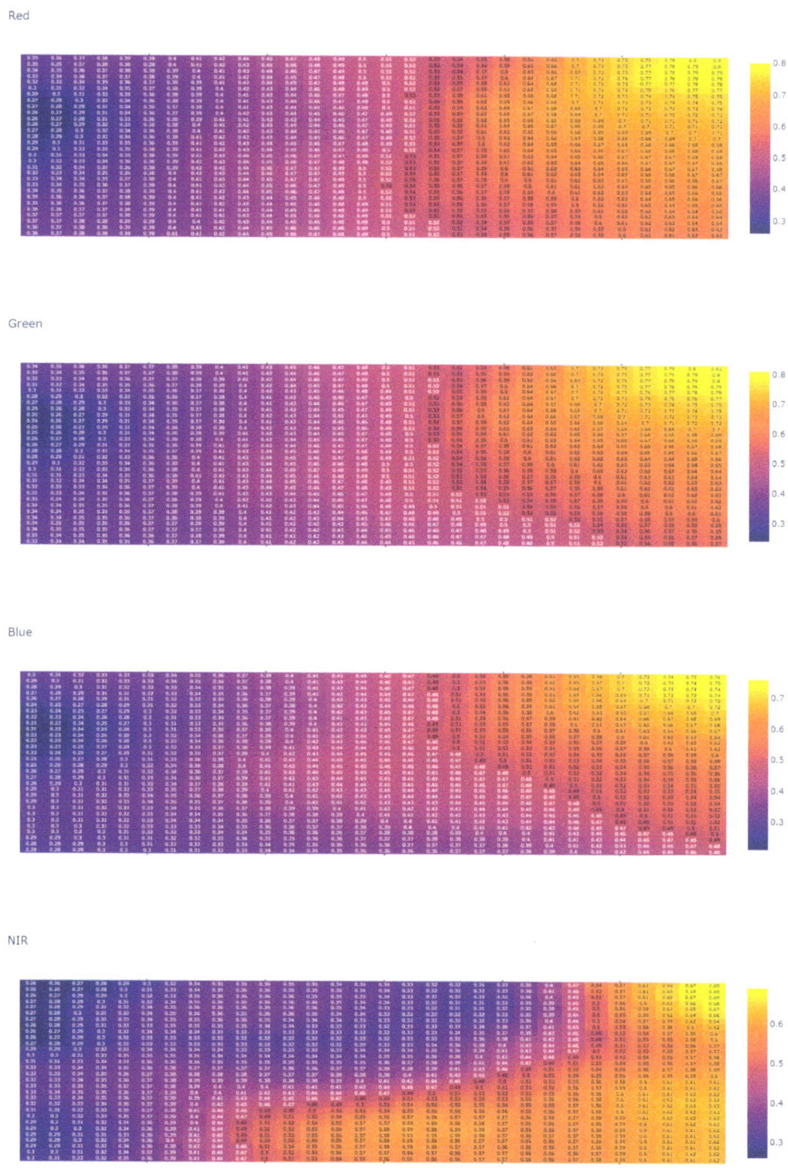

Figure A5. *Cont.*

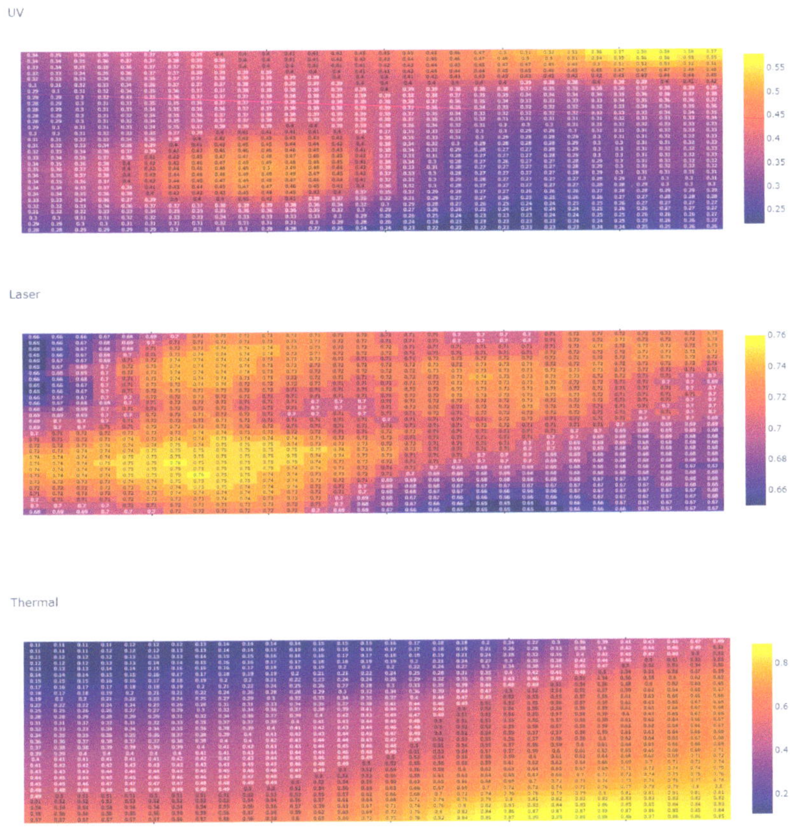

Figure A5. SOM codebooks for voxel 10 cm.

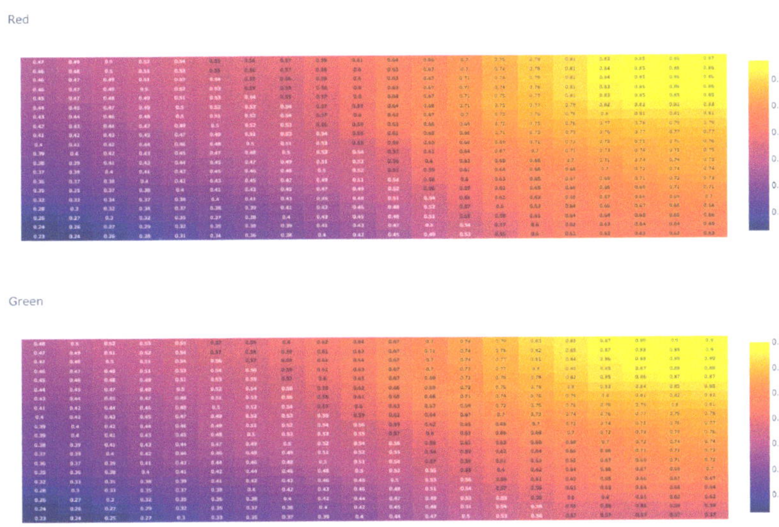

Figure A6. *Cont.*

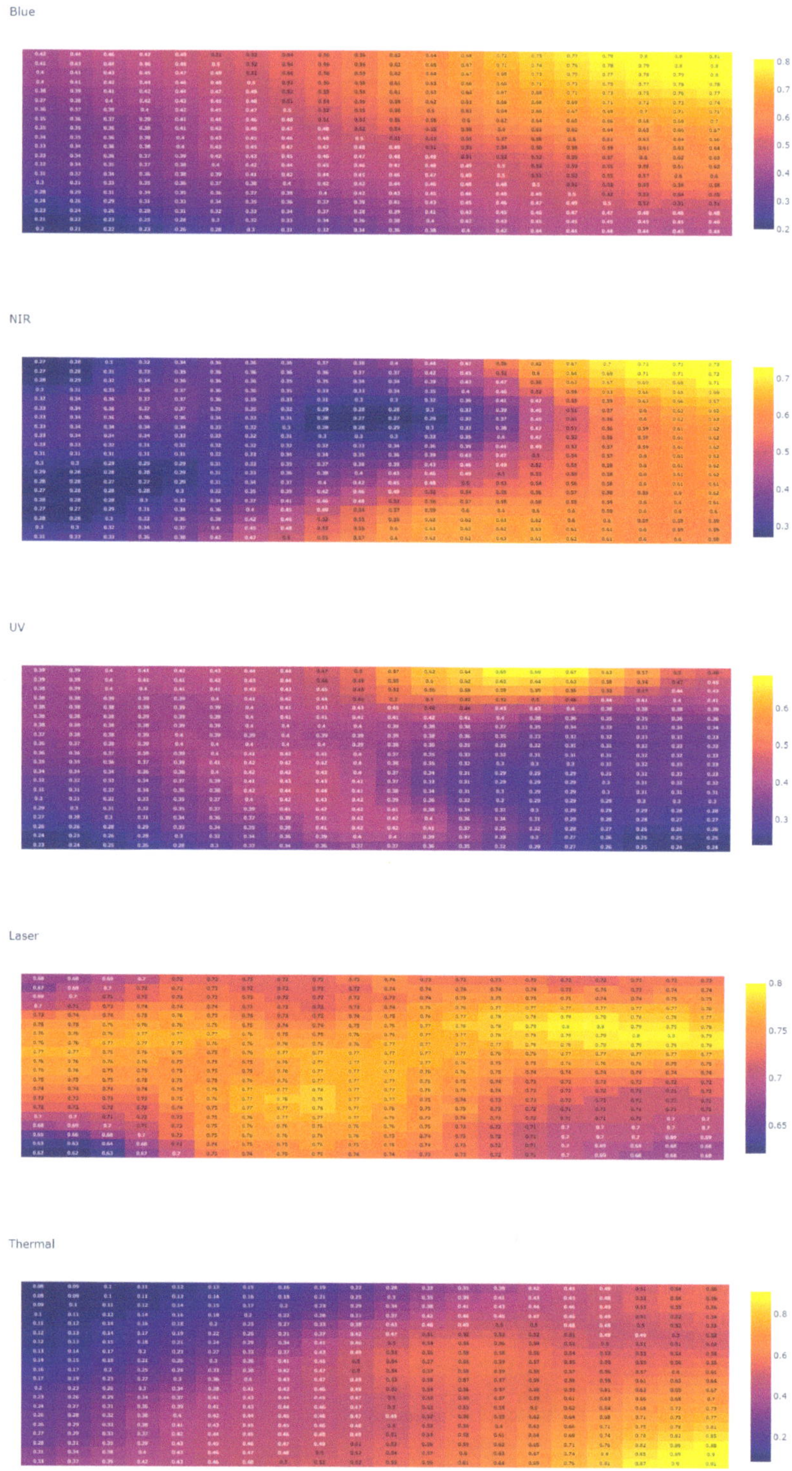

Figure A6. SOM codebooks for voxel 25 cm.

References

1. Del Pozo Aguilera, S. Multispectral Imaging for the Analysis of Materials and Pathologies in Civil Engineering, Constructions and Natural Spaces. Ph.D. Thesis, Universidad de Salamanca, Salamanca, Spain, 2016.
2. Torres-Martínez, J.A.; Sánchez-Aparicio, L.J.; Hernández-López, D.; González-Aguilera, D. Combining Geometrical and Radiometrical Features in the Evaluation of Rock Art Paintings. *Digit. Appl. Archaeol. Cult. Herit.* **2017**, *5*, 10–20. [CrossRef]
3. Crespo, C.; Armesto, J.; González-Aguilera, D.; Arias, P. Damage Detection on Historical Buildings Using Unsupervised Classification Techniques. *Int. Arch. Photogramm. Remote Sens. Spat. Inf. Sci. ISPRS Arch.* **2010**, *38*, 184–188.
4. Moropoulou, A.; Labropoulos, K.C.; Delegou, E.T.; Karoglou, M.; Bakolas, A. Non-Destructive Techniques as a Tool for the Protection of Built Cultural Heritage. *Constr. Build. Mater.* **2013**, *48*, 1222–1239. [CrossRef]
5. Oreni, D.; Cuca, B.; Brumana, R. Three-Dimensional Virtual Models for Better Comprehension of Architectural Heritage Construction Techniques and Its Maintenance over Time. In *Lecture Notes in Computer Science (Including Subseries Lecture Notes in Artificial Intelligence and Lecture Notes in Bioinformatics)*; Springer: Berlin/Heidelberg, Germany, 2012; Volume 7616 LNCS, pp. 533–542. [CrossRef]
6. Turco, M.L.; Mattone, M.; Rinaudo, F. Metric Survey and Bim Technologies to Record Decay Conditions. *Int. Arch. Photogramm. Remote Sens. Spat. Inf. Sci. ISPRS Arch.* **2017**, *42*, 261–268. [CrossRef]
7. González-Jorge, H.; Lagüela, S.; Krelling, P.; Armesto, J.; Martínez-Sánchez, J. Single Image Rectification of Thermal Images for Geometric Studies in Façade Inspections. *Infrared Phys. Technol.* **2012**, *55*, 421–426. [CrossRef]
8. Dias, I.; Flores-Colen, I.; Silva, A. Critical Analysis about Emerging Technologies for Building's Façade Inspection. *Buildings* **2021**, *11*, 53. [CrossRef]
9. Masiero, A.; Costantino, D. TLS for detecting small damages on a building façade. *Int. Arch. Photogramm. Remote Sens. Spat. Inf. Sci.* **2019**, *XLII-2/W11*, 831–836. [CrossRef]
10. Kim, N.; Choi, Y.; Hwang, S.; Park, K.; Yoon, J.S.; Kweon, I.S. Geometrical Calibration of Multispectral Calibration. In Proceedings of the 2015 12th International Conference on Ubiquitous Robots and Ambient Intelligence, URAI 2015, Goyangi, Republic of Korea, 28–30 October 2015; pp. 384–385. [CrossRef]
11. Wang, Q.; Kim, M.K.; Sohn, H.; Cheng, J.C. Surface Flatness and Distortion Inspection of Precast Concrete Elements Using Laser Scanning Technology. *Smart Struct. Syst.* **2016**, *18*, 601–623. [CrossRef]
12. Wang, Q.; Kim, M.K.; Cheng, J.C.; Sohn, H. Automated Quality Assessment of Precast Concrete Elements with Geometry Irregularities Using Terrestrial Laser Scanning. *Autom. Constr.* **2016**, *68*, 170–182. [CrossRef]
13. Chiabrando, F.; Lo Turco, M.; Rinaudo, F. Modeling the Decay in an Hbim Starting from 3d Point Clouds. A Followed Approach for Cultural Heritage Knowledge. *Int. Arch. Photogramm. Remote Sens. Spat. Inf. Sci. ISPRS Arch.* **2017**, *42*, 605–612. [CrossRef]
14. Yaagoubi, R.; Miky, Y. Developing a Combined Light Detecting And Ranging (LiDAR) and Building Information Modeling (BIM) Approach for Documentation and Deformation Assessment of Historical Buildings. *MATEC Web Conf.* **2018**, *149*, 02011. [CrossRef]
15. Previtali, M.; Barazzetti, L.; Redaelli, V.; Scaioni, M.; Rosina, E. Rigorous Procedure for Mapping Thermal Infrared Images on Three-Dimensional Models of Building Façades. *J. Appl. Remote Sens.* **2013**, *7*, 073503. [CrossRef]
16. Lerma, C.; Mas, Á.; Gil, E.; Vercher, J.; Peñalver, M.J. Pathology of Building Materials in Historic Buildings. Relationship Between Laboratory Testing and Infrared Thermography. *Mater. Constr.* **2014**, *64*, e009. [CrossRef]
17. Mercuri, F.; Cicero, C.; Orazi, N.; Paoloni, S.; Marinelli, M.; Zammit, U. Infrared Thermography Applied to the Study of Cultural Heritage. *Int. J. Thermophys.* **2015**, *36*, 1189–1194. [CrossRef]
18. Sfarra, S.; Marcucci, E.; Ambrosini, D.; Paoletti, D. Infrared Exploration of the Architectural Heritage: From Passive Infrared Thermography to Hybrid Infrared Thermography (HIRT) Approach. *Mater. Constr.* **2016**, *66*, e094. [CrossRef]
19. Luib, A. Infrared thermal imaging as a non-destructive investigation method for building archaeological purposes. *Int. Arch. Photogramm. Remote Sens. Spat. Inf. Sci. ISPRS Arch.* **2019**, *42*, 695–702. [CrossRef]
20. Del Pozo, S.; Herrero-Pascual, J.; Felipe-García, B.; Hernández-López, D.; Rodríguez-Gonzálvez, P.; González-Aguilera, D. Multispectral Radiometric Analysis of Façades to Detect Pathologies from Active and Passive Remote Sensing. *Remote Sens.* **2016**, *8*, 80. [CrossRef]
21. Del Pozo, S.; Rodríguez-Gonzálvez, P.; Sánchez-Aparicio, L.J.; Muñoz-Nieto, A.; Hernández-López, D.; Felipe-García, B.; González-Aguilera, D. Multispectral Imaging in Cultural Heritage Conservation. *Int. Arch. Photogramm. Remote Sens. Spat. Inf. Sci. ISPRS Arch.* **2017**, *42*, 155–162. [CrossRef]
22. Armesto-González, J.; Riveiro-Rodríguez, B.; González-Aguilera, D.; Rivas-Brea, M.T. Terrestrial Laser Scanning Intensity Data Applied to Damage Detection for Historical Buildings. *J. Archaeol. Sci.* **2010**, *37*, 3037–3047. [CrossRef]
23. Naranjo, J.M.; Parrilla, Á.; de Sanjosé, J.J. Geometric Characterization and Interactive 3D Visualization of Historical and Cultural Heritage in the Province of Cáceres (Spain). *Virtual Archaeol. Rev.* **2018**, *9*, 1–11. [CrossRef]
24. Antón, D.; Carretero-Ayuso, M.J.; Moyano-Campos, J.; Nieto-Julián, J.E. Laser Scanning Intensity Fingerprint: 3D Visualisation and Analysis of Building Surface Deficiencies. In *New Technologies in Building and Construction*; Bienvenido-Huertas, D., Moyano-Campos, J., Eds.; Springer Nature Singapore: Singapore, 2022; Volume 258, pp. 207–223. [CrossRef]
25. Batur, M.; Yilmaz, O.; Ozener, H. A Case Study of Deformation Measurements of Istanbul Land Walls via Terrestrial Laser Scanning. *IEEE J. Sel. Top. Appl. Earth Obs. Remote Sens.* **2020**, *13*, 6362–6371. [CrossRef]

26. Bayarri, V.; Prada, A.; García, F.; Díaz-González, L.M.; De Las Heras, C.; Castillo, E.; Fatás, P. Integration of Remote-Sensing Techniques for the Preventive Conservation of Paleolithic Cave Art in the Karst of the Altamira Cave. *Remote Sens.* **2023**, *15*, 1087. [CrossRef]
27. Xu, Y.; Tong, X.; Stilla, U. Voxel-Based Representation of 3D Point Clouds: Methods, Applications, and Its Potential Use in the Construction Industry. *Autom. Constr.* **2021**, *126*, 103675. [CrossRef]
28. Poux, F.; Billen, R. Voxel-Based 3D Point Cloud Semantic Segmentation: Unsupervised Geometric and Relationship Featuring vs Deep Learning Methods. *ISPRS Int. J. Geo-Inf.* **2019**, *8*, 213. [CrossRef]
29. Raimundo, J.; Lopez-Cuervo Medina, S.; Aguirre De Mata, J.; Prieto, J.F. Multisensor Data Fusion by Means of Voxelization: Application to a Construction Element of Historic Heritage. *Remote Sens.* **2022**, *14*, 4172. [CrossRef]
30. Alikhodja, N.; Zeghlache, H.; Bousnina, M. Remote Sensing Method (TLS) in Architectural Analysis and Constructive Pathology Diagnosis. *Res. Sq.* **2023**. [CrossRef]
31. Herrero-Tejedor, T.R.; Maté-González, M.Á.; Pérez-Martín, E.; López-Cuervo, S.; López De Herrera, J.; Sánchez-Aparicio, L.J.; Villanueva Llauradó, P. Documentation and Virtualisation of Vernacular Cultural Heritage: The Case of Underground Wine Cellars in Atauta (Soria). *Heritage* **2023**, *6*, 5130–5150. [CrossRef]
32. Suchocki, C.; Błaszczak-Bąk, W.; Janicka, J.; Dumalski, A. Detection of Defects in Building Walls Using Modified OptD Method for Down-Sampling of Point Clouds. *Build. Res. Inf.* **2021**, *49*, 197–215. [CrossRef]
33. Li, J.; Liu, P.R.; Liang, Z.X.; Wang, X.Y.; Wang, G.Y. Three-Dimensional Geological Modeling Method of Regular Voxel Splitting Based on Multi-Source Data Fusion. *Yantu Lixue Rock Soil Mech.* **2021**, *42*, 1170–1177. [CrossRef]
34. Goodbody, T.R.; Tompalski, P.; Coops, N.C.; Hopkinson, C.; Treitz, P.; van Ewijk, K. Forest Inventory and Diversity Attribute Modelling Using Structural and Intensity Metrics from Multi-Spectral Airborne Laser Scanning Data. *Remote Sens.* **2020**, *12*, 2109. [CrossRef]
35. Inano, R.; Oishi, N.; Kunieda, T.; Arakawa, Y.; Yamao, Y.; Shibata, S.; Kikuchi, T.; Fukuyama, H.; Miyamoto, S. Voxel-Based Clustered Imaging by Multiparameter Diffusion Tensor Images for Glioma Grading. *Neuroimage Clin.* **2014**, *5*, 396–407. [CrossRef] [PubMed]
36. Li, Y.; Xie, H.; Shin, H. 3D Object Detection Using Frustums and Attention Modules for Images and Point Clouds. *Signals* **2021**, *2*, 98–107. [CrossRef]
37. Wang, X.; Xu, D.; Gu, F. 3D Model Inpainting Based on 3D Deep Convolutional Generative Adversarial Network. *IEEE Access* **2020**, *8*, 170355–170363. [CrossRef]
38. Poux, F.; Neuville, R.; Nys, G.A.; Billen, R. 3D Point Cloud Semantic Modelling: Integrated Framework for Indoor Spaces and Furniture. *Remote Sens.* **2018**, *10*, 1412. [CrossRef]
39. Kuang, H.; Wang, B.; An, J.; Zhang, M.; Zhang, Z. Voxel-FPN: Multi-Scale Voxel Feature Aggregation for 3D Object Detection from LIDAR Point Clouds. *Sensors* **2020**, *20*, 704. [CrossRef]
40. Gebhardt, S.; Payzer, E.; Salemann, L.; Fettinger, A.; Rotenberg, E.; Seher, C. Polygons, Point-Clouds, and Voxels, A Comparison of High-Fidelity Terrain Representations. In Proceedings of the Fall Simulation Interoperability Workshop 2009, 2009 Fall SIW, Orlando, FL, USA, 21–25 September 2009; pp. 357–365.
41. Arayici, Y.; Counsell, J.; Mahdjoubi, L.; Nagy, G.; Hawas, S.; Dweidar, K. *Heritage Building Information Modelling*; Taylor & Francis Ltd.: London, UK, 2021.
42. Castellazzi, G.; D'Altri, A.; Bitelli, G.; Selvaggi, I.; Lambertini, A. From Laser Scanning to Finite Element Analysis of Complex Buildings by Using a Semi-Automatic Procedure. *Sensors* **2015**, *15*, 18360–18380. [CrossRef]
43. Bitelli, G.; Castellazzi, G.; D'Altri, A.; De Miranda, S.; Lambertini, A.; Selvaggi, I. Automated Voxel Model from Point Clouds for Structural Analysis of Cultural Heritage. *Int. Arch. Photogramm. Remote Sens. Spat. Inf. Sci. ISPRS* **2016**, *XLI-B5*, 191–197. [CrossRef]
44. Zhang, C.; Jamshidi, M.; Chang, C.C.; Liang, X.; Chen, Z.; Gui, W. Concrete Crack Quantification Using Voxel-Based Reconstruction and Bayesian Data Fusion. *IEEE Trans. Ind. Inform.* **2022**, *18*, 7512–7524. [CrossRef]
45. Wang, Y.; Xiao, Y.; Xiong, F.; Jiang, W.; Cao, Z.; Zhou, J.T.; Yuan, J. 3DV: 3D Dynamic Voxel for Action Recognition in Depth Video. In Proceedings of the IEEE Computer Society Conference on Computer Vision and Pattern Recognition, Seattle, WA, USA, 13–19 June 2020; pp. 508–517. [CrossRef]
46. Brock, A.; Lim, T.; Ritchie, J.M.; Weston, N. Generative and Discriminative Voxel Modeling with Convolutional Neural Networks. *arXiv* **2016**, arXiv:1608.04236. [CrossRef].
47. Li, B. 3D Fully Convolutional Network for Vehicle Detection in Point Cloud. In Proceedings of the 2017 IEEE/RSJ International Conference on Intelligent Robots and Systems (IROS), Vancouver, BC, Canada, 24–28 September 2017; pp. 1513–1518. [CrossRef]
48. Liu, J.; Mills, S.; McCane, B. Variational Autoencoder for 3D Voxel Compression. In Proceedings of the 2020 35th International Conference on Image and Vision Computing New Zealand (IVCNZ), Wellington, New Zealand, 25–27 November 2020; pp. 1–6. [CrossRef]
49. Huang, B.C.; Feng, Y.C.; Liang, T.Y. A Voxel Generator Based on Autoencoder. *Appl. Sci.* **2022**, *12*, 10757. [CrossRef]
50. Rezaei, M.; Yang, H.; Meinel, C. Voxel-GAN: Adversarial Framework for Learning Imbalanced Brain Tumor Segmentation. In *Brainlesion: Glioma, Multiple Sclerosis, Stroke and Traumatic Brain Injuries*; Crimi, A., Bakas, S., Kuijf, H., Keyvan, F., Reyes, M., Van Walsum, T., Eds.; Springer International Publishing: Cham, Switzerland, 2019; Volume 11384, pp. 321–333. [CrossRef]

51. Kleineberg, M.; Fey, M.; Weichert, F. Adversarial Generation of Continuous Implicit Shape Representations. *arXiv* **2020**, arXiv:2002.00349. [CrossRef].
52. Zhang, W.; Ma, Y.; Zhu, D.; Dong, L.; Liu, Y. MetroGAN: Simulating Urban Morphology with Generative Adversarial Network. In Proceedings of the 28th ACM SIGKDD Conference on Knowledge Discovery and Data Mining, Washington, DC, USA, 14–18 August 2022; pp. 2482–2492. [CrossRef]
53. Kniaz, V.V.; Remondino, F.; Knyaz, V.A. Generative adversarial networks for single photo 3D reconstruction. *Int. Arch. Photogramm. Remote Sens. Spat. Inf. Sci. ISPRS* **2019**, *XLII-2/W9*, 403–408. [CrossRef]
54. Bastonero, P.; Donadio, E.; Chiabrando, F.; Spanò, A. Fusion of 3D Models Derived from TLS and Image-Based Techniques for CH Enhanced Documentation. *Int. Arch. Photogramm. Remote Sens. Spat. Inf. Sci. ISPRS* **2014**, *2*, 73–80. [CrossRef]
55. Berra, E.; Gibson-Poole, S.; MacArthur, A.; Gaulton, R.; Hamilton, A. Estimation of the Spectral Sensitivity Functions of Un-Modified and Modified Commercial off-the-Shelf Digital Cameras to Enable Their Use as a Multispectral Imaging System for UAVs. *Int. Arch. Photogramm. Remote Sens. Spat. Inf. Sci. ISPRS Arch.* **2015**, *40*, 207–214. [CrossRef]
56. Yeshwanth Kumar, A.; Noufia, M.A.; Shahira, K.A.; Ramiya, A.M. Building information modelling of a multi storey building using terrestrial laser scanner and visualisation using potree: An open source point cloud renderer. *Int. Arch. Photogramm. Remote Sens. Spat. Inf. Sci. ISPRS* **2019**, *XLII-2/W17*, 421–426. [CrossRef]
57. Acorsi, M.G.; Gimenez, L.M.; Martello, M. Assessing the Performance of a Low-Cost Thermal Camera in Proximal and Aerial Conditions. *Remote Sens.* **2020**, *12*, 3591. [CrossRef]
58. Zoreda, L.C.; Mier, M.F.; Pérez, A.R.; Arnanz, A.M.; López, P.; Rosado, R.M.; Uzquiano, P. Notas Sobre El Complejo Productivo de Melque (Toledo): Prospección del Territorio y Análisis de Carbono 14, Polínicos, Carpológicos y Antracológicos y de Morteros; In *Archivo Español de Arqueología*; Technical Report; Consejo Superior de Investigaciones Cientifica: Madrid, Spain, 1999.
59. Ortega, R.; Vanacker, V.; Sanjurjo-Sánchez, J.; Miralles, I. Human-Landscape Interactions during the Early and High Medieval Period in Central Spain Based on New Estimates of Sediment Yield from the Melque Agricultural Complex. *Geoarchaeology* **2017**, *32*, 177–188. [CrossRef]
60. Zoreda, L.C. Un Canal de Transmisión de Lo Clásico En La Alta Edad Media Española: Arquitectura y Escultura de Influjo Omeya En La Península Ibérica Entre Mediados Del Siglo VIII e Inicios Del X (I). *Al-Qantara Rev. Estud. Árabes* **1994**, 321–350.
61. Boehm, J.; Werl, B.; Schuh, H. Troposphere Mapping Functions for GPS and Very Long Baseline Interferometry from European Centre for Medium-Range Weather Forecasts Operational Analysis Data. *J. Geophys. Res. Solid Earth* **2006**, *111*, 2005JB003629. [CrossRef]
62. Velasco-Gómez, J.; Prieto, J.F.; Molina, I.; Herrero, T.; Fábrega, J.; Pérez-Martín, E. Use of the Gyrotheodolite in Underground Networks of Long High-Speed Railway Tunnels. *Surv. Rev.* **2016**, *48*, 329–337. [CrossRef]
63. Foley, J.D. *Computer Graphics: Principles and Practice*; Addison Wesley: Boston, MA, USA, 1990.
64. Zhou, Q.Y.; Park, J.; Koltun, V. Open3D: A Modern Library for 3D Data Processing. *arXiv* **2018**, arXiv:1801.09847. [CrossRef].
65. García-Tejedor, Á.J.; Nogales, A. An Open-Source Python Library for Self-Organizing-Maps. *Softw. Impacts* **2022**, *12*, 100280. [CrossRef]
66. Ruiz-Varona, A.; Lacasta, J.; Nogueras-Iso, J. Self-Organizing Maps to Evaluate Multidimensional Trajectories of Shrinkage in Spain. *ISPRS Int. J. Geo-Inf.* **2022**, *11*, 77. [CrossRef]
67. Forest, F.; Lebbah, M.; Azzag, H.; Lacaille, J. A Survey and Implementation of Performance Metrics for Self-Organized Maps. *arXiv* **2020**, arXiv:2011.05847. [CrossRef].
68. Bauer, H.U.; Pawelzik, K.; Geisel, T. A Topographic Product for the Optimization of Self-Organizing Feature Maps. In *Advances in Neural Information Processing Systems*; Moody, J., Hanson, S., Lippmann, R., Eds.; Morgan-Kaufmann: Burlington, MA, USA, 1991; Volume 4.
69. Ortiz, R.; Ortiz, P.; Martín, J.M.; Vázquez, M.A. A New Approach to the Assessment of Flooding and Dampness Hazards in Cultural Heritage, Applied to the Historic Centre of Seville (Spain). *Sci. Total. Environ.* **2016**, *551–552*, 546–555. [CrossRef]

Disclaimer/Publisher's Note: The statements, opinions and data contained in all publications are solely those of the individual author(s) and contributor(s) and not of MDPI and/or the editor(s). MDPI and/or the editor(s) disclaim responsibility for any injury to people or property resulting from any ideas, methods, instructions or products referred to in the content.

heritage

MDPI

Article

Mortar Characterization of Historical Masonry Damaged by Riverbank Failure: The Case of *Lungarno Torrigiani* (Florence)

Sara Calandra [1,2], Teresa Salvatici [1,*], Elena Pecchioni [1], Irene Centauro [1] and Carlo Alberto Garzonio [1]

1 Department of Earth Sciences, University of Florence, 50121 Florence, Italy; sara.calandra@unifi.it (S.C.); elena.pecchioni@unifi.it (E.P.); irene.centauro@unifi.it (I.C.); carloalberto.garzonio@unifi.it (C.A.G.)
2 Department of Chemistry Ugo Schiff, University of Florence, 50019 Florence, Italy
* Correspondence: teresa.salvatici@unifi.it; Tel.: +39-33750088

Abstract: The research of structural masonry associated with geo-hydrological hazards in Cultural Heritage is a multidisciplinary issue, requiring consideration of several aspects including the characterization of used materials. On 25 May 2016, loss of water from the subterranean pipes and of the aqueduct caused an Arno riverbank failure damaging a 100 m long portion of the historical embankment wall of *Lungarno Torrigiani* in Florence. The historical masonry was built from 1854–1855 by Giuseppe Poggi and represents a historical example of an engineering approach to riverbank construction, composed of a scarp massive wall on foundation piles, with a rubble masonry internal core. The failure event caused only a cusp-shaped deformation to the wall without any shattering or toppling. A complete characterization of the mortars was performed to identify the technologies, raw materials and state of conservation in order to understand why the wall has not collapsed. Indeed, the mortars utilized influenced the structural behavior of masonry, and their characterization was fundamental to improve the knowledge of mechanical properties of civil architectural heritage walls. Therefore, the aim of this research was to analyze the mortars from mineralogical–petrographic, physical and mechanical points of view, to evaluate the contribution of the materials to damage events. Moreover, the results of this study helped to identify compatible project solutions for the installation of hydraulically and statically functional structures to contain the riverbank.

Keywords: historical buildings; riverbank failure; mineralogical–petrographic, physical and mechanical characterization; mortars; stone materials

Citation: Calandra, S.; Salvatici, T.; Pecchioni, E.; Centauro, I.; Garzonio, C.A. Mortar Characterization of Historical Masonry Damaged by Riverbank Failure: The Case of *Lungarno Torrigiani* (Florence). *Heritage* **2023**, *6*, 3820–3834. https://doi.org/10.3390/heritage6050203

Academic Editors: Daniela Fico and Daniela Rizzo

Received: 29 March 2023
Revised: 19 April 2023
Accepted: 20 April 2023
Published: 22 April 2023

Copyright: © 2023 by the authors. Licensee MDPI, Basel, Switzerland. This article is an open access article distributed under the terms and conditions of the Creative Commons Attribution (CC BY) license (https://creativecommons.org/licenses/by/4.0/).

1. Introduction

The construction of riverbanks to defend cities began in ancient times. The reasons for construction were linked, on the one hand to the recurrence of disastrous floods and on the other to the expansion of anthropized zones. The current morphology of the Arno river in Florence is the result of typical urbanization of cities that develop along rivers, with the construction of retaining walls and the reduction of the river section [1,2]. The proximity of a river to civil structures and historical center provides a strong reason for detailed stability evaluation [3,4], considering the damage that the failure of the riverbanks could cause. *Lungarno Torrigiani* is a section of Florentine riverbank, located on the left side, 200 m long and near *Ponte Vecchio*. Like most of the Florentine riverbanks, it dates back to specific urban redevelopment, started between 1854–1855 (the realization of the "*Lungarno Nuovo*") and completed during the "*Piano Particolareggiato*" interventions of the Florence Capital (1866–1872).

On 25 May 2016 a riverbank failure damaged a portion of the 100 m long historical embankment wall of *Lungarno Torrigiani* [5].

The displacement of the wall towards the river was about 3 m, with an extension of about 80 m, forming a chasm with a maximum depth of 3.5 m [6]. However, the failure exceptionally did not lead to the collapse of the riverbank, and the structure and the materials used in manufacture remained standing [7].

For this reason, comprehensive study of the riverbank was carried out, focusing on the construction technique and materials used in the original project. Mineralogical–petrographic, physical and mechanical characteristics of mortar were identified. These properties of the building materials are fundamental to evaluate their durability, decay phenomena and service life [8–10].

The analysis related to historical masonry walls is a task to be approached from several aspects. It is indeed necessary to obtain information about the inner core of the structural elements, identifying the mechanical properties of materials that can change greatly from point to point. In addition, the type of mortar used in the realization of its masonry can also influence the performance of the wall, due to the weathering products that can form and which can affect the durability of the structure [11,12]. Therefore, the masonry should be able to accommodate some degree of movement from creep or thermal effects without cracking, and must be sufficiently strong to develop appropriate adhesion between the elements.

Therefore the aim of this work was to study the mortars of the damaged wall of *Lungarno Torrigiani* from mineralogical–petrographic, physical and mechanical points of view, in order to identify and to explain their overall characteristics, understanding how the walls resisted the collapse; the results were compared with other two *Lungarni* of Florence, *Lungarno degli Acciauoli* and *Lungarno delle Grazie* as described in the paper of Calandra et al., 2022 [13], in order to understand the differences between the various riversides.

2. Historical and Area Framework

Lungarno Torrigiani is located on the hydrographical left side of the Arno river in Florence, between *Ponte Vecchio* and *Ponte alle Grazie* (Figure 1a). The construction was started during the years 1854–1855 and continued with all the work for the new urban infrastructures (*Piano Particolareggiato*) until 1872 [14].

The general working program was developed by Luigi Del Sarto, while the architect and director of the works was Giuseppe Poggi. The plan involved the complete transformation of the city; it was divided into ten parts, six on the hydrographical right side and four on the hydrographical left side of the Arno river [15].

In terms of preventing the damage of collapse and flooding, Giuseppe Poggi designed the new Arno riverbanks. In particular, *Lungarno Torrigiani* was built expanding the left bank of the Arno river and demolishing some facade building for their alignment; in fact, the new *Lungarno* cut through the land from *Ponte Vecchio* to *Ponte alle Grazie* (Figure 1b).

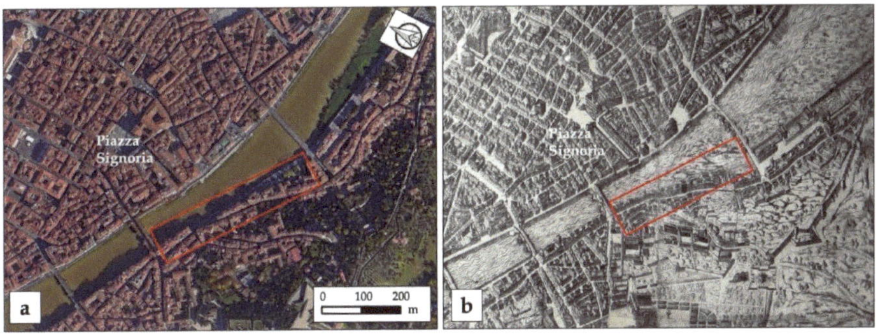

Figure 1. In the red square the study area (**a**) today, (**b**) in 1594 by Stefano Buonsignori [16]. (adapted with permission from https://commons.wikimedia.org/wiki/File:Mappa_del_buonsignori,_dettaglio_03.JPG#/media/File:Pianta_del_buonsignori,_1594,_00.JPG, 2013, Sailko, accessed on 28 March 2023).

According to the Poggi project, by the Budini and Gattai construction company (active in many construction sites in Florence [17]), the retaining wall of *Lungarno Torrigiani* riverbank is composed of a massive scarp with four rows of wood piles driven 1.20 m under the medium level of the river (Figure 2). The masonry is made up of a vertical foundation wall of 0.75 m, a massive scarp wall of 7.17 m and a brick railing of 1.20 m high from the road surface complete the riverbank.

Figure 2. Construction techniques of the *Lungarno Torrigiani* riverbank (modified from original drawings preserved in the municipal historical archive of Florence).

The wall is constructed with an external layer and an internal core using the rubble masonry building method. The former consists of a cortical wall constructed of various squared stones, quarried near Florence and therefore widely used in the city's buildings (mainly Pietra Serena sandstone, and in other stretches of the Lungarno also with Pietraforte sandstone and Pietra Alberese marly limestone, as reported by the municipal historical archive of Florence). The latter consists of mortar made of coarse aggregate (mostly stone elements and, in some levels, also bricks) bound together by a binder filling the interior.

The behavior during the failure of the *Lungarno Torrigiani* highlighted an extremely compact, supportive, rigid wall system. In fact, there were three main breaks, corresponding to clear fractures, the central one in correspondence with the movement towards the river and two "lateral hinges". This compactness is certainly due to the construction techniques of specialized workers, very active during the entire period from and after Florence as the Capital of Italy. According to this technique, all the elements of the masonry (stone blocks, fillings, binders and coatings) were carried out simultaneously.

Historical research shows that in 1944 during World War II, a section of *Lungarno Torrigiani* near the *Ponte Vecchio* was partially destroyed. The Nazi high command had ordered the German commander, Field Marshal Albert Keisselring, to blow up Florence's bridges, but to save any he deemed "culturally significant". During the operation called *Feuerzauber* or *magic fire*, only the *Ponte Vecchio* was saved, but not the buildings lining the approaching streets to the *Ponte Vecchio*, in order to block the way of the Allies. Figure 3a shows the photo after the Florence liberation, where only a part of the *Lungarno Torrigiani* close to the *Ponte Vecchio* was destroyed.

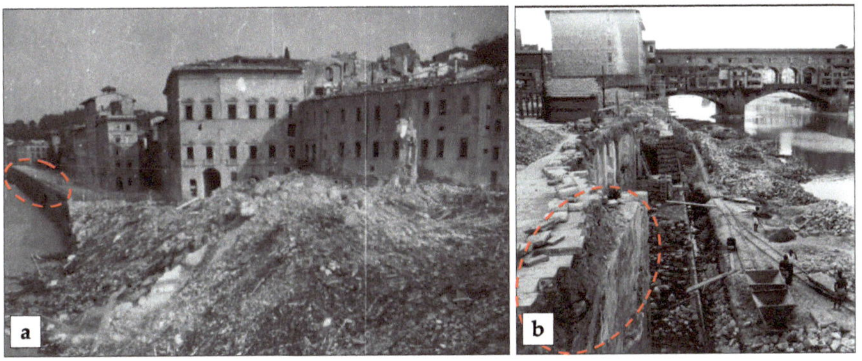

Figure 3. (**a**) Image of *Lungarno Torrigiani* close to the *Ponte Vecchio* after the liberation in the 1944: the condition of the riverbank; (**b**) the reconstruction of the *Lungarno* close to the *Ponte Vecchio*, in 1947 (photo Fondo Berti, ISRT). In red are highlighted the walls of *Lungarno Torrigiani* that remained unharmed.

The reconstruction of this part of the *Lungarno Torrigiani* (which does not include the area object of this research, as visible in Figure 3a,b) was carried out considering the reconstruction plan "*Città sul fiume, piano di ricostruzione della zona intorno al Ponte Vecchio*" proposed by Edoardo Detti, Riccardo Gizdulich, Rolando Pagnini and Danilo Santi in 1947 [18,19]. Figure 3b shows reconstruction works throughout the post-war period.

3. Materials and Methods

3.1. Sampling

The sampling was carried out from an area affected by fractures, cracks (Figure 4a) and detachments in the low part of the massive scarp wall, at the height of almost 1 m. The damaged section of the riverbank, where the sampling masonry was carried out, is located on a portion that remained unharmed from destruction during World War II. The samples shown in Figure 4b were collected from bedding mortars (LT1-A), jointing mortars (LT2-B) and core wall mortars (LT3, 4, 5, 6-C).

Figure 4. Study area and sampling position of *Lungarno Torrigiani* riverbank: (**a**) damaged area by embankment failures, on which an assay was carried out to study the masonry and perform sampling; (**b**) the assay and position of mortars sampling from bedding, jointing and from the core of masonry.

The mineralogical, petrographic and microchemical characterization was performed on 2 samples of mortar, belonging to bedding and jointing masonry, and 4 samples from core masonry. The physical and mechanical properties were determined on mortar sampling from the core masonry.

3.2. Mineralogical–Petrographic and Microchemical Analyses

The mineralogical–petrographic observations (OM) were performed with video camera (5 megapixels resolution), and analysis software AxioVision (V1), on the thin sections (30 μm) of the mortars [20]; this kind of analysis allows for highlighting the texture of the binder and the finer and medium coarse of the aggregate.

Mineralogical semi-quantitative characterizations were conducted with X-ray diffraction (XRPD) on bulk powder samples, using a diffractometer (Philips PW 1050/37 coupled with Philips X'Pert PRO for data acquisition and interpretation system); detection limit was 4% [21–23]. The operating conditions were: 40 kV-20 mA, Cu anode, graphite monochromator and 2°/min goniometry speed in 5–70° θ range.

A SEM-EDS electronic microscope (ZEISS EVO MA 15) with W filament and an analytical system in the dispersion of energy EDS/SDD, Oxford Ultimax 40 (40 mm^2 with resolution 127 eV @5.9 keV) was used to determine semi-quantitative microchemical and morphological characterizations.

The measurements were carried out on thin sections (after carbon-metallized pretreatment) in binder and lumps areas. The operating conditions were: 15 kV acceleration potential, 500 pA beam current, 9–8.5 mm working distance; 20 s live time as acquisition rate useful to archive at least 600,000 cts, on Co standard and process time 4 for point analyses; and 500 μs pixel dwell time for maps acquisition with 1024 × 768 pixel resolution.

The microanalysis was performed using Aztec 5.0 SP1 software employing the XPP matrix correction scheme developed by [24]. The process used purchased standard elements for calculations, allowing "standard-less" quantitative analysis. With numerous analyses of a Co metallic standard, the tracking of constant analytical conditions (i.e., filament emission) was recorded.

3.3. Physical and Mechanical Analysis

Samples of mortar of the core wall from the masonry embankment were used for the physical analysis to calculate the water absorption coefficients by capillarity and for total immersion, together with the open porosity and apparent density.

The water absorption by capillarity tests were performed based on the European standard [25] on four prismatic modules obtained by cutting the samples with dimensions of about 4 cm × 4 cm × 4 cm. The modules were dried in an oven at 60 °C until constant mass was reached. Then, each module was placed on a humid filter on footholds in the tank in contact with the water, allowing the samples to adsorb water by upward capillary forces. After a fixed time (5, 10, 20 30, 60, 90, 120,180, 240, 360, 1440, 2880 min) the samples were weighed, removing the water on the surface which was not absorbed using a wet cloth [26–28]. The test allowed us to estimate the water absorption by capillarity of the mortars.

The capillary water absorption coefficient C is calculated by linear regression, using at least five successive aligned points, considering the slope of the linear section of the curve obtained and plotting the mass difference versus the square root of time per area (A) (1).

$$C = \frac{(M_2 - M_1)}{A\sqrt{t_f - t_i}} \tag{1}$$

where A equals the area of the sample, M_1 and M_2 are the sample mass at different time instant and t_i and t_f the initial and final time, respectively.

In order to better characterize the mortar physical properties, the samples were also tested for total water absorption, apparent density and open porosity. In addition, the mineralogical and petrographic analyses to identify mortar composition, structure, texture and pore system were important for testing the water absorption (described in Section 3.2).

The water absorption by total immersion tests provided the absorption of water during the time of totally immersed samples in water, based on the UNI EN 1097-6 [29]. The samples were obtained by cutting the four specimens used for capillarity absorption determination. Initially the specimens were dried in the oven until reaching a constant mass. Subsequently, they were saturated with water at a temperature of about 20 °C. The obtained results allowed us to determine the imbibition coefficient, according to Equation (2):

$$W = \frac{(M_s - M_d)}{M_d} * 100 \tag{2}$$

where M_s is the sutured mass and M_d the dry mass.

The total quantity of water absorbed is related to the total open porosity, after saturation, and the hydrostatic mass was determined according to the UNI EN 1936 [30] standard. Thanks to the hydrostatic balance of the Mettler Toledo XS 204, with a maximum weighing of 220 g and a resolution of 0.1 mg, hydrostatic mass, apparent density and open porosity were defined as:

$$\rho = \frac{M_s}{M_s - M_h}(\rho_0 - \rho_L) + \rho_L \tag{3}$$

$$P = \frac{M_s - M_d}{M_s - M_h} * 100 \tag{4}$$

where M_h is the hydrostatic weight, ρ_0 is the density of the auxiliary liquid and ρ_L is the air density.

In order to understand the mechanical behaviors of the samples, the ultrasonic velocity test was carried out on the same test samples (4 cm × 4 cm × 4 cm) for physical analysis. The cubic specimens, after the physical analysis, were again dried in a stove at a temperature of 60 °C until a constant weight was reached. In this way it was possible to compare all values and to correlate the results. The ultrasonic tests were performed through an IMG 5200 CSD ultrasonic instrument characterized by two 50 kHz transducers. The measurements were carried out in the direct transmission approach with two transducers placed on the opposite faces of samples in three dimensions. For each position, three measures were carried out in order to minimize errors; the result is the mean of nine measures for each sample. The calculated parameter is the velocity of the first ultrasonic wave able to travel through the material (Vp); based on its value it is possible to detect the occurrence of internal defects and inhomogeneities of the mortar itself.

4. Results and Discussion

4.1. Mineralogical–Petrographic and Microchemical Data

The petrographic observations of the mortar of the bedding, of the jointing and of the core wall highlight similar characteristics, showing a binder made with an air hardening calcitic lime of light brown color, anisotropic and homogeneous appearance and with micritic and microsparitic texture; small dark impurities are widespread in the binder (Figure 5a,b). The aggregate is always well distributed, from sub-angular to sub-rounded in shape, with two-dimensional distribution (size from 200–300 μm to 500–600 μm); sometimes fragments of millimetric size are also present.

Figure 5. (**a**) Image under PLM, xpl: mortar with an air hardening calcitic lime and aggregate characterized of different rock fragments (sandstone, marly limestone, pelite, quartzite); (**b**) image under PLM, ppl: small dark impurities are widespread in the binder of the mortar.

The composition mainly consists of sand in mono and polycrystalline quartz grains (Figure 6a) beyond calcite fragments, rare gypsum (Figure 6b), plagioclases and micas; besides sandstone, carbonate, marly limestones, mica schists, pelites and quartzitic rock fragments widespread in a binder showing small dark impurities (Figure 6c–e), rare *cocciopesto* is also present (Figure 6a). Lumps in the form of unmixed lime and of underburnt marly limestone remains are present, as well as a higher concentration of small dark impurities in the binder (Figure 6d–f). The macro porosity visible in thin sections is medium–low; there are also fractures and the binder/aggregate ratio is 1/3.

Pietra Alberese, a marly limestone from the Ligurian series (Monte Morello Formation, of Eocene age), is most likely the marly limestone found in the aggregate.

The locality near Florence where this marly formation exhibits the typical outcrop is where the term "Monte Morello Formation" originates. This stone was mainly employed to utilize the lime in the area from Florence to Pistoia.

The confirmation of the Pietra Alberese use in the past as a raw material for the production of lime is usually testified by "ghosts" of underburnt rock fragments in the historical mortars. Such inclusions show the same petrographic features of this marly limestone [31]. In the samples analyzed (Figure 6d–f) both fragments of Pietra Alberese limestone and lumps of the underburnt limestone are identified. This also justifies the presence, within the binder, of neoformation phases in the form of small dark impurities (Figures 5 and 6). These consist of not-hydrated calcium silicates and aluminates as product of the reaction (in burning), between the calcium oxide coming from the dissociation of calcite and the amorphous silicate compound coming from the destruction of the clay minerals present in the marly limestone (Pietra Alberese) [32]. The use of Pietra Alberese in which a variable amount of clay minerals is present, from 7% to 26%, proves the presence of two varieties: *sasso alberese* with a lower content of clay minerals and lighter color and of *sasso porcino* with a higher content of clay minerals and darker color; both varieties can give rise to a partial hydraulicity in the mortar. The microscopic appearance of the *Lungarno Torrigiani* binders seem to confirm the use of *sasso alberese* [31].

The mineralogical semi-quantitative analyses on X-ray diffraction confirm the petrographic observations. The data in all samples show a prevalence of quartz and calcite, while plagioclases and k-feldspars (feldspars) with micas and clay minerals are in lower amounts (Table 1).

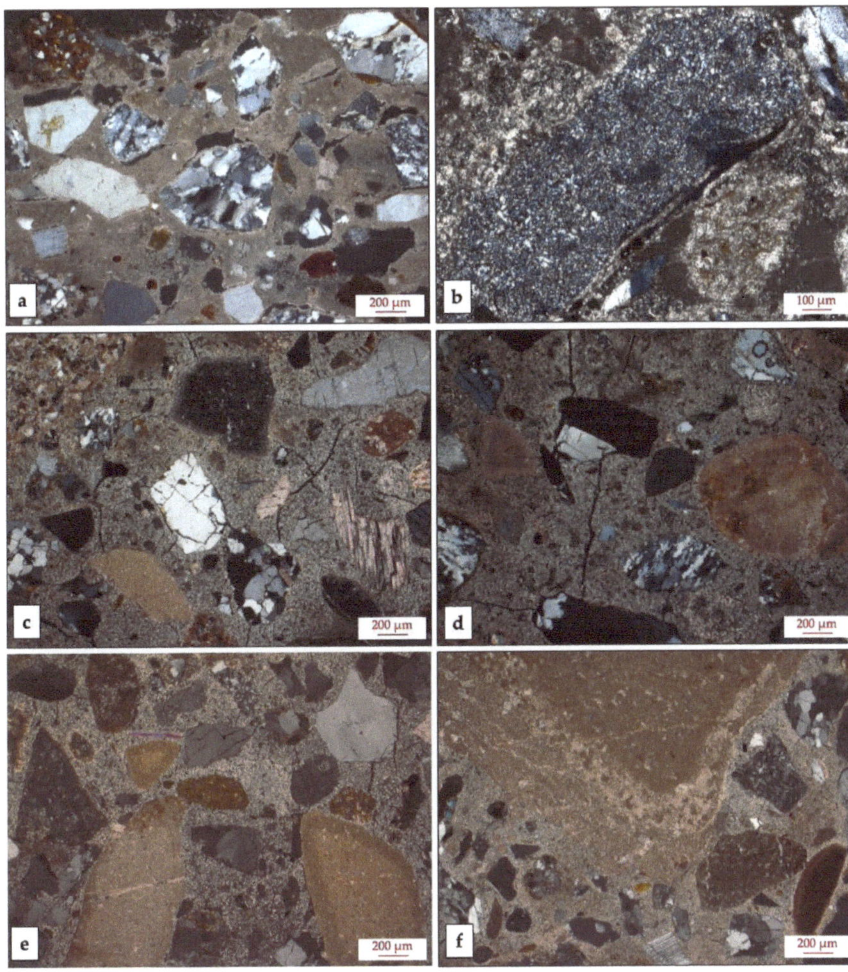

Figure 6. Images under PLM, xpl; (**a**) LT1-A, mortar sample with an air hardening calcitic lime and aggregate constituted in prevalence of polycrystalline quartz, calcite and quartzitic fragments; (**b**) LT2-B, mortar showing a fragment of gypsum; (**c**) LT3-C, mortar with characteristic aggregate of two different sizes and quartzitic and calcitic composition; (**d**) LT4-C, mortar with quartzite aggregate, a lump in form of an underburnt marly limestone remains and small dark impurities in the binder; (**e**) LT5-C, mortar with prevalence of marly limestone fragments; (**f**) LT6-C, mortar showing mainly a lump in form of an underburnt marly limestone remains.

Table 1. XRPD results of the analysis of *Lungarno Torrigiani* mortar samples.

Sample	Quartz	Calcite	Feldspars	Micas	Clay Minerals
LT1-A	xxx	xx	x	tr	tr
LT2-B	xxx	xx	x	tr	tr
LT3-C	xxx	x	x	tr	tr
LT4-C	xxx	xx	x	tr	tr
LT5-C	xxx	xx	x	tr	tr
LT6-C	xxx	xx	tr	tr	tr

Semi-quantitative amount xxx = high amount; xx = medium amount; x = low amount; tr = in traces.

The SEM-EDS microchemical and morphological studies highlighted the characteristics of the binder and the lumps [33]. The presence of hydraulicity is often linked to the appearance of the binder and the lumps, showing small dark impurities that can be easily recognized by the polarized microscope, as underlined above. These impurities could be referred to as non-hydrated relics of belite (C_2S). Therefore, SEM-EDS microchemical analyses of the unburnt lumps and binder areas were carried out (Figures 7 and 8). The analysis on the unburnt lumps shows high values of Ca, and a variable amount of Si and Al, due to the presence of clay minerals in the original marly limestone (Figure 7). Such microchemical variability is reflected also on the composition of the binder, made by burning a marly limestone. Indeed, the binder shows a varying composition and a medium hydraulicity index about HI 0.20. The binder presents areas with Si and Al in higher amounts, due to the pollution given by the finest aggregate fraction (quartz and feldspars) widespread and areas with Ca prevailing due to a bad mixing. The punctual microchemical analysis in the binder sample LT6-C (core wall) confirms that Ca is 90% and Si 6% in spectrum 1, while in spectrum 2 Ca is prevailing (79%) with Si (14%) (Figure 8).

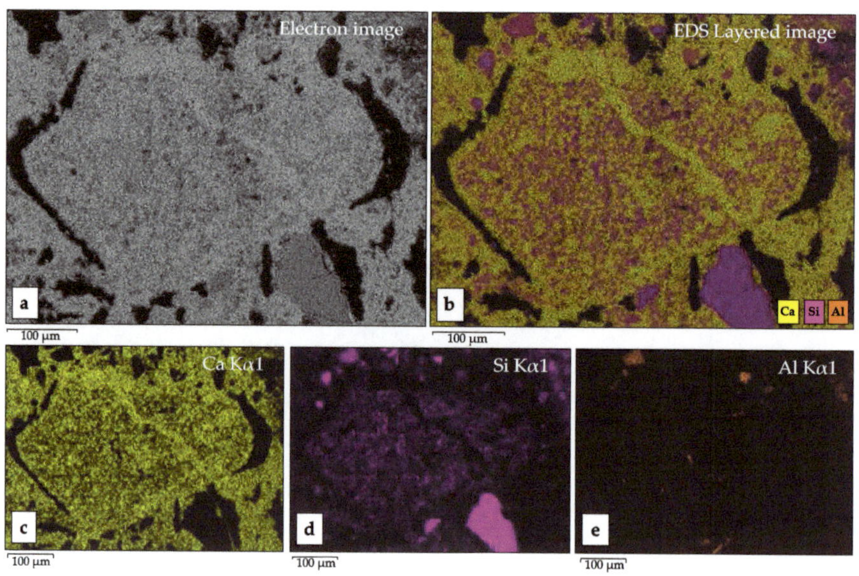

Figure 7. (**a**) BS image of a detail of a binder portion and of the unburnt limestone (sample LT1-A) in which areas with different gray tones can be observed; (**b**) SEM-EDS map layered of the previous area; (**c**) SEM-EDS map of Ca; (**d**) SEM-EDS map of Si; (**e**) SEM-EDS map of Al.

Nonetheless, it is possible to highlight that the core sample LT6-C has areas in which the binder has a higher content of Si and Al (Figure 8b–d), confirming the petrographic observations showing a higher concentration of small dark impurities in the binder.

4.2. Physical and Mechanical Data

Figure 9a shows the physical analysis results of the four core wall samples.

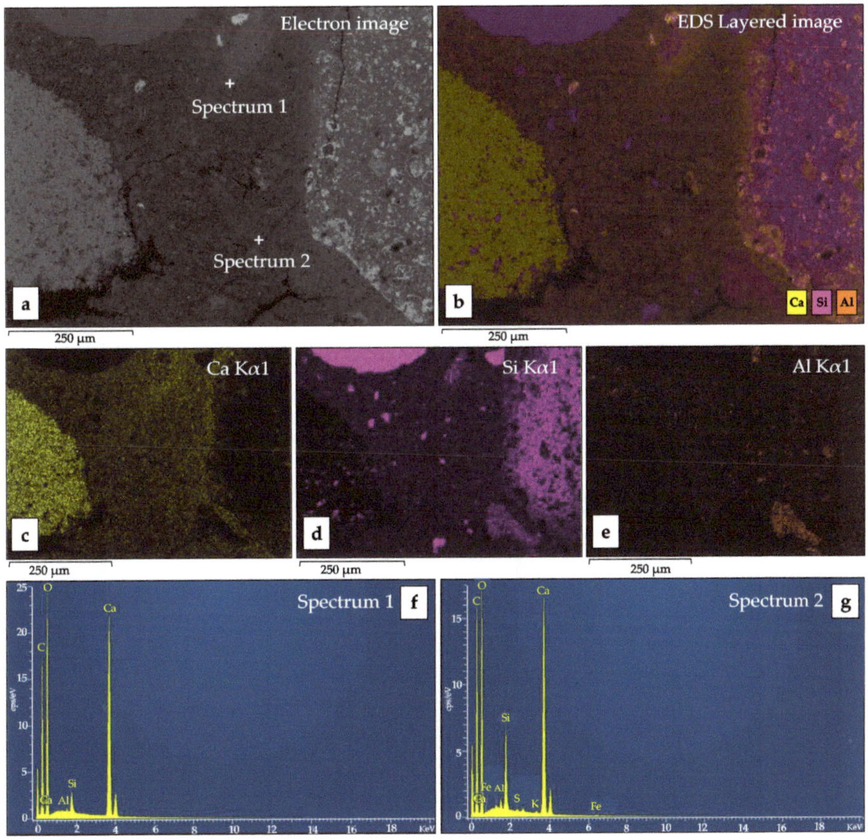

Figure 8. (**a**) BS image of a detail of the binder portion (sample LT6-C) in which areas with different gray tones can be observed and punctual microchemical analysis of the sample; (**b**) SEM-EDS layered image of the previous area; (**c**) SEM-EDS map of Ca; (**d**) SEM-EDS map of Si; (**e**) SEM-EDS map of Al; (**f**,**g**) EDS spectra of punctual analysis of binder.

The average capillarity absorption coefficient (C) is about 27.75 $g/m^2s^{0.5}$, corresponding to an amount of water absorbed of about 15% typical of mortars. All the samples demonstrated a remarkably swift imbibition during the initial 60 min of the test, as the largest porosities became saturated through contact immediately [13,34–36]. In addition, IC, P and ρ showed typical values for the mortar [37]. There is a good correlation between the physical data and macroscopic observation. The mortar presented well sorted coarse aggregates from centimeter to millimeter size mixed with finer binder portions.

The ultrasonic velocity results are affected by the pores, fine portion/coarse aggregate ratio and aggregate dimensions. The presence of aggregates elevates velocity values, whereas a high porosity level in the sample results in lower ultrasonic velocities [38]. The pores in the fine portion of mortar and the cracks between the binder and aggregates cause discontinuity in the mortar, to which correspond both low ultrasonic velocities and low mechanical characteristics. The samples exhibited average Vp values of about 2300 m/s that include mortar and stone aggregates in good state of conservation, with quite good mechanical properties. The results of Vp for each examined sample show a good correlation with the macroscopic consideration on fine portion/coarse aggregate ratio, grading and content of aggregates (Figure 9b). In particular, sample LT3-C, characterized by the presence of coarse aggregate and less fine fraction, showed higher Vp values than sample LT6-C, which appeared more degraded with less presence of the coarse aggregate. In addition, if physical test results are also taken into account, sample LT3-C was characterized by lower porosity than the other three. Mortars with lower porosity levels show high density (ρ) and Vp values, whereas mortars with higher porosity levels and lower bulk density are expected to have lower ultrasonic propagation velocities.

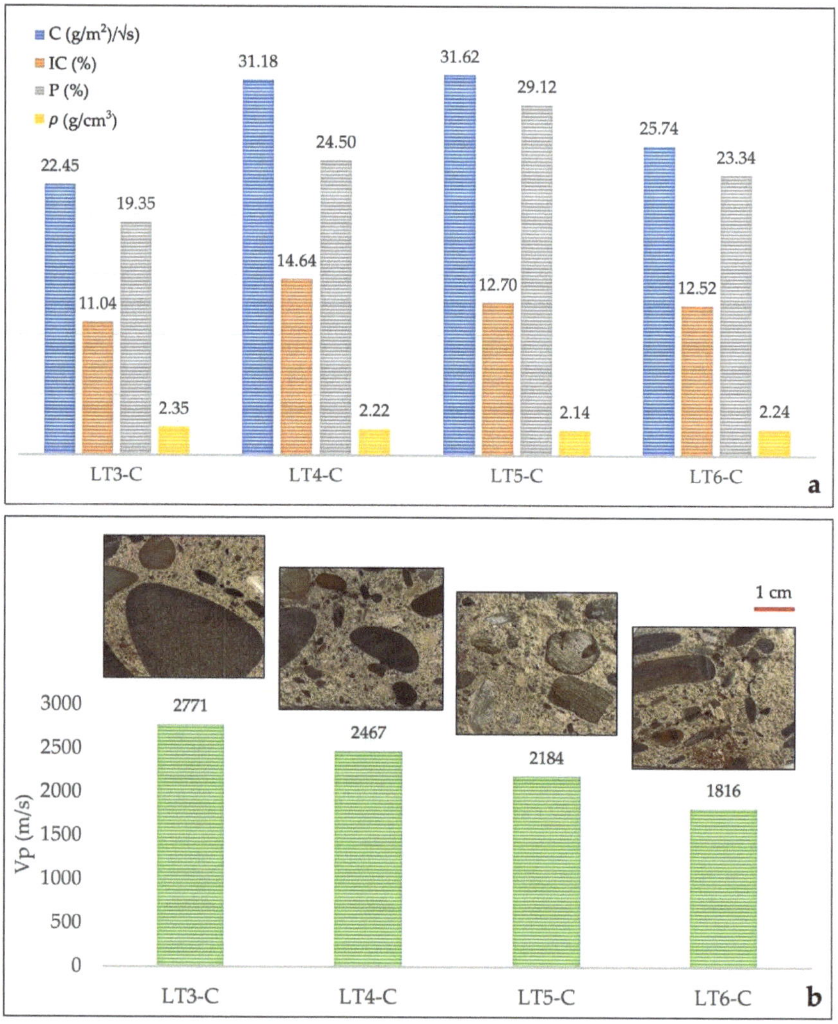

Figure 9. Results of physical and mechanical analyses. (**a**) The values reported are capillary water absorption coefficient (C), open porosity (P, or porosity accessible to the water), total imbibition coefficient (IC) and density (ρ). (**b**) The Vp mean value and macroscopic images of each examined sample.

5. Discussion about the Comparison between the *Lungarno Torrigiani* Mortars with *Lungarno degli Acciaiuoli* and *Lungarno delle Grazie*

It is interesting to compare the results obtained from analyzing the mortar in the inner part of the *Lungarno Torrigiani* rubble masonry with the information gathered in a study by Calandra et al. in 2022 [13]. The latter investigated the chemical, mineralogical–

petrographic and physico-mechanical properties of core samples taken from the stonework of two other riverbanks in Florence (*Lungarno degli Acciaiuoli* and *Lungarno delle Grazie*).

All riverbanks of Florence were in fact expanded by Giuseppe Poggi over the course of the 1800s for the redevelopment of the city. In *Lungarno degli Acciaiuoli* and *Lungarno delle Grazie*, four typologies of mortar were identified; two ancient (named Type X and Type Y in [13]) and two modern mortars. Based on a macroscopic description, the mortar closer to *Lungarno Torrigiani* mortar is the one named Type Y [13], described as a historic mortar with centimeter-sized coarse aggregates. Such mortar corresponds to the deepest portion in the two riverbanks masonry stratigraphy. The microscopic characterization results, however, suggest that the fine portions of the two ancient mortars (Type Y and Type X) have the same composition and differ only by macroscopical aspects.

From the comparison between *Lungarno degli Acciaiuoli*, *Lungarno delle Grazie* and *Lungarno Torrigiani* petrographic observations, differences were observed, even though the raw material used to produce lime remained constant (Pietra Alberese). *Lungarno degli Acciaiuoli* and *Lungarno delle Grazie* samples showed a darker colored binder with heterogeneous appearance (areas with low birefringence were identified) and micritic/microsparitic texture. This is typical of mortars obtained from burning of Pietra Alberese with higher clay minerals (*sasso porcino*). On the contrary, the *Lungarno Torrigiani* mortars presented a lighter colored binder with micritic/microsparitic texture, but homogeneous appearance typical of mortars obtained from burning of Pietra Alberese with lower clay minerals amount (*sasso alberese*). The different binder characteristics of *Lungarno degli Acciaiuoli* and *Lungarno delle Grazie* may therefore be due to a higher hydraulicity (obtained burning *sasso porcino*) [31]. In addition, the endoscopic investigations of *Lungarno degli Acciaiuoli* and *Lungarno delle Grazie* have shown a storm drain running parallel to the *Lungarni*. Such a structure could allow more water into the masonry; this could change the optical features of the binder, making it darker.

Comparing the results of the physical analysis (Figure 10), it is possible to highlight that the mortars of the *Lungarno Torrigiani* have similar porosity and imbibition coefficient values to the mortars of type Y from *Lungarno degli Acciaiuoli* and *Lungarno delle Grazie*. While the capillarity absorption and the ultrasonic velocity are slightly lower, such distinction could be due to higher hydraulic behavior of *Lungarno degli Acciaiuoli* and *Lungarno delle Grazie* mortars, confirming the use of a different variety of Pietra Alberese with higher content of clay minerals.

In any case, the findings indicate that all riverbanks exhibit favorable mineralogical–petrographic and physico-mechanical properties. The mortars display both high compactness and cohesion levels.

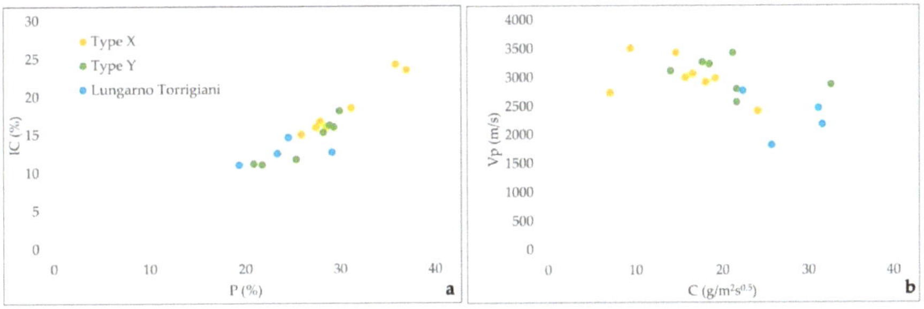

Figure 10. Comparison between physical and mechanical data of *Lungarno Torrigiani* mortar and Type Y and Type X of *Lungarno degli Acciaiuoli* and *Lungarno delle Grazie* mortars. (**a**) P (%) vs. IC (%) and (**b**) C ($g/m^2s^{0.5}$) vs. Vp (m/s).

6. Conclusions

In general terms, it can be assessed that the original mortar used in the *Lungarno Torrigiani* historical embankment wall has been created by burning a marly limestone

(Pietra Alberese, probably variety *sasso alberese*), and is able to provide a partial hydraulicity to the mortars and give a good cohesion with coarse and fine aggregates utilized in the manufacturing. The partial hydraulicity is confirmed both by the presence of not-hydrated calcium silicates and aluminates, visible in small dark impurities widespread in the binder and by the results of the microchemical analysis highlighting binder areas with a medium hydraulicity index. It is interesting to underline that the petrographic and microchemical analyses carried out on the internal core show a greater thickening of dark small impurities and a higher amount of Si and Al in the binder, confirming a greater resistance of this part of the masonry. Therefore, the numerous lumps (both underburnt fragments of Pietra Alberese and of unmixed lime) present in the mixture prove a historical production technique of the mortars. The centimeter/millimeter-sized coarse aggregates are constituted mainly by sandstone and marly limestone stone fragments (most likely Pietra Serena, Pietraforte and Pietra Alberese of local origin), while the finer fraction is constituted mainly by quartzitic sand, also in this case from the Florentine area, probably the Arno valley. These aggregates are always well distributed. The macroporosity is medium–low; it is interesting to underline that the core samples (LT3, 4, 5, 6-C) show a lower porosity and appear more compact.

The mortars show good physical properties: low porosity (P = 24%) that ensure a low imbibition coefficient and low capillarity absorption coefficient (IC = 12.72% and $C = 27.75\ g/m^2s^{0.5}$), with good density ($\rho = 2.23\ g/cm^3$) and good mechanical properties, in agreement with the characteristic described by [36]. A correlation between petrographic observation (macroporosity, type of binder, composition of aggregate, etc.) and physical and mechanical data has been observed, confirming the good quality of the mortars.

It is important to underline the excellent quality of the materials used and the type of traditional manufacturing, combined with the construction method used by the Budini and Gattai construction company. The rubble masonry, in which the wall is created with an outer layer and an internal core built gradually at the same time (the first consisting in a cortical wall made with squared stone (sandstone) and the inner filled with mortar) allowed it to avoid the catastrophic event [17]. Indeed, during the failure event of 25 May 2016, the *Lungarno Torrigiani* riverbank did not collapse into the Arno river.

Author Contributions: Conceptualization E.P., S.C., T.S., I.C. and C.A.G.; methodology, T.S., E.P. and S.C.; investigation, S.C., E.P. and T.S.; data curation, S.C., E.P. and T.S.; writing—original draft preparation, S.C., E.P., I.C. and T.S.; writing—review and editing, S.C., E.P., I.C. and T.S.; supervision, C.A.G. All authors have read and agreed to the published version of the manuscript.

Funding: This research received no external funding.

Data Availability Statement: Not applicable.

Acknowledgments: The authors are grateful to MEMA Centro di Servizi di Microscopia Elettronica e Microanalisi (DST-UNIFI) for the technical support with the SEM-EDS analysis.

Conflicts of Interest: The authors declare no conflict of interest.

References

1. Morelli, S.; Segoni, S.; Manzo, G.; Ermini, L.; Catani, F. Urban planning, flood risk and public policy: The case of the Arno River, Firenze, Italy. *Appl. Geogr.* **2012**, *34*, 205–218. [CrossRef]
2. Morelli, S.; Pazzi, V.; Tofani, V.; Raspini, F.; Bianchini, S.; Casagli, N. Reconstruction of the slope instability conditions before the 2016 failure in an Urbanized District of Florence (Italy), a UNESCO world heritage site. In *Understanding and Reducing Landslide Disaster Risk*; Sendai Landslide Partnerships (Kobe-City, Japan); Kyoto Landslide Commitment 5th: Kyoto, Japan, 2021; Volume 1, pp. 449–455.
3. Morelli, S.; Battistini, A.; Catani, F. Rapid assessment of flood susceptibility in urbanized rivers using digital terrain data: Application to the Arno river case study (Firenze, northern Italy). *Appl. Geogr.* **2014**, *54*, 35–53. [CrossRef]
4. Morelli, S.; Pazzi, V.; Tanteri, L.; Nocentini, M.; Lombardi, L.; Gigli, G.; Casagli, N. Characterization and geotechnical investigations of a riverbank failure in Florence, Italy, UNESCO World Heritage Site. *J. Geotech. Geoenviron. Eng.* **2020**, *146*, 05020009. [CrossRef]
5. Dapporto, S.; Rinaldi, M.; Casagli, N.; Vannocci, P. Mechanisms of riverbank failure along the Arno River, Central Italy. *Earth Surf. Process. Landf.* **2003**, *28*, 1303–1323. [CrossRef]

6. Asioli, C.; Bertero, A.; Cribari, F.; Agostini, C.; Chiarugi, M.; Galli, O.; Tognotti, M.; Spinelli, P.; Aiello, E.; Micheloni, M. Intervento di messa in sicurezza in somma urgenza del tratto interessato da dissesto sul Lungarno Torrigiani a Firenze: Progetto ed esecuzione delle opere di consolidamento, conservazione e ripristino del muro spondale dislocate. In Proceedings of the XXVI Convegno Nazionale di Geotecnica, Rome, Italy, 20–22 June 2017; Associazione Geotecnica Italiana (AGI): Rome, Italy, 2017.
7. Pazzi, V.; Lotti, A.; Chiara, P.; Lombardi, L.; Nocentini, M.; Casagli, N. Monitoring of the vibration induced on the Arno masonry embankment wall by the conservation works after the May 25, 2016 riverbank landslide. *Geoenviron. Disasters* **2017**, *4*, 6. [CrossRef]
8. Ponce-Antón, G.; Arizzi, A.; Cultrone, G.; Zuluaga, M.C.; Ortega, L.A.; Agirre Mauleon, J. Investigating the manufacturing technology and durability of lime mortars from Amaiur Castle (Navarre, Spain): A chemical–mineralogical and physical study. *Constr. Build. Mater.* **2021**, *299*, 123975. [CrossRef]
9. Sardella, A.; Canevarolo, S.; Marrocchino, E.; Tittarelli, F.; Bonazza, A. Investigation of Building Materials Belonging to the Ruins of the Tsogt Palace in Mongolia. *Heritage* **2021**, *4*, 2494–2514. [CrossRef]
10. Columbu, S.; Lisci, C.; Sitzia, F.; Lorenzetti, G.; Lezzerini, M.; Pagnotta, S.; Raneri, S.; Legnaioli, S.; Palleschi, V.; Gallello, G.; et al. Mineralogical, petrographic and physical-mechanical study of Roman construction materials from the Maritime Theatre of Hadrian's Villa (Rome, Italy). *Measurement* **2018**, *127*, 264–276. [CrossRef]
11. Franzoni, E.; Santandrea, M.; Gentilini, C.; Fregni, A.; Carloni, C. The role of mortar matrix in the bond behavior and salt crystallization resistance of FRCM applied to masonry. *Constr. Build. Mater.* **2019**, *209*, 592–605. [CrossRef]
12. Haneefa, K.M.; Rani, S.D.; Ramasamy, R.; Santhanam, M. Microstructure and geochemistry of lime plaster mortar from a heritage structure. *Constr. Build. Mater.* **2019**, *225*, 538–554. [CrossRef]
13. Calandra, S.; Salvatici, T.; Centauro, I.; Cantisani, E.; Garzonio, C.A. The Mortars of Florence Riverbanks: Raw Materials and Technologies of Lungarni Historical Masonry. *Appl. Sci.* **2022**, *12*, 5200. [CrossRef]
14. Cresti, C.; Zangheri, L. *Architetti e Ingegneri nella Toscana dell'800*; UNEDIT: Florence, Italy, 1978.
15. Poggi, G. *Sui Lavori per L'ingrandimento di Firenze. Relazione di G.P. (1864–1877)*; Barbera, G., Ed.; Fiorentinagas—Giunti: Florence, Italy, 1882.
16. Del Badia, I. Pianta topografica della città di Firenze di don Stefano Bonsignori dell'anno 1584. In Proceedings of the Atti del III Congresso Geografico Italiano, Florence, Italy, 12–17 April 1898; pp. 570–577.
17. Fei, S.; Spini, G. *Nascita e Sviluppo di Firenze Città Borghese*; Editrice: Alberobello, Italy, 1971.
18. Fantozzi Micali, O. *Piani di Ricostruzione e Città Storiche 1945–1955*; Alinea: Florence, Italy, 1998.
19. Gabellini, P.; Ragghianti, R. Edoardo Detti 1913–1984. In *Le Sculture di Paolo Borghi Omaggio agli Urbanisti del Novecento*; Ministero dei Lavori Pubblici: Rome, Italy, 2001.
20. *UNI EN 12407*; Natural Stone Test Methods—Petrographic Examination. Ente Nazionale Italiano di Normazione: Milan, Italy, 2007.
21. *UNI EN 13925-1*; Non-Destructive Testing—X-Ray Diffraction from Polycrystalline and Amorphous Material—Part 1: General Principles. Ente Nazionale Italiano di Normazione: Milan, Italy, 2006.
22. *UNI EN 13925-2*; Non-Destructive Testing—X-Ray Diffraction from Polycrystalline and Amorphous Materials—Part 2: Procedures. Ente Nazionale Italiano di Normazione: Milan, Italy, 2006.
23. *UNI-EN 13925-3*; Non-Destructive Testing—X-Ray Diffraction from Polycrystalline and Amorphous Materials—Part 3: Instruments. Ente Nazionale Italiano di Normazione: Milan, Italy, 2006.
24. Pouchou, J.L.; Pichoir, F. Quantitative Analysis of Homogeneous or Stratified Microvolumes Applying the Model "PAP". In *Electron Probe Quantitation*; Springer: Boston, MA, USA, 1991; pp. 31–75. [CrossRef]
25. *UNI EN 1015-18:2002*; Methods of Test for Mortar for Masonry—Part 18: Determination of Water Absorption Coefficient due to Capillary Action of Hardened Mortar. Ente Nazionale Italiano di Normazione: Milan, Italy, 2002.
26. Martys, N.S.; Ferraris, C.F. Capillarity transport in mortars and concrete. *Cem. Concr. Res.* **1997**, *27*, 747–760. [CrossRef]
27. Lanzón, M.; García-Ruiz, P.A. Evaluation of capillary water absorption in rendering mortars made with powdered waterproofing additives. *Constr. Build. Mater.* **2009**, *23*, 3287–3291. [CrossRef]
28. Cunha, S.; Aguiar, J.; Ferreira, V.; Tadeu, A. Mortars based in different binders with incorporation of phase-change materials: Physical and mechanical properties. *Eur. J. Environ. Civ. Eng.* **2015**, *19*, 1216–1233. [CrossRef]
29. *UNI-EN 1097-6*; Tests for Mechanical and Physical Properties of Aggregates—Part 6: Determination of Particle Density and Water Absorption. Ente Nazionale Italiano di Normazione: Milan, Italy, 2013.
30. *UNI EN 1936*; Natural Stones Test Methods: Determination of Real Density and Apparent Density and of Total and Open Porosity. Ente Nazionale Italiano di Normazione: Milan, Italy, 2007.
31. Fratini, F.; Cantisani, E.; Pecchioni, E.; Pandeli, E.; Vettori, S. Pietra Alberese: Building Material and Stone for Lime in the Florentine Territory (Tuscany, Italy). *Heritage* **2020**, *3*, 1520–1538. [CrossRef]
32. Pecchioni, E.; Fratini, F.; Cantisani, E. *Atlas of the Ancient Mortars in Thin Section under Optical Microscope*, 2nd ed.; Nardini: Florence, Italy, 2020; p. 78.
33. Cantisani, E.; Fratini, F.; Pecchioni, E. Optical and electronic microscope for minero-petrographic and microchemical studies of lime binders of ancient mortars. *Minerals* **2022**, *12*, 41. [CrossRef]
34. Veiga, R.; Velosa, A.; Magalhaes, A. Experimental applications of mortars with pozzolanic additions: Characterization and performance evaluation. *Constr. Build. Mater.* **2009**, *23*, 318–327. [CrossRef]

35. Aragon, G.; Aragon, A.; Santamaria, A.; Esteban, A.; Fiol, F. Physical and mechanical characterization of a commercial rendering mortar using destructive and non-destructive techniques. *Constr. Build. Mater.* **2019**, *224*, 835–849. [CrossRef]
36. Cantisani, E.; Calandra, S.; Barone, S.; Caciagli, S.; Fedi, M.; Garzonio, C.A.; Vettori, S. The mortars of Giotto's Bell Tower (Florence, Italy): Raw materials and technologies. *Constr. Build. Mater.* **2021**, *267*, 120801. [CrossRef]
37. Moropoulou, A.; Bakolas, A.; Anagnostopoulou, S. Composite materials in ancient structures. *Cem. Concr. Compos.* **2005**, *27*, 295–300. [CrossRef]
38. Arizzi, A.; Cultrone, G. The influence of aggregate texture, morphology and grading on the carbonation of non-hydraulic (aerial) lime-based mortars. *Q. J. Eng. Geol. Hydrogeol.* **2013**, *46*, 507–520. [CrossRef]

Disclaimer/Publisher's Note: The statements, opinions and data contained in all publications are solely those of the individual author(s) and contributor(s) and not of MDPI and/or the editor(s). MDPI and/or the editor(s) disclaim responsibility for any injury to people or property resulting from any ideas, methods, instructions or products referred to in the content.

heritage

MDPI

Article

The Flight of Saint Mary Magdalene—A Case Study of the Dismantling, Repositioning and Restoration of a Votive Aedicule and Wall Painting in Nardò, Lecce, Italy

Fabrizio Ghio [1,*], Enrica Marcella Stefanelli [2] and Enrico Ampolo [3]

1 Independent Researcher, 73100 Lecce, Italy
2 Independent Researcher, 73100 Lecce, Italy; margi.stefa@libero.it
3 Independent Researcher, 73100 Lecce, Italy; enrico.ampolo@apiup.it
* Correspondence: fabrizio_ghio@yahoo.it

Abstract: The work carried out as part of construction on a ring road around Nardò (Lecce, South Italy) involved a votive aedicule attached to the perimeter wall of a rural building that had been scheduled for demolition. Provision was made for the movement and relocation of the building, dedicated to St. Mary Magdalene, to an adjacent area, in addition to the restoration of the pictorial decoration itself. The intervention was designed to conserve the building as testimony of its use for future generations, given its value and sense of identity for the local community. The project was largely conservatory, thus guaranteeing the continuity of its still active devotional function. In this sense, the restoration of the building was also associated with an educational aim of raising awareness of the importance of eco-compatible care of the landscape. So, this project highlights the importance of social, as much as environmental, sustainability.

Keywords: restoration; heritage; rural landscape; aedicule; wall paintings; St. Mary Magdalene

Citation: Ghio, F.; Stefanelli, E.M.; Ampolo, E. The Flight of Saint Mary Magdalene—A Case Study of the Dismantling, Repositioning and Restoration of a Votive Aedicule and Wall Painting in Nardò, Lecce, Italy. *Heritage* **2023**, *6*, 3429–3447. https://doi.org/10.3390/heritage6040182

Academic Editor: Austin Nevin

Received: 30 December 2022
Revised: 8 March 2023
Accepted: 27 March 2023
Published: 29 March 2023

Copyright: © 2023 by the authors. Licensee MDPI, Basel, Switzerland. This article is an open access article distributed under the terms and conditions of the Creative Commons Attribution (CC BY) license (https://creativecommons.org/licenses/by/4.0/).

1. Introduction

Work was carried out during the construction works of a ring road that involved a votive aedicule leaning against the perimeter wall of a rural building that had been scheduled for demolition (the works were carried out in 2019 by Nicolì S.p.A Company). The project was organized by the Province of Lecce (Dario Corsini, engineer, manager of the Public Works Sector; Anna Maria Riccio, engineer, Responsible for the Procedure), in collaboration with the Superintendence of Archaeology, Fine Arts and Landscape for the provinces of Brindisi and Lecce (Luisa Rosato, conservator restorer; Antonio Zunno, architect official) and the Municipality of Nardò (Nicola D'Alessandro, engineer, Executive; Gianluca Manieri, engineer, technical instructor; Cosimo Pellegrino, engineer, functional area manager).

1.1. The Rural Building and the Aedicule

The small rural building, devoid of architectural value that housed the Magdalene aedicule (Figure 1) in the countryside around Nardò, northwest of the ancient village was at the crossroads of a historic road system. The road connects the town to the sea, close to Sant'Isidoro (Figure 2) [1–7], passing through the town's San Francesco gate to the Castle of *Agnano*, on to Masseria Corsari and Villaggio Resta. Even though the area is now besieged by the expansion of the urban periphery, the building and its aedicule still retains its rural characteristics and retains the toponym La Maddalena, referring directly to Saint Mary Magdalene (Figure 3).

Figure 1. The rural building, from the southeast; on the left, the Magdalene aedicule.

Figure 2. Satellite image of the intervention area (in red).

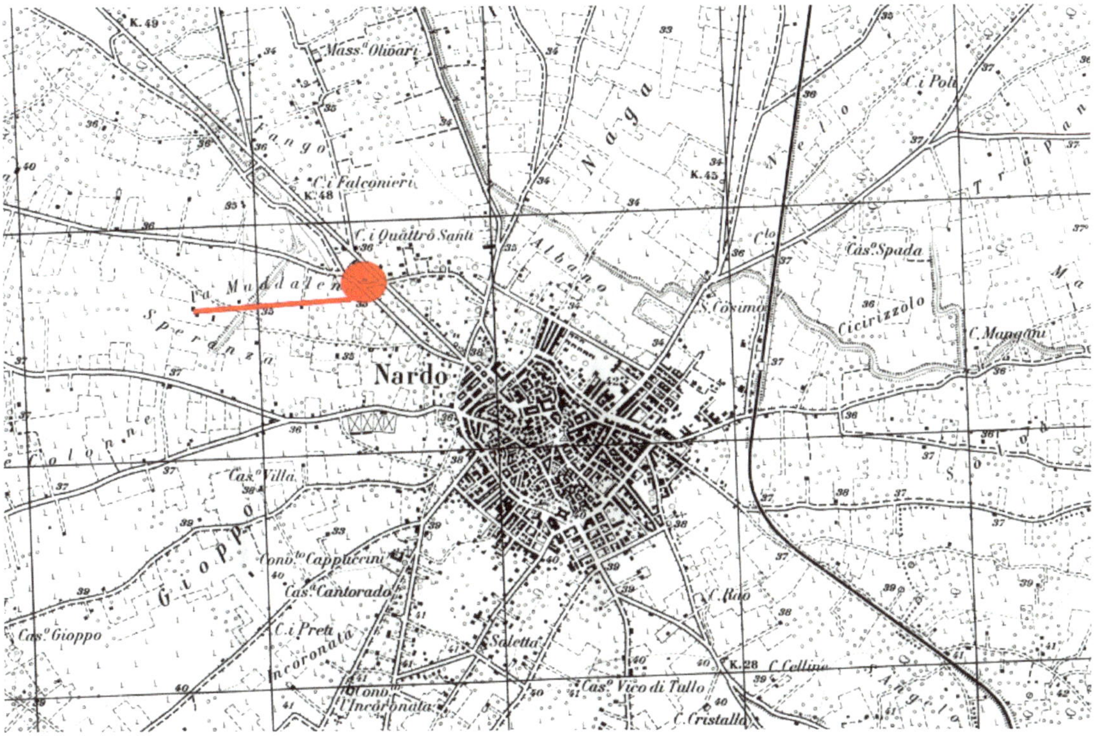

Figure 3. The intervention area (in red) on the IGM map, 1:25.000 scale (F 214IV SW—Nardò), 1947.

The aedicule, entirely stuccoed and decorated, sits on a base resembling an altar table finished at the top with a moulded cornice. On the front are fragmentary traces of a now illegible epigraph. Flanked by two lateral jambs, it ends at the top with an arch surmounted by a further cornice protected by a short roofing overhang made with opus signinum (Figure 4). It leans against a plastered curtain wall framed by two pilasters (the one on the left rounded to accommodate passing vehicles), delimited at the top by a cornice surmounted by a parapet protected by a layer of Lecce stone paving.

This type of aedicule (*cuneddhra* or *conella* in the local dialect) is a common sight in the Salento countryside. It is a votive aedicule, meaning, as the Latin origin of the term suggests, "house of the god" (*aedicula*, diminutive of aedes). However simple, these constructions signal the widespread need to mark religious devotion and to "adopt" a superior entity that would in confidence listen, protect and overcome moments of difficulty. These votive constructions come in many shapes and sizes but always retain the original function of small chapels, whether they are to be found in the countryside, in town or inside residential or religious buildings.

In the countryside, they constitute points of reference or resting places for pilgrims and wayfarers. They host pictorial or sculptural representations, evocative expressions of popular art and devotion free from imposed canons. These constructions will usually be built by local, and unnamed, artisans. It is precisely this widespread humble tradition both of construction and devotion that gives these aedicules their value. They create a strong identifying aspect of the local landscape that has been heavily affected by the continuous presence of man. This centuries-old relationship with the environment has produced a classic cultural landscape. However, over the last few decades, it has lost its vitality following the crisis in the agricultural economy and has led to a progressive degradation of the rural landscape [8–10].

Figure 4. Votive aedicule.

1.2. The Pictorial Decoration

The "pregnant Mary Magdalene" takes center stage on a high relief of the southern portal—attributed to Placido Boffelli and decorated with panels of religious and mythological subjects—of the seventeenth-century Church of the Crucifix (*Chiesa del Crocifisso*) just outside Muro Leccese (Figure 5a,b) [11,12]. She is clearly identifiable by the soft drapery of her dress, which highlights the roundness of the maternal womb, the result of a carnal as well as a spiritual relationship with Christ. As the local Salentine saying has it: "The Magdalene went to the Holy vineyard and came back pregnant" (*A Maddalena sciu alla vigna e turnau prena*). This typology of representation is also associated with the being patron of the "*pulandre*" (Greek dialect, for women, such as prostitutes, who have had many men in their lives). The theme of the pregnant Magdalene can also be found in a number of popular Christian legends of the French tradition.

(a) (b)

Figure 5. Muro Leccese, church of the Crucifix. Southern portal (**a**) and detail of the left jamb (**b**). Photo by O. Caroppo. http://naturalizzazioneditalia.altervista.org (accessed on 18 March 2023).

Indeed, the cult of the Magdalene, particularly venerated in the south of France, was introduced to the Kingdom of Naples at the end of the 13th century by the Angevins. Also, the stories of Mary in Jacopo da Varazze's *Golden Legend* refer to her as settling in Provence.

Artistic evidence [13] of the popular devotion to the Magdalene can be found throughout Salento from the Middle Ages to the modern era. Of note are the cave church in Manduria (*San Pietro Mandurino*); the *Torre di Belloluogo*, Lecce (*Cappella della Maddalena*) [14]; the castle at Copertino (*Cappella della Maddalena*) [15,16]; the cathedral at Nardò; the basilica at Galatina (*Santa Caterina d'Alessandria*) and the cave church (*Sant'Anna*); the church at Soleto (*Santo Stefano*), Melpignano (*Cappella di San Michele Arcangelo*); two churches in Sogliano Cavour (*Santa Maria Annunziata* and *San Lorenzo*); the main church, Uggiano (*chiesa matrice*); Nociglia (*Santa Maria de Itri*); Castiglione di Andrano (*Cappella della Maddalena*) and the main church (*chiesa matrice*) and the cave church at Santa Eufemia Tricase (*Cripta della Madonna del Gonfalone*).

The painting here is enclosed in a minimal architecture. It is a poor and simple reworking of the canons of the local iconographic tradition. Mary is depicted in the center of the aedicule, kneeling with her long hair loose on her shoulders, her head bowed and her gaze turned to heaven. She holds the Crucifix with her left hand, while her right hand is delicately placed on her abdomen as if to protect it. In addition to the wooden Crucifix, we note the ointment jar and the skull which represent the iconographic attributes of the classical tradition. In the background, we see a sky with clouds enlivened by two cherubs and a cluster of houses, which could refer to the town of Nardò (Figure 6a,b).

(**a**) (**b**)

Figure 6. Central panel, general (**a**) and particular (**b**).

In the intrados of the summit arch, a dove with a radiated crown, a symbol of the Holy Spirit, is depicted (Figure 7); in the left jamb, there is an angelic figure while the one on the right houses the effigy of the patron Saint of Nardò, *San Gregorio Armeno* (Saint Gregory the Armenian), in the robes of a bishop in a gesture of benediction.

Figure 7. The Holy Spirit.

1.3. *Analysis of the State of Conservation of the Pictorial Decoration and of the Support [17]*

The pictorial decoration appeared in a very bad state of conservation. It is a "mixed" painting, with "dry" parts—lime paint, probably tempera—on single-layer plaster laid on the masonry face in regular ashlars of local tuff (Pliocene biocalcarenite). Execution appears simple and without the canonical rough coat-*arriccio-intonachino* stratification, which would have been typical of better-quality wall paintings, especially when using the *a fresco* or *mezzo fresco* techniques.

Adhesive qualities of the plaster to the masonry support were compromised in several points due to the poor grip qualities of the underlying stone, favoring its detachment with the consequent formation of pockets and visible areas of layer lifting, which entailed the loss of parts of the paintings (lacunas), concentrated above all in the intradossal portion of the vault. The plaster, in addition to a dense network of micro and macro cracks, has suffered widespread biological attack (biological patina) and surface deposits.

Generally abraded due to exposure to atmospheric agents, the pictorial film presented lacunas and was fragile and partially or totally missing, mainly in the lower part of the painting, with the greatest loss in the lower area of the image of Mary Magdalene and *San Gregorio Armeno* (Figure 8a,b).

(**a**) (**b**)

Figure 8. Details highlighting the poor state of conservation: St. Mary Magdalene (**a**); *San Gregorio Armeno* (**b**).

Despite the extensive losses, the artifact is legible as a whole, and no evidence of any signature or date was found. Considering the stylistic and architectural style, the quality and the technique, the aedicule that houses it and on the basis of comparison with similar creations, it can be presumed to conform to the local tradition of the 18–19th century.

2. The Intervention

2.1. Methodological Premise

The new road infrastructure was already well underway, and the dilapidated building was close to collapse. So, it was clearly necessary to work on the votive aedicule with immediate effect. Restoration and relocation to the edge of the new road layout was clearly necessary and was carried out at all times following Superintendence procedures [18,19].

Preliminarily, an accurate and in-depth investigation was conducted. This was not limited to the analysis of the building alone, but took into consideration the wider rural context, in line with the principles of conservation, to enhance the area through redeployment where compatible. Considering the lack of historical documentation regarding the aedicula, the origin and history of the building and its transformations over time were reconstructed through comparisons with what we know about similar structures of the time. In particular, analogies were made regarding locally popular construction systems and techniques, building materials and architectural styles. Analysis was also made of the state of conservation of the materials and the constituent elements [20–24].

The first hypothesis, the detachment of the fresco and its musealisation at the Castromediano museum in Lecce, was discarded, given the inappropriacy of separating the sacred image from its aedicule. Instead, it was important to preserve the entire votive aedicule for its cultural value as evidence of a rooted tradition in the Salento countryside.

The transfer of an artifact or the detachment of a wall painting from its original site was common practice up until the 1970s. It does however involve very complex operations to be carried out only when all other possibilities of intervention are to be excluded and on careful analysis case by case.

Below, we note a number of some removals in the Salento area. In 2015, during the restoration works of the small church of the Assunta in Botrugno, the following were removed: the main altar in Lecce stone (18th century) and a wooden altarpiece with the canvas depicting the Virgin of the Assumption and Saints Carlo Borromeo and Gaetano Thiene. This particular removal allowed new historical–archaeological investigations to be carried out as well as the restoration of the original 14th-century fresco on the apse wall [25].

In 2008, during the restoration work on the crypt of *Santa Maria della Grotta* in Ortelle, following stratigraphic investigations, a portion of a recent and rather amateurish mural painting was detached. The work, depicting the Medici Saints Cosimo and Damiano and two unidentified bishop saints, was located in the lunette of one of the apses. This rough painting was covering an older one attributable to the same 18th-century fresco painter of the other apses, where the subjects are always the Santi Medici Holy Unmercenaries, i.e., medical practitioners who refused (mercenary) payment. In this painting, it is clearly written who the saints are: Nicholas and Liborio [26].

Returning to the aedicule discussed here, the solution proposed was disassembly, anastylosis of the building, and relocation to a bordering area, compatible with safety standards and to the new road layout. In consideration of the inseparable relationship between the aedicule, the historical road system and the desire to maintain an active devotional function, an unused area of municipal property was found a short distance from the original position along Via dei Bernardini (Figure 9a,b).

Reconstruction was limited to ensure minimal visual impact and to conform harmoniously within the rural landscape setting. In this sense, chromatic solutions compatible with the building and its surroundings were adopted, avoiding strong contrasts and favoring the colors in the local area.

The flooring of the area was reduced to the bare minimum and made with draining material (rammed earth and gravel) to avoid further waterlogging of the soil. New planting was introduced including indigenous herbaceous species from local ecotypes to avoid the uprooting of any trees or shrubs.

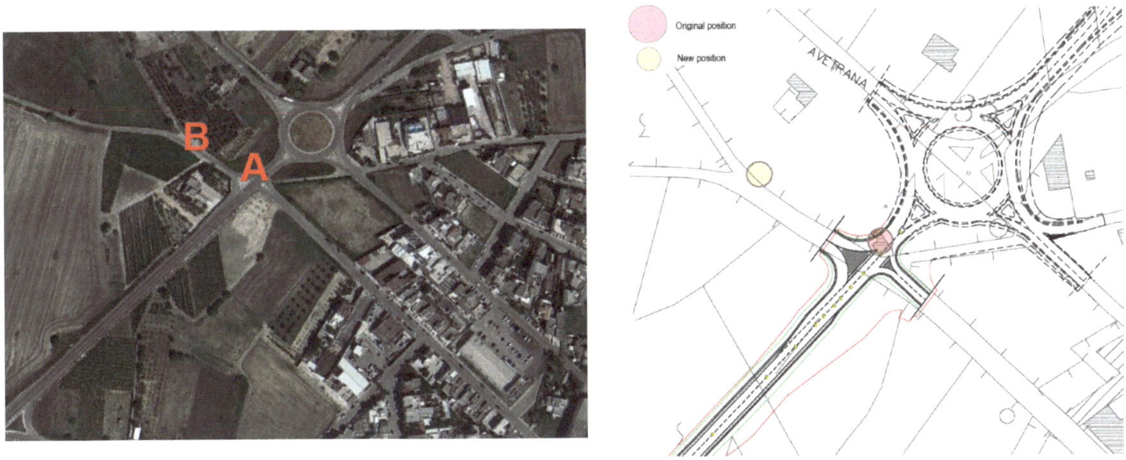

(**a**) (**b**)

Figure 9. Satellite image (**a**) of the original location of the building (A) and of the new location (B). Project plan on and aerophotogrammetric map (**b**).

As mentioned above, the original plan involved disassembly and anastylosis. However, the company engaged for the project suggested removing the entire building in one piece, using a crane for lifting, and a particular vehicle for transport to the new site. This was an innovative solution that significantly reduced execution times and also simplified the degree of intervention.

2.2. The Operational Phases

2.2.1. Preliminary Investigations

To begin with, preliminary diagnostic investigations were carried out [27–32]. These were aimed at identifying and determining the exact microclimatic conditions, stone and restoration techniques and materials for the intervention.

Samples were painstakingly taken from the vault and the walls at points without pictorial decoration. Mineralogical and petrographic analyses of the material were then made to determine the characterization and behavior of the material and its resistance to the action of degradation agents. The soluble salts in these samples were also analyzed to determine the extent of their action on the structure.

Very small stratigraphic pieces were made by hand with a scalpel to verify the existence of previous applications of pictorial decoration below those legible today. None were found.

A further preventive assessment was made to provide further information to help decide how to organize the restoration phases. Tests were conducted to identify the most suitable techniques and products for biological disinfestation, to ascertain the application timing of the most appropriate methodologies and products for the neutralization of residues in relation to the state of the paintings and to determine the execution technique as well as the conditions of the plaster and stone support. The tests were carried out in the areas affected by biological attack using numerous application pieces with varying concentrations and times of application of a number of products.

At the end of this investigation of monitoring, data research and testing, it was possible to draw up a project protocol of the work to be carried out, summarized in the following paragraphs.

2.2.2. Scaffolding Work (Figure 10a,b)

Prior to the detachment phase and transport of the aedicule, necessary temporary shoring, cambering and slinging of the entire masonry body of the building was carried out. This involved both vertical and horizontal wooden shoring to prevent sagging or structural movements, using mainly wooden beams, with bracing and wedging of sections. A scaffolding cage was made of carefully placed scaffolding tubes and wooden planks incorporating the entire masonry body of the votive aedicule ready for transfer to the new location. Prior to this, the entire fresco was protected with support straps and the entire pictorial surface was covered to prevent any danger of detachment of the painted surfaces or of the support plaster during handling and transport. Finally, the entire masonry envelope was protected by an insulating membrane to separate the masonry structure from the scaffolding itself.

(a)

(b)

Figure 10. (**a**,**b**) Scaffolding work.

2.2.3. Detachment and Handling

Prior to the transfer, a portion of the wall the aedicule was leaning against was cut vertically. Disassembly of overhanging portions of the wall was carefully carried out with both electric and manual equipment. The operation of detachment from the entire body of the building also included the assembly of scaffolding and work surfaces and their subsequent disassembly when the operation was completed.

We then proceeded with the handling of the product and transport to the reassembly location using a mobile crane (Figure 11a–c).

2.2.4. Reassembly of the Aedicule (Figure 12a,b)

First, the relocation site was prepared. This involved the construction of special masonry work on the foundation and, for the construction of the foundations, the mechanical excavation of the earth up to the rock bank. The foundation itself was composed of blocks of local stone and mortar, using traditional techniques. Traditional mortar was used to help reposition the aedicule. The new back structure was built following the shape of the original building using the same technique as above. Stone constraint elements were also applied for the rear support of the building to give it solidity and for safety reasons. To top it off, a level in local stone was put in place. Finishing consisted of the comparison and final patination of the building using materials similar to the original (local stone facing, lime-based mortars and plasters, whitewashing and coloring with natural pigments) and applied with traditional techniques.

(a) (b) (c)

Figure 11. (**a**–**c**) Handling.

2.2.5. Restoration of Pictorial Surfaces [33–35]

The first fundamental interventions, prior to the dismantling of the votive aedicule, consisted of ensuring the safety of the portions of the painted plaster at risk of falling. The cracks and gaps were filled with mortar based on hydraulic lime and river sand, while pre-consolidation was performed with injections of PLM premixed mortar. The cohesion defects of the pictorial film were resolved, and surface deposits were removed. Three layers of cotton gauze were then applied with Paraloid B72 acrylic resin diluted with trichloroethane. This operation was aimed at protecting the surface of the painting to prevent parts of the pictorial film and of the preparatory layers or of the support from falling (Figure 13).

After the reassembly in the new site, the layers of gauze were carefully removed with the aid of brushes and solvent along with the subsequent elimination of PARALOID residues from the entire painted surface with cotton wool and solvent (Figure 14a,b).

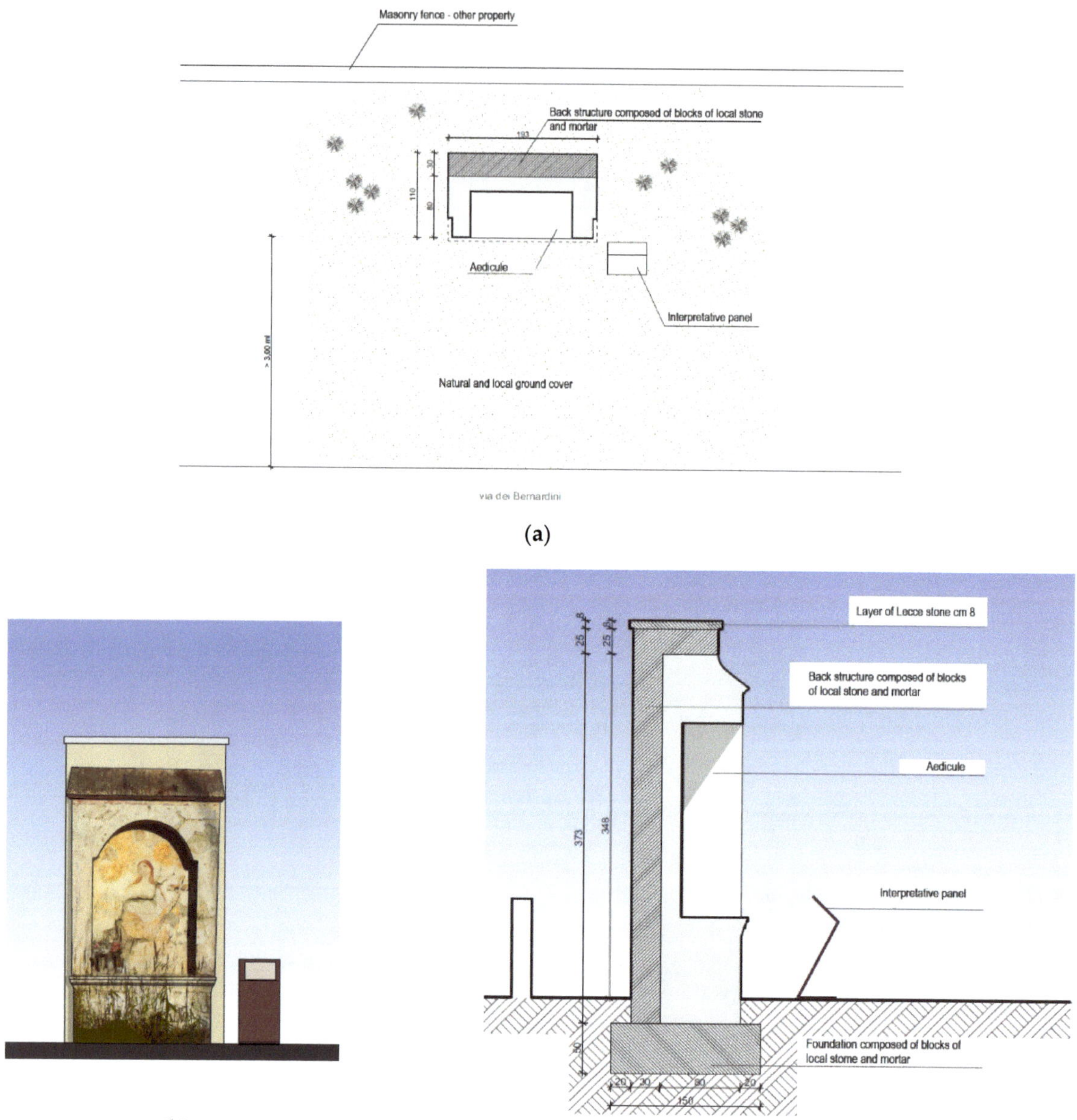

Figure 12. Project drawings: plan (**a**), elevation (**b**), section (**c**).

The next phase saw the nebulization of BIOTIN T, a biocidal product, over the entire surface to preserve and protect the surface from microbiological attack (Figure 15). Biodeteriogen residues were removed with soft brushes, and then neutralization was carried out by washing with deionized water.

Figure 13. The aedicule covered by protective gauze.

(a)

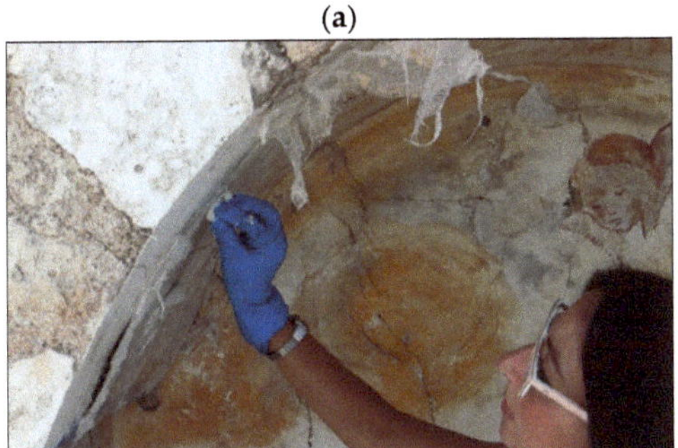

(b)

Figure 14. Bandage removal (**a**,**b**).

Figure 15. Nebulization of the biocide.

We then proceeded with the reinforcement. For the deep filling of the plaster that had detached from the stone support, we injected a water-based calcium hydroxide mortar, PML 33, with demineralized water injections and alcohol. We then worked on the more superficial detachments between the plaster and the painted plaster, injecting acrylic resin (ACRIL 33) in a 5% aqueous solution (Figure 16a–c).

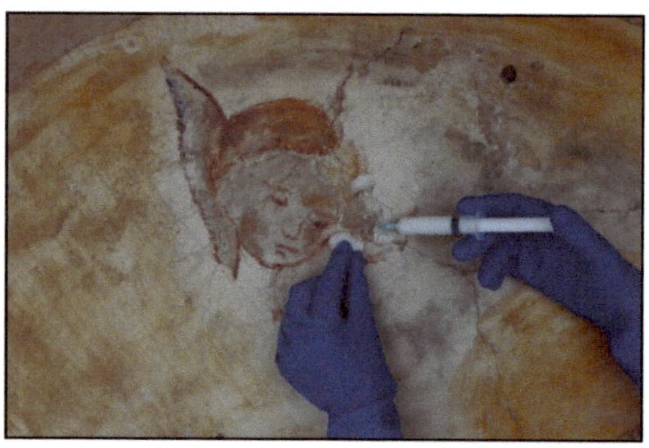

(**a**)

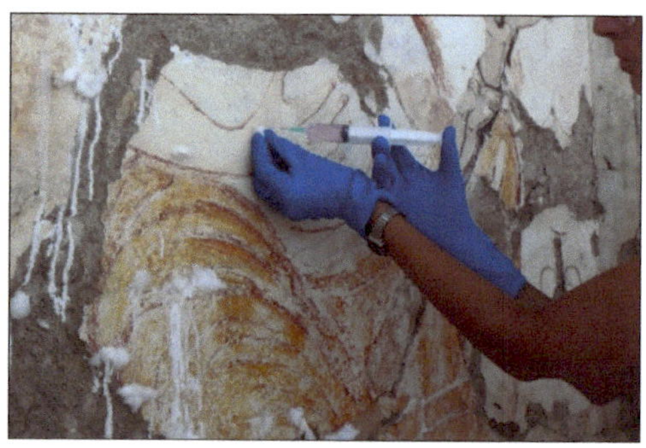

(**b**)

(**c**)

Figure 16. Consolidation operations and application with a syringe (**a**–**c**).

Subsequently, the grouting of the gaps and fractures was carried out, levelled with hydraulic lime mortar, using a variety of grain-sized river sand according to gap depth; yellow *cocciopesto* plaster was added to match the original chromatic tone (Figure 17a–c).

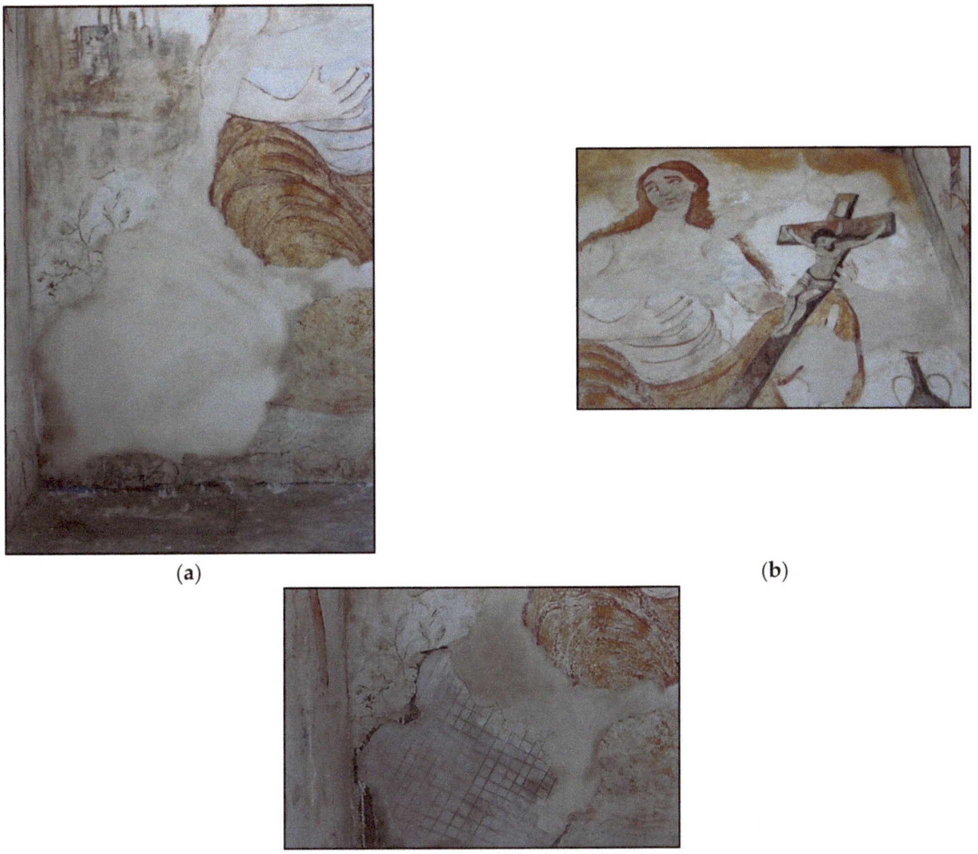

(**a**) (**b**) (**c**)

Figure 17. (**a–c**) Plaster grouting.

The pictorial intervention, aimed at reducing the visual disturbance caused by the evident gaps and abrasions,. was strictly limited to the area to be integrated and always in full respect of the original pictorial layout (Figure 18a–c). For the reintegration, carried out with an undertone veiling technique, pure CTS pigments were used, bound with very diluted solutions of acrylic resin to ensure low alterability over time.

As part of the final report, photographic documentation and chronological documentation of the entire work (pre, during and after restoration) was provided (Figure 19a,b).

2.2.6. External Arrangement of the Area

After the spreading of topsoil, to reduce environmental impact to the minimum, native and local ground cover was planted to greenify the area (*Sedum* spp., *Lobularia marittima*, *Cynodon dactylon*).

An interpretative panel was also installed to secure the memory of the site and of the work that had been carried out.

(a)

(b)

(c)

Figure 18. Pictorial integration of *San Gregorio Armeno* (**a**) of the face of Mary Magdalene (**b**) and of an angel (**c**).

(**a**) (**b**)

Figure 19. The aedicule before (**a**) and after (**b**) intervention.

3. Conclusions

A distinctive feature of Puglian "minor" cultural heritage in rural areas consists of constructions, including votive aedicules, where local activities took place, such as craftwork or gatherings for social and spiritual life. Once knowledge of this aspect of rural heritage is lost, conservation is neglected, leading to impairment or irreparable degradation. The same level of degradation can occur through uninformed restoration incompatible with the identity of the structures themselves and/or the landscapes that had originally welcomed them.

The intervention, based on environmental sustainability and socio-cultural compatibility, was therefore designed to restore the building for cultural heritage, to ensure not only its physical and functional survival, but also that of its testimonial value, as a mark of identity for the community in this particular area of Nardò, thus guaranteeing the continuity of its function as an actively used place of devotion. In particular, the work carried out made it possible to preserve the original architectural elements and surface finishes, as well as to restore the landscape in harmony with the rural context. This was made possible through the use of traditional materials and construction techniques, and conserving the morpho-typological, typological-constructive, material and geometric identity of the building.

Specifically, for the recovery of the building and its harmonious insertion in the landscape context, the indications of the current Regional Territorial Landscape Plan were taken into account, especially the detailed *Guidelines for the recovery, maintenance and reuse of buildings and rural goods*. This guide provides information and cognitive support for complex regional rural building heritage sites by directing building recovery, restoration and restructuring operations, focusing on the problem of reuse and destinations compatible with the architectural and construction characteristics of the artifacts and their landscaped surroundings. All the work done respected the fundamental principles of restoration: recognisability, reversibility, compatibility and minimal intervention, in line with the rules of the 1972 Restoration Charter.

For the intervention, sustainable environmental technologies were used, based on the use of traditional construction techniques. These techniques, the result of working with the area's limited resources, have for centuries provided more than satisfactory results for the inhabitants.

To conclude, the case study appears important, in our opinion, to turn the spotlight on what is ostensibly a minor heritage, humble and locally popular. It is, however, important and worthy of being preserved due to its high testimonial value, which today is particularly threatened through neglect, abandonment, vandalism, urban sprawl and soil erosion. In this sense, the restoration of the building is also associated with a wider educational function of not only raising awareness of local history and traditions, but also of eco-compatible care of the local landscape.

The flight of St. Mary Magdalene: time lapse of the movement.

Author Contributions: Conceptualization, E.A., F.G. and E.M.S.; methodology, E.A., F.G. and E.M.S.; validation, E.A., F.G. and E.M.S.; formal analysis, E.A., F.G. and E.M.S.; investigation, E.A., F.G. and E.M.S.; resources, E.A., F.G. and E.M.S.; data curation, E.A., F.G. and E.M.S.; writing—original draft preparation, F.G. and E.M.S.; writing—review and editing, E.A., F.G. and E.M.S. All authors have read and agreed to the published version of the manuscript.

Funding: The intervention was carried out as part of the Nardò Ring Road Project—3rd Lot, financed by the Province of Lecce.

Data Availability Statement: Data sharing is not applicable for this project.

Acknowledgments: The authors thank the Province of Lecce, the municipality of Nardò (LE) and the Superintendence of Archaeology, Fine Arts and Landscape for the provinces of Brindisi and Lecce.

Conflicts of Interest: The authors declare no conflict of interest.

References and Note

1. Vetere, B. (Ed.) *Città e Monastero: I Segni Urbani di Nardò (Secc. XI–XV)*; Congedo: Galatina, Italy, 1986.
2. Gaballo, M.; De Cupertinis, G. (Eds.) *Nardò Nostra. Studi in Onore di don Salvatore Leonardo*; Congedo: Galatina, Italy, 2000.
3. Mennonna, M. *Nardò dalle Origini alla Metà del '900*; Congedo: Galatina, Italy, 2003.
4. Cazzato, M.; Nardò. *Insediamenti del Salento dall'Antichità all'Età Moderna*; Cazzato, V., Guaitoli, M., Eds.; Congedo: Galatina, Italy, 2005; pp. 123–126.
5. Giuri, P.; Giuri, F. La città di Nardò nel XIX secolo: Dall' "intra moenia" all' "extra moenia". *Kronos* **2007**, 3, 77–81.
6. Mastrangelo, M.V. L'antica viabilità nel territorio neritino. *Spicilegia Sallentina* **2009**, 5, 47–51.
7. Giuri, P. Nardò. In *Lecce e il Salento/1 i Centri Urbani, le Architetture e il Cantiere Barocco*; Cazzato, V., Cazzato, M., Eds.; De Luca: Rome, Italy, 2015; pp. 291–315.
8. Basile, A. Momenti di religiosità popolare nel Salento: Le edicole votive. *Sallentum Riv. Quadrimestrale Cult. Attività Salentina* **1982**, *V*, 1.
9. Costantini, A. *Le Edicole Votive nell'Area della "Cupa"*; Cartografica Rosato: Lecce, Italy, 2005.
10. Costantini, A. Presenze Barocche nell'Edilizia Domestica e nell'Architettura Rurale. In *Lecce e il Salento/1 i Centri Urbani, le Architetture e il Cantiere Barocco*; Cazzato, V., Cazzato, M., Eds.; De Luca: Rome, Italy, 2015; pp. 81–96.
11. Patella, S. Muro Leccese. In *Lecce e il Salento/1 i Centri Urbani, le Architetture e il Cantiere Barocco*; Cazzato, V., Cazzato, M., Eds.; De Luca: Rome, Italy, 2015; pp. 286–291.
12. Caroppo, O. La Maddalena Raffigurata Incinta a Muro Leccese. Available online: http://naturalizzazioneditalia.altervista.org (accessed on 18 March 2023).

13. Mosco, M. (Ed.) *La Maddalena tra Sacro e Profano. Da Giotto a De Chirico, Catalogo della Mostra (Florence, Palazzo Pitti, 24 May–7 September 1986)*; Mondadori: Milan, Italy; Florence, Italy, 1986.
14. Cassiano, A. Momenti di Arte Gotica nel Salento. In *Ricerche e Studi in Terra d'Otranto*; Vetrugno, P.A., Ed.; Congedo: Cellino San Marco, Italy, 1988; Volume 3, pp. 39–70.
15. Ortese, S. Una Committenza del Balzo Orsini Chiaromonte nella Cappella della Maddalena a Copertino e Alcune Note Sulla Pittura Tardogotica del Salento. In *Un Principato Territoriale nel Regno di Napoli?: Gli Orsini del Balzo Principi di Taranto (1399–1463), Atti del Convegno di Studi (Lecce, 20–22 Ottobre 2009)*; Petracca, L., Vetere, B., Eds.; Ist. Storico per il Medioevo: Rome, Italy, 2013; pp. 577–588.
16. Ortese, S. *Pittura Tardogotica nel Salento*; Congedo: Galatina, Italy, 2014; pp. 229–246.
17. The Terminology Used for the State of Conservation and the Definitions of Deterioration of the Building Refers to the "NORMAL lexicon C.N.R./I.C.R. 1/88". Available online: www.inforestauro.org/wptest/wp-content/uploads/2012/02/biblio_norma_06.pdf (accessed on 18 March 2023).
18. Weyer, A.; Roig Picazo, P.; Pop, D.; Cassar, J.-A.; Özköse, A.; Vallet, J.M.; Srša, I. (Eds.) *EwaGlos. European Illustrated Glossary of Conservation Terms for Wall Paintings and Architectural Surfaces*; Michael Imhof Verlag: Oetersberg, Germany, 2015; Volume 17.
19. With a note dated 26 January 2015, the Superintendence for Architectural and Landscape Heritage for the Provinces of Lecce, Brindisi and Taranto "given the advanced state of execution of the road infrastructure and the dilapidated building, close to collapse, invites proposals to recover the votive aedicule, and carry out work of restoration and possible relocation to the side of the road, subject to the acquisition of authorization by this Superintendence pursuant to art. 21 of Legislative Decree 42/2004, as provided for by art. 50 of the same decree". The Superintendence of Archaeology, Fine Arts and Landscape for the Provinces of Brindisi Lecce and Taranto, with a note dated 4 December 2017 "does not agree with the proposal for the detachment of the fresco and its musealisation at the Castromediano museum of Lecce" observing "that it is the entire votive aedicule that takes on cultural value, as evidence of typical construction in the Salento countryside. Therefore, this Superintendence believes that the sacred image cannot be separated from the structure that houses it. Furthermore, the opportunity to re-locate the aedicule, must be on an area bordering the new road layout compatible with current safety standards".
20. De Venere, C.; Di Capua, M.G.; Lorusso, A. *Restauri in Puglia 1971–1981*; Schena: Fasano, Italy, 1983; Volume 1.
21. Di Capua, M.G. *Restauri in Puglia, Catalogo della Mostra (Bari, Castello Svevo, March–October 1982)*; Schena: Fasano, Italy, 1983; Volume 2.
22. Lombardo, M. (Ed.) *Tecnologie per i Beni Culturali*; TorGraf: Galatina, Italy, 2007.
23. Tosco, C. *I Beni Culturali. Storia, Tutela e Valorizzazione*; Il Mulino: Bologna, Italy, 2014.
24. Capanna, A.; Mele, G. (Eds.) *Rilievo Progetto Riuso*; Maggioli Editore: Sant'Arcangelo di Romagna, Italy, 2017.
25. Ortese, S. *Botrugno Chiesa dell'Assunta. La Parete Absidale Dopo il Disvelamento degli Affreschi*; Congedo: Galatina, Italy, 2016.
26. Ortese, S. *Ortelle Cripta di Santa Maria della Grotta. Storia e Restauri*; Lupo: Copertino, Italy, 2009.
27. Dipartimento Geomineralogico, Università degli Studi di Bari. *Metodologie Analitiche Mineralogiche e Petrografiche*; Digilabs: Bari, Italy, 2002.
28. Sajeva, S. *Pitture Murali. I Degradi di Origine Meccanica, Manuale per Restauratori ed Ingegneri*; DEI: Rome, Italy, 2014.
29. Osama, A.; Danila, A.; Ezzeldin, K.M.; Akram, T.; Ashraf, S. Experimental Investigations and Microstructural Characterization of Construction Materials of Historic Multi-Leaf Stone-Masonry Walls. *Heritage* **2021**, *4*, 2390–2415. [CrossRef]
30. Visco, G.; Serra, A.; Buccolieri, G.; Manno, D.; Calcagnile, L.; Quarta, G.; Buccolieri, A. Mortar Characterization and Radiocarbon Dating as Support for the Restoration Work of the Abbey of Santa Maria di Cerrate (Lecce, South Italy). *Heritage* **2022**, *5*, 4161–4173. [CrossRef]
31. Fico, D.; Rizzo, D.; Casciaro, R.; Esposito Corcione, C. Historically Accurate Reconstruction of the Materials and Conservation Technologies Used on the Facades of the Artistic Buildings in Lecce (Apulia, Italy). *Materials* **2022**, *15*, 3658. [CrossRef] [PubMed]
32. Fioretti, G.; Clausi, M.; Eramo, G.; Longo, E.; Monno, A.; Pinto, D.; Tempesta, G. A Non-Invasive and Sustainable Characterization of Pigments in Wall Paintings: A Library of Apulian Colors. *Heritage* **2023**, *6*, 1567–1593. [CrossRef]
33. Mora, P.; Mora, l.; Philippot, P. *La Conservazione delle Pitture Murali*; Editrice Compositori: Bologna, Italy, 1999.
34. Guarini, I.; Martignano, G.; Mascalchi, M.; Minerva, B.; Patrizi, M.G.; Siano, S. Pulitura Laser degli Affreschi della Cappella "Madonna dei Farauli" a Sternatia. In *Aplar 4. Applicazioni Laser nel Restauro, atti del Convegno (Roma, Museo Nazionale Romano di Palazzo Massimo alle Terme, 14–15 giugno 2012)*; Il Prato Publishing House S.r.l.: Padova, Italy, 2013; pp. 155–168.
35. Minerva, B. Il restauro degli affreschi della chiesa "Madonna dei farauli" a Sternatia (Le). Un nuovo metodo di pulitura con tecnologia laser. *Kronos* **2013**, *15*, 165–174.

Disclaimer/Publisher's Note: The statements, opinions and data contained in all publications are solely those of the individual author(s) and contributor(s) and not of MDPI and/or the editor(s). MDPI and/or the editor(s) disclaim responsibility for any injury to people or property resulting from any ideas, methods, instructions or products referred to in the content.

heritage

Article

In Situ Investigation of the Medieval Copper Alloy Door in Troia (Southern Italy)

Giovanni Buccolieri [1], Antonio Serra [1], Giorgio Giuseppe Carbone [1], Vito Nicola Iacobellis [2], Alfredo Castellano [1], Lucio Calcagnile [1] and Alessandro Buccolieri [1,*]

[1] CEDAD (CEnter of Applied Physics, DAting and Diagnostics)—Dipartimento di Matematica e Fisica "Ennio De Giorgi", University of Salento, 73100 Lecce, Italy
[2] Polo Museale della Puglia, via Pier l'Eremita 25/b, 70122 Bari, Italy
* Correspondence: alessandro.buccolieri@unisalento.it

Abstract: This work describes experimental results concerning the chemical composition of the alloy patina and inlays from the medieval copper alloy door of the cathedral of Troia (southern Italy), dating back to 1127 CE. The analyses were conducted in situ with no sampling or sample preparation required, using a portable energy-dispersive X-ray fluorescence (ED-XRF) instrument. The compositional results show that the two door leaves were made using a binary alloy of copper and lead, while the nails, lion protomes, and handles have a different chemical composition. Moreover, the analyses revealed uniform concentrations of chlorine, probably due to cleaning treatments during the restoration. It is important to emphasise that the obtained results are the only ones related to this valuable masterpiece analysed.

Keywords: medieval door; copper alloy; patina; ED-XRF; portable apparatus

Citation: Buccolieri, G.; Serra, A.; Carbone, G.G.; Iacobellis, V.N.; Castellano, A.; Calcagnile, L.; Buccolieri, A. In Situ Investigation of the Medieval Copper Alloy Door in Troia (Southern Italy). *Heritage* **2023**, *6*, 2688–2700. https://doi.org/10.3390/heritage6030142

Academic Editors: Daniela Fico and Daniela Rizzo

Received: 22 December 2022
Revised: 28 February 2023
Accepted: 1 March 2023
Published: 2 March 2023

Copyright: © 2023 by the authors. Licensee MDPI, Basel, Switzerland. This article is an open access article distributed under the terms and conditions of the Creative Commons Attribution (CC BY) license (https://creativecommons.org/licenses/by/4.0/).

1. Introduction

The final decades of the Western Roman Empire were the time when the majority of historical bronze doors were made. The plainness of the ornamental plant or animal designs that embellish these doors is what makes them distinctive. The Middle Ages saw a resurgence in bronze door production, with many doors being installed across Europe. Sadly, a significant number of these doors have been lost because of environmental agents' eroding effects, and many more since the valuable metal from which they were manufactured was repurposed for other applications.

There are currently just about thirty medieval copper alloy doors in Europe that can be visited, and, in particular, in Italy, it is possible to admire them in Amalfi (1060 CE), Monte Cassino (1066 CE), Rome (1070 CE), Monte Sant'Angelo (1076 CE), Atrani (1087 CE), Salerno (1085 CE), Benevento (1150–1151 CE), Canosa di Puglia (1111–1118 CE), Trani (1111–1118 CE), Troia (with two: 1119 CE and 1127 CE), Monreale (1140 CE), and Venice (with two: 1080 CE and 1112 CE) [1–8].

Such great works of art were mostly created in Constantinople and brought to Italy or were manufactured by Italian artisans using byzantine doors as inspiration.

Although commonly considered "bronze doors", byzantine doors were constructed using orichalcum, a form of brass composed of a copper alloy with a low zinc percentage. Since brass oxidizes to a greenish hue over time, similar to how bronze does, these doors were likely given the name "bronze". They include illustrations of sacred icons and were distinguished by priceless goldsmiths' embellishments done with inlay and niello techniques.

Byzantine doors are composed of panels, which are tiny modular orichalcum pieces that are nailed to a wooden core and used as the doors' structural building blocks. The creation of controllable, lightweight doors that were simple to move was made possible through this building technology.

In the Middle Ages, lost-wax casting was frequently employed for producing bronze (or brass) doors. In this technique, a chalk model of the artifact was covered in wax and buried under a layer of refractory soil. This type of mould was then baked in an oven. The dissolving of the wax induced by the temperature left a tiny gap between the chalk model and the refractory soil that was filled with molten metal, producing the final piece. For single-part doors, this process worked quite well, but it was costly and time-consuming for multi-part doors like the bronze ones used in Italy.

Regarding byzantine doors, some research shows that their leaves may have been cast using the so-called sand or mould casting process. Since it did not require wax, this approach is different from lost-wax casting. Using a frame known as a casting flask, moulds for the intended object were created. The model, which was typically constructed of wood and had the shape of the object to be cast, was set in the centre of the casting flask. The model was surrounded by sand, which was then compressed to form a chamber into which the casting could be done after the model was taken out.

The chemical composition, the colour, and the thickness of the patina vary depending on both the original composition of the metal and the post-deposition environment (air, soil, and water) [9]. The chemical composition of the deposition environment, and its temperature, humidity, and pH contribute to the formation of different patinas.

Moreover, it is important to evaluate the phenomenon of surface enrichment of certain elements, such as, for example, tin, which often appears in higher percentages in surface measurements. Cleaned copper has been shown to provide similar results between destructive methods and portable energy-dispersive X-ray fluorescence (ED-XRF) [10]. Some success has also come with the use of ED-XRF through the implementation of Monte Carlo simulations, as a way to circumvent the effects of the patina [11]. The issue with the implementation of Monte Carlo simulations is that it currently still requires knowledge of the cross-section of a corroded alloy [12].

There is another qualitative method that is often considered, that is, the evaluation of the ratio Sn-Kα/Sn-Lα of the X-ray intensities. If there is a uniform distribution of tin between the surface and the bulk of the alloy, it is expected that the analysis is free from any corrosive patinas. In these ways, the artefacts are cleaned to remove the patina, or the object is sectioned to analyse the bulk of the alloy.

Very often, however, destructive sampling of objects of antiquity is not possible due to the historical–artistic value of the object and its integrity [10]. Usually, the need for destructive sampling is assessed beforehand by a team of experts (researchers, restorers, and superintendents) to see if this is really necessary.

With this in view, portable ED-XRF has found wide use over the past decades as an in situ, non-invasive, fast, low-cost surface technique. Of course, the advantages and disadvantages of such equipment must be evaluated on a case-by-case basis, precisely because of the experimental difficulties that patinas create on the materials to be analysed.

It is clear that, for bronze samples that require specifically objective results, energy-dispersive X-ray fluorescence could be defined as not valid, but it is a very useful technique for the preliminary analysis of bronze artefacts and/or when it is not possible to perform destructive analyses on historical–artistic manufacts [13].

The objective of this research is to use portable energy-dispersive X-ray fluorescence to perform an in situ, non-destructive investigation of the medieval door of the Cathedral of Troia (Figure 1), on which very little information is available in the literature [14–16]. In particular, our aim was to investigate the patinas of these precious medieval doors using the portable ED-XRF instrument we had available. Moreover, we would also like to underline that the superintendent did not allow destructive analyses.

This study is the result of several years of investigation by the authors of this manuscript, who analysed different doors in Apulia, including those in Canosa di Puglia [17] and Monte Sant'Angelo [18], in an effort to ascertain the composition of these priceless examples of medieval art and aid future research to clarify the intricate processes involved in their manufacture.

Figure 1. Picture of the medieval door called *"Freedom Door"* in Troia (southern Italy) with the measurement points.

2. Materials and Methods

2.1. Description of the Analysed Door

Troia (Foggia, Apulia, Southern Italy) is considered the heir of an ancient city known as Aecae, and from 1031 CE it became a bishopric and one of the most important religious centres in southern Italy, culturally and politically linked to the Church of Rome. The actual cathedral of Troia stands on the site of a primitive church dedicated to the Blessed Virgin Mary Assumed into Heaven.

The work for its construction began under the episcopate of Girardo of Piacenza and in 1106 CE was refinanced by bishop William II, known as William the Norman, who decided to seal the importance of this sacred place by commissioning a bronze door to the master foundryman Oderisius of Benevento. Soon before the cathedral was completed in 1119 CE, the door was installed on the main entrance and was renamed the *"Door of Prosperity"* to honour the euphoric atmosphere created by the church and its bishop, who acted as a crucial mediator between the papacy and the Norman barons.

In 1127 CE, only eight years after the execution of the main door, bishop William II again commissioned Oderisius of Benevento to craft a bronze door for the side of the cathedral, which is the subject of this study.

The door, measuring 284 cm in height and 158 cm in width, is divided into twenty-four panels framed in bolted bronze frameworks and divided into six bands of four panels each, which narrate an important page in the life of the city of Troia: the liberation from Norman domination. For this reason, the studied manufact was called the "*Freedom Door*".

Starting from the left (Figure 2), the first band of panels has an epigraphic inscription in Latin that states:

PRINCEPS PATRONU(M) PETRE TROIA(M) SUSCIPE DONUM QUA(M) L(A)ETABUN DUS GUILELM(US) DONO SECUNDUS

which can be translated as:

O Peter, prince (of the apostles) and patron, receive as a gift the city of Troia, which I, William the Second, gladly offer to you.

Figure 2. Picture of the first twelve upper panels of the studied door.

Next to this phrase is in fact depicted the bishop William II, who commissioned the door, towering in the act of donating the "*Civitas Trojana*" to the Church of Rome symbolised in the next panels by the inlays depicting St. Peter and St. Paul.

On the left outer edge of the second and third panels' bands is engraved the phrase:

HIC FUIT NOVAE TROIAE

which means that this was of the new Troia.

To introduce, starting from the second band there is a series of eight panels engraved with eight flexuous and harmonious stylised figures with a bicorne mitre surrounded by a halo, which represent, in reverse chronological order, the first eight bishops who succeeded one another in the city of Troia:

GUILIELMUS PRIMUS EPISCOPUS, HIC OCTAVUS (William the First, eighth bishop);
UBBERTUS EPISCOPUS, HIC SEPTIMUS (Hubertus, seventh bishop);
GIRARDUS EPISCOPUS, HIC SESTUS (Girard, sixth bishop);
GUALTIERUS EPISCOPUS, HIC QUINTUS (Gualtiero, fifth bishop);

STEPHANUS EPISCOPUS, HIC QUARTUS (Stephen, fourth bishop);
IOHANNES EPISCOPUS, HIC TERTIUS (John, third bishop);
ANGELUS EPISCOPUS, HIC SECUNDUS (Angel, second bishop);
ORIANUS EPISCOPUS, PRIMUS EPISCOPUS (Horian, first bishop).

It is important to highlight that the panel with the inlay Horian, first bishop of Troia, is the only documented evidence of his existence.

In the fourth band, it is possible to read an engraving with the author's signature (Figure 3):

FACTOR PORTARUM FUIT ODERISIUS HARUM BENEVENTANUS

That means that the creator of this door was Oderisius of Benevento.

Figure 3. Photograph of the four leonine protomes of the door. On the lower frame, it is possible to read the engraving *"Factor portarum fuit Oderisius harum beneventanus"*, which represents the signature of Oderisius of Benevento, builder of the bronze door.

The fourth band also contains four lion protomes holding clapper rings between their teeth. The style of these lions is simpler than those of the *"Door of Prosperity"*, but Oderisius did not fail to give them sufficient vigour to make them stand out, breaking up the uniformity of the composition. Some of the protomes have gaps, behind which the wooden support of the bronze door can be glimpsed. As an example, in Figure 4, two gaps are shown around the third leonine protome.

The last two bands are engraved with a Latin inscription (Figure 5):

AEQUITATIS MODERATOR/LIBERATOR PATRIAE/DOMINUSGUI/LIELMUS SECU(n)UDUS/DEI GRATIA/VENERABILIS HUJUSTROJANAE/SEDIS E(pisco)PU(u)S/NONUS/HAS ETIAM/PORTAS AENEAS/DE PROPRIO ERARIO/LARG(e)DISPEN/SATOR FIERI IUSSIT/ANNO INCAR/NATIO(n)IS D(omi)NICAE MCXXVII ET/CIVITATIS HUJUS CONDITIO(n)E/ANNO C(entesimo) AT(que) OCTAVO PONTIFICAT(us) VERO/D(omi)NI HONORII P(apae)/SEC(un)DI TERTIO ITEM/PONTIFCATUSD(omi)NI/GUILIELMI EPI(scopi)/SEC(un)DI XXI/INDIC(tione) V ANNO/Q(uo) GUILIELM(us) T(er)T(iu)/N(o)RMANN(o)R(um) DUX/SALERNI(s)OBIIT MORTE COMUNI/(Tun)C TROJANUS/POP(u)L(u)S PRO LIBER/TATE TUENDA ARCEM SUB/VERTITETUR/BEM VALLO/MURISQ(ue)MUNIVIT

which can be translated as:

The moderator of equity, liberator of the homeland, William II by the grace of God, ninth bishop of this Trojan See, magnanimous dispenser, also had these bronze doors made with his own money in the year of the incarnation of the Lord 1127 CE, the eighth hundredth from the foundation of this city, the third of the pontificate of Pope Honorius II, the twenty-first of the pontificate of bishop William himself, the fifth indiction, in which year William, third duke of the Normans, died a natural death in Salerno. Then the people of Troia, to defend their freedom, pulled down the castle and fortified the city with walls and a moat.

Figure 4. Pictures of the gaps near the third leonine protome, behind which the wooden support of the bronze door can be seen.

Figure 5. Photo of the bottom of the door.

This story engraved on the "*Freedom Door*" proudly and accurately summarises several historical events in the city: the year the gate was built (in 1127 CE) and who commissioned it, William II; the year Troia was founded (in 1019 CE); the year Pope Honorius II assumed the papal throne (in 1124 CE); the year William II took office as bishop (in 1106 CE); and the demolition of the castle on the death of Duke William III in 1127 CE to defend the freedom of the Trojan people.

A further constructive detail is that the neighbouring panels have complementary engravings, probably made to facilitate the correct assembly of the door. It is important to underline that these engravings are present in various parts of the door under the frame. An example of such engravings is shown in Figure 6.

Figure 6. Picture of three complementary engravings between two neighbouring panels, inscribed to facilitate the assembly of the door.

As stated on the door, the last restoration dates to 1951 CE [19], but oxidation on the bronze doors, clearly visible to the naked eye, is causing irreparable damage to the panels made by Master Oderisius. The experimental results obtained in this study could therefore represent a valuable resource in the now-necessary restoration of this priceless artefact.

2.2. *ED-XRF Analysis*

Energy-dispersive X-ray fluorescence (ED-XRF) analysis was performed using a portable instrument designed at the University of Salento [20,21]. It is composed of an X-ray tube produced by MOXTEK® (Orem, UT, USA) with a palladium anode operating at 1–40 kV voltage and 0–100 µA current, and a Si-PIN detector produced by AMPTEK® (Bedford, MA, USA), model XR_100CR, thermoelectrically cooled, with a beryllium window of 25 µm. It has a resolution of 150 eV at 5.9 keV, a resolution of about 250 eV in the range 10–15 keV, and a pocket multi-channel analyser produced by AMPTEK®, model MCA8000A, interfaced with a laptop. The diameter of the X-ray beam is an ellipse, whose axes are equal to 2 mm and 3 mm.

Cu, Pb, Zn and Fe were quantitatively determined at high energy: 20 kV voltage, 3 µA current, with an acquisition time of 60 s. Sn, S and Ag were determined at low energy: 6 kV voltage, 40 µA current, with an acquisition time of 60 s.

The calibration was carried out by analysing five standards, consisting of CuO, $CuCO_3$, Cu/Sn (80/20% *w*/*w*), Pb_3O_4, Zn, Fe, $CuCl_2$, $CuSO_4$, and SnO_2 in different ratios. The concentration of each standard was chosen in order to reproduce the chemical composition of the investigated patinas.

All chemical compounds were analytical grade and purchased from Sigma-Aldrich® (St. Louis, MO, USA), except the Cu/Sn (80/20% *w*/*w*) standard, which was purchased from Goodfellow® (Huntingdon, UK).

Each standard was prepared by mixing the compounds in different weight percentages. In particular, the chemical compounds were weighted using an analytical balance KERN® model ABT 100–5M (KERN & SOHN GmbH: Balingen, Germany), subsequently mixed and homogenised in an agate mortar for ten minutes, and finally compressed at 200 bar for ten minutes. The homogeneity of the elements in the standard meets the requirements for ED-XRF quantitative analysis [22].

Table 1 shows the quantitative chemical composition of the five analysed standards for ED-XRF analysis.

Table 1. Chemical composition of the standards used for calibration in ED-XRF analysis.

Standard	Cu	Pb	Sn	Zn	Fe
			(% wt)		
1	73.5	12.3	2.2	4.8	0.7
2	65.0	8.4	4.9	7.5	1.1
3	58.1	5.9	8.4	12.4	2.2
4	51.5	3.6	10.7	18.4	2.9
5	48.9	2.5	11.8	21.2	3.3
Standard deviation	*0.5*	*0.5*	*0.3*	*0.3*	*0.2*

The samples analysed were assumed to have "infinite thickness" [23–25] and the quantitative results are expressed in terms of weight percentage (% wt).

From a visual examination, the door does not show any corrosive phenomena in progress and the surface patina appears compact and homogeneous in the various regions. Before performing the ED-XRF analysis, cleaning of the metal surface was performed with Japanese paper and deionised water.

Both the calibration and analytical samples were analysed under the same instrumental conditions. For all the spectra obtained, the intensities of the different signals were determined by integrating the different signals using Gaussian curves.

The calibration curves were obtained by reporting the intensity of the signals as a function of the concentrations of the element under investigation.

Figure 1 shows the locations of all the measurement points, while Table 2 shows their brief description. The twenty-nine points were chosen to have enough measurements on both leaves. In particular, the choice of the various areas was performed to characterize as much as possible of the patinas of the alloys (points n. 1–11), the frame (points n. 12–13), the regions added in 1951 CE by the ICR (points n. 14–17), the different engravings (points n. 18–22), the different ornaments of the door (points n. 23–28), and an evident welding (point n. 29).

Table 2. Brief description of the analysed measurement points and their composition obtained using ED-XRF.

Sample	Description	Cu	Pb	Sn	Zn	Fe	Cl	S	Note
					(% wt)				
1	Alloy (dark)	63.8	16.5			3.9	13.7		
2	Alloy (dark)	68.5	13.0				16.7		
3	Alloy	63.6	12.2		9.6	2.5	11.5		
4	Alloy (dark)	62.8	16.5			0.9	12.8		
5	Alloy	65.2	14.2				15.1		
6	Alloy	65.0	15.8				11.8		
7	Alloy	62.8	13.7				13.8		
8	Alloy	66.0	13.6				14.5		
9	Alloy	65.0	14.5				12.1		
10	Alloy	64.1	15.0				13.2		
11	Alloy	69.6	10.2			3.5	11.2		
12	Frame	65.4	14.5			2.0	12.2		
13	Frame	66.0	14.0				13.1		
14	Addition	78.8	2.9				17.8		
15	Addition (rest 1951)	70.5	3.4	6.8	2.8	2.7	11.9		
16	Rawplug	41.4	5.2		38.2	6.4	8.4		
17	Addition (rest 1951)	73.2	2.9		2.1	2.9	18.8		
18	Inlay	48.8	13.1				5.2	5.3	Ag trace
19	Inlay	63.2	6.0	3.0	7.9		9.3	7.2	Ag trace
20	Inlay	45.1	12.7			0.7	11.9		Ag trace
21	Inlay	20.0	3.1				17.9		Ag
22	Inlay	10.0	1.1				15.4		Ag
23	Leonine protome	62.4	17.5	4.8			10.9		
24	Leonine protome	63.6	22.5	7.9			4.5		
25	Handle	70.8	6.1	6.4	7.7		7.2		
26	Handle	70.4	8.0	3.3	7.9	2.0	6.8		
27	Nail	70.3	2.0	13.2		2.0	12.4		
28	Nail	72.9	2.5	6.9		4.3	11.2		Ti
29	Welding	10.0	60.5	12.2					
	Detection limit	*2.0*	*1.0*	*2.0*	*1.0*	*0.5*	*3.0*	*3.0*	

3. Results and Discussion

The analysed door includes two wooden-core copper-based leaves covered with metal-framed panels fixed with nails. Table 2 shows a brief description of the analysed measurement points and their chemical composition obtained using portable energy-dispersive X-ray fluorescence.

Regarding the chemical composition, the two wings show no significant differences.

The panels' patina (samples n. 1–11) and the frames (n. 12 and n. 13) consist of a binary alloy of copper and lead with a Cu/Pb ratio for the thirteen samples of $4.7 \pm 0.8\%$ wt. The percentage of tin detected was always below the detection limit. The quantity of zinc was also always below the detection limit of the instrument, except for sample n. 3 where the zinc percentage was 9.6% wt.

The chemical composition of some areas (n. 1, n. 3, n. 4 and n. 11) differs from that typical of the door due to the presence of iron at between 0.9% wt and 3.9% wt. These are original areas, and the high iron content may suggest dripping. Other areas of the door were indeed restored (n. 14–17). For example, at point n. 14 there was certainly a gap that was filled during restoration by casting of an alloy consisting of copper (78.8% wt) and lead (2.9% wt). In contrast, at point n. 16, a repair was carried out during the restoration using a square dowel consisting of copper and zinc in a 1:1 ratio and with traces of lead (5.2% wt) and iron (6.4% wt).

As reported in the "*Bollettino dell'Istituto Centrale del Restauro, 1952*" these areas are the result of reconstruction processes during the 1951 CE restoration, which used a quaternary alloy composed of copper in high concentration, tin, lead, zinc, and small quantities of iron [19].

The analysed inlays depicting the bishops of Troia were mainly made of copper, lead, and silver. In particular, inlays n. 18, n. 19, and n. 20 are made of copper with a high lead content and traces of silver, while agemine n. 21 and n. 22 are made of silver with low copper and lead content. It is important to emphasise that the amount of silver has not been quantified, since the precious metal is not a constituent of the alloy of the panels but was used in the form of foils in the inlaying process to fill the grooves of the engravings. Therefore, the terms trace or high content, referring to silver, only indicate a qualitative estimation of the metal present in the ageminature, taking into account that most of the silver was removed over the centuries. Furthermore, very often, the dimensions of the agemine are smaller than the emission spot of the tube; therefore, the emission signal is not only due to the chemical composition of the silver, but also to the surface of the alloy patina.

Sample n. 19, somewhat darker in colour, showed the presence of tin (3% wt) and zinc (7.9% wt).

The protomes (n. 23 and n. 24) and handles (n. 25 and n. 26) were made of a different patina than the one used to make the door moulds and the door brim. For the protomes, this is a ternary alloy of copper, lead, and tin, while for the handles it is a quaternary alloy consisting of copper, lead, tin, and zinc.

The fasteners analysed (n. 27 and n. 28) are made of copper (70.3% wt–72.9% wt) and tin (6.9% wt–13.2% wt) and lower concentrations of lead (2% wt–2.5% wt) and iron (2.0% wt–4.3% wt). Therefore, the nails seem to be made of brass alloyed with iron. A lower amount of lead is understandable, since lead reduces to some extent the mechanical properties of the alloy, which is important for their functions.

All the regions analysed show the presence of chlorine in relatively high concentrations, ranging from 4.5% wt to 18.8% wt, except for region n. 29, in which is evident a solder showing high concentrations of lead (60.5% wt), as well as copper (10.0% wt) and tin (12.2% wt).

The chlorine concentration in the different investigated regions may be due to atmospheric corrosion; indeed, during wet and dry cycles, chlorides could concentrate in the surface, causing chlorine-rich patinas [26]. A further cause of the widespread presence of chlorine on the surface can be attributed to the reagents used to patinate the pieces crafted to repair the door during the 1951 CE restoration.

The chemical composition of the alloy patina was compared with that of the alloy used to make eight medieval bronze doors in Italy, as shown in Table 3. The table shows the average concentration values of copper, lead, zinc, and tin in nine Italian doors and, for further comparison among them, the ratio between the concentration of copper and the concentration of lead.

The patina analysed on the "*Freedom Door*" in Troia turned out to consist of a binary alloy of copper (65.2% wt) and lead (14.1% wt). Moreover, it has a copper-to-lead ratio of 4.6, which is very different from other manufacts.

Therefore, it is clear that the chemical composition of the alloy studied is quite different from that of the other doors; in fact, the remaining medieval doors are made up of ternary or quaternary alloys. For example, the Amalfi door has a ternary alloy consisting of copper (62% wt), lead (19% wt), and zinc (17% wt) and a copper/lead ratio equal to 3.3.

The Monte Cassino door has a quaternary alloy with the highest concentration of copper (80% wt) among the eight doors, in addition to lead (3% wt), zinc (10% wt), and tin (7% wt). The door in Monte Cassino also has the highest copper/lead ratio value (26.7).

The door in Rome S. Paul's outside the walls has a quaternary alloy with a high concentration of copper (73.5% wt), in addition to lead (8.48% wt), zinc (17.9% wt), and tin (0.12% wt) and a copper/lead ratio equal to 8.7. These values are very similar to the door in Monte Sant'Angelo; in fact, it has a quaternary alloy with a high concentration of copper (70.2% wt), in addition to lead (11.2% wt), zinc (14.2% wt), and tin (0.8% wt) and a copper/lead ratio equal to 6.3.

The door in Venice, S. Clemente, has a quaternary alloy with a high concentration of copper (72.4% wt), in addition to lead (8.6% wt), zinc (16.8% wt), and tin (2.2% wt) and a copper/lead ratio equal to 8.4.

The door in Salerno has a quaternary alloy with a high concentration of copper (77% wt), in addition to lead (11% wt), zinc (4.5% wt), and tin (3.5% wt) and a copper/lead ratio equal to 7.0.

The central door in Venice has a quaternary alloy with a high concentration of copper (78% wt), in addition to lead (8% wt), zinc (9% wt), and tin (5% wt) and a copper/lead ratio equal to 9.8.

The door in Canosa di Puglia has a quaternary alloy with a high concentration of copper (69.6% wt), in addition to lead (10.2% wt), zinc (0.3% wt), and tin (8.5% wt) and a copper/lead ratio equal to 6.8.

Table 3. Comparison between the chemical compositions of alloys obtained in this work and those of other Italian medieval copper-alloy doors.

Italian Medieval Door	Cu	Pb	Zn	Sn	Cu/Pb	Reference
	(% wt)					
Amalfi (1060 CE)	62	19	17	Traces	3.3	[3]
Monte Cassino (1066 CE)	80	3	10	7	26.7	[1]
Rome, S. Paul's outside the Walls (1070 CE)	73.5	8.48	17.9	0.12	8.7	[1]
Monte Sant'Angelo (1076 CE)	70.2	11.2	14.2	0.8	6.3	[18]
Venice, S. Clemente (1080 CE)	72.4	8.6	16.8	2.2	8.4	[1]
Salerno (1085 CE)	77	11	4.5	3.5	7.0	[3]
Venice, central door (1112 CE)	78	8	9	5	9.8	[1]
Canosa di Puglia (1111–1118 CE)	69.6	10.2	0.3	8.5	6.8	[17]
Troia, "Freedom Door" (1127 CE)	*65.2*	*14.1*	*<1.0*	*<2.0*	*4.6*	*This work*

4. Conclusions

The experimental results obtained have shown that the two wings and the frame of the medieval door in Troia have the same patina chemical composition. In these experimental situations, it is possible to correlate the patina's composition with the composition of the alloy because the same substrates are compared, and it can be assumed that the environmental conditions have been the same for the areas compared over the centuries.

Moreover, the results show that the door studied has a very different composition from the other Italian medieval doors investigated. In fact, the copper/lead ratio in the studied manufact, equal to 4.6, is different from eight medieval copper doors analysed in Italy.

The entire structure is uniformly covered in a patina of chlorine, probably added during the 1951 CE restoration, or it could be due to atmospheric phenomena; in fact, during wet and dry cycles, chlorides can concentrate in the surface of the manufact causing chlorine-rich patinas.

Even the leonine protomes, handles, nails, and parts added during the restoration are different from the original alloy. The few remaining agemine were also analysed, and the results showed that they are of two distinct types: copper-based agemine and silver-based agemine.

The data obtained from this work are the only date relating to this valuable manufact and may be of help for future research in the field of medieval copper alloy doors still present in Europe.

Author Contributions: Conceptualization, G.B., A.S., G.G.C., V.N.I., A.C., L.C. and A.B.; methodology, G.B., G.G.C., V.N.I. and A.B.; validation, G.B., A.S., G.G.C., V.N.I., A.C., L.C. and A.B.; formal analysis, G.B., V.N.I. and A.B.; investigation, G.B., A.S., V.N.I. and A.B.; resources, G.B., A.S., G.G.C., V.N.I., A.C., L.C. and A.B.; data curation, G.B., A.S., G.G.C., V.N.I., A.C., L.C. and A.B.; writing—original draft preparation, G.B., A.S., G.G.C., V.N.I., A.C., L.C. and A.B.; writing—review and editing, G.B., A.S., G.G.C., V.N.I., A.C., L.C. and A.B. All authors have read and agreed to the published version of the manuscript.

Funding: This research received no external funding.

Data Availability Statement: Not applicable.

Conflicts of Interest: The authors declare no conflict of interest.

References

1. Matthiae, G. *Le Porte Bronzee Bizantine in Italia*; Officina Edizioni: Rome, Italy, 1971.
2. Bloch, H. Origin and Fate of the Bronze Doors of Abbot Desiderius of Monte Cassino. *Dumbart. Oaks Pap.* **1987**, *41*, 89–102. [CrossRef]
3. Banti, O. La Porta Di Bonanno Nel Duomo Di Pisa e Le Porte Bronzee Medioevali Europee: Arte e Tecnologia: Atti Del Convegno Internazionale Di Studi. In Proceedings of the Opera Della Primaziale Pisana, Pontedera, Pisa, Italy, 6–8 May 1993; Bandecchi & Vivaldi: Pontedera, Pisa, Italy, 1993; p. 326.
4. Cadei, A. La porta del mausoleo di Boemondo a Canosa tra Oriente e Occidente. In *Le Porte del Paradiso: Arte e Tecnologia Bizantina tra Italia e Mediterraneo*; Iacobini, A., Ed.; Milion; Campisano: Rome, Italy, 2009; pp. 429–471. ISBN 978-88-88168-43-2.
5. Brenk, B. Bronzi della Sicilia normanna: Le porte del duomo di Monreale. In *Le Porte del Paradiso: Arte e Tecnologia Bizantina tra Italia e Mediterraneo*; Iacobini, A., Ed.; Milion; Campisano: Rome, Italy, 2009; pp. 471–491. ISBN 978-88-88168-43-2.
6. Davì, G. *Le Porte Bronzee Del Duomo Di Monreale*; Centro di Cultura: Cefalù, Italy, 1987; pp. 102–106.
7. Leisinger, H. *Bronzi Romanici. Porte Di Chiese Nell'Europa Medioevale*; Silvana Editoriale d'Arte: Milan, Italy, 1956.
8. Angelucci, S. Il Rapporto Tra Materia, Tecnica e Forma Nelle Porte Bizantine d'Italia. In *Storia Dell'arte Marciana: Sculture, Tesoro, Arazzi, Proceedings of the Atti del Convegno Internazionale di Studi, Venice, Italy, 11–14 October 1994*; Polacco, R., Alcouffe, D., Eds.; Marsilio: Venice, Italy, 1997; ISBN 978-88-317-6852-8.
9. Emmitt, J.; McAlister, A.; Armstrong, J. Pitfalls and Possibilities of Patinated Bronze: The Analysis of Pre-Roman Italian Armour Using PXRF. *Minerals* **2021**, *11*, 697. [CrossRef]
10. Holmqvist, E.; Wessman, A.; Mänttäri, I.; Lahaye, Y. Lead Isotope and Geochemical Analyses of Copper-Based Metal Artefacts from the Iron age Water Burial in Levänluhta, Western Finland. *J. Archaeol. Sci. Rep.* **2019**, *26*, 101854. [CrossRef]
11. Brunetti, A.; Golosio, B.; Melis, M.G.; Mura, S. A High-Quality Multilayer Structure Characterization Method Based on X-ray Fluorescence and Monte Carlo Simulation. *Appl. Phys. A* **2015**, *118*, 497–504. [CrossRef]
12. Manso, M.; Schiavon, N.; Queralt, I.; Arruda, A.M.; Sampaio, J.M.; Brunetti, A. Alloy Characterization of a 7th Century BC Archeological Bronze Vase—Overcoming Patina Constraints Using Monte Carlo Simulations. *Spectrochim. Acta Part B At. Spectrosc.* **2015**, *107*, 93–96. [CrossRef]
13. Pearce, M. The Curse of the PXRF: The Negative Consequences of the Popularity of Handheld XRF Analysis of Copper-Based Metal Artefacts. *Metalla* **2019**, *24*, 81–85. [CrossRef]
14. Weinryb, I. *The Bronze Object in the Middle Ages*; Cambridge University Press: Cambridge, UK, 2016; ISBN 978-1-316-53902-6.
15. Belli D'Elia, P. Le Porte Della Cattedrale Di Troia. In *Le Porte di Bronzo Dall'antichità al Secolo XIII*; Giubileo 2000; Edipuglia: Bari, Italy, 1990; pp. 341–357. ISBN 978-88-7228-252-6.
16. Belli D'Elia, P. Art. Oderisio Da Benevento. In *Enciclopedia Dell'arte Medievale*; Istituto della Enciclopedia Italiana: Rome, Italy, 1997; Volume VIII, pp. 791–792. Available online: https://www.treccani.it/enciclopedia/oderisio-da-benevento_%28Enciclopedia-dell%27-Arte-Medievale%29/ (accessed on 12 December 2022).
17. Buccolieri, G.; Castellano, A.; Iacobelli, V.N.; Carbone, G.G.; Serra, A.; Calcagnile, L.; Buccolieri, A. Non-Destructive In Situ Investigation of the Study of a Medieval Copper Alloy Door in Canosa Di Puglia (Southern Italy). *Heritage* **2022**, *5*, 145–156. [CrossRef]
18. Buccolieri, A.; Castellano, A.; Iacobellis, V.N.; Paladini, F.; Buccolieri, G. ED-XRF Analysis of the Mediaeval Copper-Based Door in Monte Sant'Angelo (Southern Italy). *Archaeol. Anthropol. Sci.* **2021**, *13*, 95. [CrossRef]
19. Cagiano De Azevedo, M. *Restauri a Porte Di Bronzo—Porta Laterale Del Duomo Di Troia*; Bollettino dell'Istituto Centrale del Restauro; Ministero della Pubblica Istruzione: Rome, Italy, 1952; p. 47.
20. Buccolieri, G.; Buccolieri, A.; Donati, P.; Marabelli, M.; Castellano, A. Portable EDXRF Investigation of the Patinas on the Riace Bronzes. *Nucl. Instrum. Methods Phys. Res. Sect. B Beam Interact. Mater. At.* **2015**, *343*, 101–109. [CrossRef]
21. Buccolieri, A.; Castellano, A.; Degl'Innocenti, E.; Cesareo, R.; Casciaro, R.; Buccolieri, G. EDXRF Analysis of Gold Jewelry from the Archaeological Museum of Taranto, Italy: EDXRF Analysis of Gold Jewelry. *X-ray Spectrom.* **2017**, *46*, 421–426. [CrossRef]

22. Li, L.; Feng, S.L.; Feng, X.Q.; Xu, Q.; Yan, L.T.; Ma, B.; Liu, L. Study on Elemental Features of Longquan Celadon at Fengdongyan Kiln Site in Yuan and Ming Dynasties by EDXRF. *Nucl. Instrum. Methods Phys. Res. Sect. B Beam Interact. Mater. At.* **2012**, *292*, 25–29. [CrossRef]
23. He, F.; Van Espen, P.J. General Approach for Quantitative Energy Dispersive X-Ray Fluorescence Analysis Based on Fundamental Parameters. *Anal. Chem.* **1991**, *63*, 2237–2244. [CrossRef]
24. Giurlani, W.; Berretti, E.; Innocenti, M.; Lavacchi, A. Coating Thickness Determination Using X-Ray Fluorescence Spectroscopy: Monte Carlo Simulations as an Alternative to the Use of Standards. *Coatings* **2019**, *9*, 79. [CrossRef]
25. Barcellos Lins, S.A.; Gigante, G.E.; Cesareo, R.; Ridolfi, S.; Brunetti, A. Testing the Accuracy of the Calculation of Gold Leaf Thickness by MC Simulations and MA-XRF Scanning. *Appl. Sci.* **2020**, *10*, 3582. [CrossRef]
26. Bernardi, E.; Chiavari, C.; Lenza, B.; Martini, C.; Morselli, L.; Ospitali, F.; Robbiola, L. The Atmospheric Corrosion of Quaternary Bronzes: The Leaching Action of Acid Rain. *Corros. Sci.* **2009**, *51*, 159–170. [CrossRef]

Disclaimer/Publisher's Note: The statements, opinions and data contained in all publications are solely those of the individual author(s) and contributor(s) and not of MDPI and/or the editor(s). MDPI and/or the editor(s) disclaim responsibility for any injury to people or property resulting from any ideas, methods, instructions or products referred to in the content.

heritage

MDPI

Article

The Version of the Composition of the Mausoleum-Khanaka Khoja Ahmed Yassawi Main Facade in Turkestan

Konstantin Samoilov [1,*], Bolat Kuspangaliyev [1], Gaukhar Sadvokasova [2] and Aizhan Akhmedova [2]

[1] T. Basenov Institute of Architecture and Civil Engineering, Satbayev University, 22 Satbayev Str., Almaty 050000, Kazakhstan
[2] International Education Corporation, Kazakh Leading Academy of Architecture and Civil Engineering/KazGASA, 28 Ryskulbekov Str., Almaty 050043, Kazakhstan
* Correspondence: samconiv@mail.ru; Tel.: +7-701-357-68-93

Abstract: The problem of preservation, and optimal demonstration of it to local residents and tourists of architectural monuments, is of constant scientific and public interest. Two concepts coexist in dialectical interaction: conservation to preserve the monument in the form in which it has come down to in our times, and restoration of the monument with the restoration of lost details. In each case, one or another decision is made, which finds both supporters and opponents. One of the aspects of this problem is the attitude to buildings that have long breaks in the history of conduct in their construction works. An interesting example of such a monument is the mausoleum-khanaka of Ahmed Yassawi in Turkestan, which remains unfinished. Given its importance for the self-determination of the culture of modern Kazakhstan, it seems appropriate to consider the planned design, which remains unrealized. For the first time, the article suggests, by way of discussion, several options proposed by the authors for solving the main facade of the mausoleum-khanaka and the shape of the central dome.

Keywords: a reconstruction; a restoration; the medieval architecture of Central Asia; mausoleum-khanaka; a graphic analysis of composition; proportions

Citation: Samoilov, K.; Kuspangaliyev, B.; Sadvokasova, G.; Akhmedova, A. The Version of the Composition of the Mausoleum-Khanaka Khoja Ahmed Yassawi Main Facade in Turkestan. *Heritage* **2023**, *6*, 1344–1377. https://doi.org/10.3390/heritage6020074

Academic Editors: Daniela Fico and Daniela Rizzo

Received: 3 January 2023
Revised: 21 January 2023
Accepted: 22 January 2023
Published: 30 January 2023

Copyright: © 2023 by the authors. Licensee MDPI, Basel, Switzerland. This article is an open access article distributed under the terms and conditions of the Creative Commons Attribution (CC BY) license (https://creativecommons.org/licenses/by/4.0/).

1. Introduction

The Medieval architecture of Kazakhstan is represented by a variety of monuments. They originally reflected the specifics of each of the historical periods in which they appeared. The mausoleum-khanaka of Ahmed Yassawi in Turkestan is one of particular importance. This building, despite the fact that it remains in an unfinished state, has great historical significance. Additionally, now, it is also considered as part of the spectrum of elements of regional identity for the modern architecture of the country [1]. Special attention is being paid to it at the present time. This is due to the fact that the city of Turkestan has become the capital of the region with the same name. In addition, in ideological terms, this city with a complex of medieval monuments is perceived as the "Spiritual capital of Kazakhstan". The studies mentioned in the literature review reflect the main part of the spectrum of available works on the topic under consideration. The works of the largest experts on the topic, published in the first half of the twentieth century, and modern studies summarizing and clarifying the existing body of knowledge are mentioned here.

The mausoleum-khanaka in Turkestan has been described and studied in detail for more than one-hundred-and-fifty years. By the beginning of this century, the scientific basis for the study of the mausoleum-khanaka consisted of more than 50 names of research materials [2]. The study of the monument is actively continuing.

One of the first published descriptions of the building was made by M.S. Bekchurin on 19 February 1866 [3]. He pointed out the location of the building in the structure of the city fortress, and briefly described the spatial composition, planning solution, functional scheme, and constructive solution. A list of state and public figures buried in the complex is given. Much attention is paid to the description of the organizational and financial aspects of the existence of the complex as an object of religious worship. The description contains data on the time of construction of the "Azret mosque"; the legend about the reasons that prompted Timur to order the construction of the complex is retold. Special attention is paid to the translation of inscriptions made in Arabic and Persian. Attention is also drawn to the emergency condition of some parts of the building. The destruction was the result of both military actions that took place at various periods and untimely routine repairs.

In 1929 and 1930, a detailed description was performed by M. E. Masson [4,5]. He spoke about Khoja Ahmed Yassavi himself; traced the history of the mausoleum and the main stages of its functioning; characterized the historical and architectural character of the mausoleum; gave a description of the architectural planning and spatial solutions of the mausoleum; and described in detail the modern appearance of the mausoleum. An assumption was made about the possible similarity in the composition of the unfinished portal of the Mausoleum in Turkestan with the portal of the Ak-Saray Palace complex in Shakhrisyabz. Special attention is paid to the sheikhs of the Mausoleum and the legends of Khoja Ahmed. These works by M. E. Masson are the first scientifically substantiated description of the monument, which became the basis for further research.

In 1950, the work of G. I. Patsevich was published [6]. It describes, in detail, the complexities of the work on the repair and restoration of the mausoleum-mosque of Khoja Ahmed Yasawi, which were carried out in 1939–1941. Thanks to this description, the specifics of the constructive solution of the building became clear, both in general and in individual nodes.

Based on previously personally conducted research and analysis of the available scientific base in 1960–1963, L. Yu. Man'kovskaya published several works on the architectural and archaeological aspects of the restoration of the mausoleum [7–9]. For the first time, the architectural features of this building became the subject of dissertation research. In the future, under the leadership of L. Yu. Man'kovskaya, work was carried out to prepare the restoration, and a project was carried out to preserve the mausoleum of Khoja Ahmed Yassawi (1968–1972). This work became the basis of a comprehensive study on the typology of architecture in Central Asia. The research carried out by L. Yu. Man'kovskaya contributed to the "clarification of the architectural phenomenon of the resolution of many functions in a single form", which revealed a kind of universal constructor, from which all types of monumental buildings were erected, while remaining specific and unique. L.Yu.Man'kovskaya was able to prove that there was a standard design. At the same time, variations were allowed on the basis of a typical solution. In addition, the architects possessed methods of flexible transformation of spatial structures. In a monograph published in 1980 [10], a typological classification was developed for the first time covering all types of human activity in medieval society and the types of buildings and structures corresponding to them: it is based on the leading feature of architectural form—the type of three-dimensional structure. The extensive factual material of many years of full-scale study of the possibility of studying the art of Central Asian architects was summarized.

At the end of the 1980s, B. T. Tuyakbayeva highlighted, at a dissertation level, the composition and content of the epigraphy of the mausoleum-khanaka [11]. This work gives the most complete description of the specifics of the architectural decoration of the monument to date.

The authors and work devoted to the direct analysis of various aspects of the form and decoration of this building are: N. B. Alliyar and K. I. Samoilov (the currently available material on the architectural and artistic solution of the mausoleum-khanaka is briefly summarized [12–14]), B. A. Baitanaev and Yu. A. Yelgin (clarified the issues of onomastics and epigraphy of the mausoleum-khanaka [15]), T. K. Basenov (basic information on the architecture of the mausoleum-khanaka is given [16]), D. Mustapayeva (the supposed features of the original mausoleum of the twelfth century on the site of the mausoleum-khanaka of the fourteenth century are described [17]—this work is important for understanding the history of the complex, which is still unfinished), N.-B. Nurmukhammedov (a somewhat romanticized description of the complex and the history of its development [18]), A. Ordabaev (the general issues of the study and restoration of the mausoleum-khanaka are described [19]—the interview describes the personal experience of many years of work on the study and restoration of the monument), Yu. A. Yelgin (the materials on the study and restoration of the mausoleum-khanaka from the mid-nineteenth century to the mid-twentieth century are summarized [20,21]—these studies form the basis for further restoration and reconstruction work), among others. These studies, conducted at different times, emphasize many aspects of the protection, study, repair, and restoration of the mausoleum-khanaka, which are important for the study of the authors of this article.

Considerations of this building in the context of the contemporary architecture of the Central Asian region is reflected in these studies: N. B. Alliyar and K. I. Samoilov (the general features of the historical and architectural context of the construction of the mausoleum-khanaka in Turkestan are considered [22]), E. Baitenov (the specifics of shaping in the memorial architecture of Kazakhstan as a whole are emphasized [23]), E. Baitenov, A. Tuyakayeva, and G. Abdrassilova (the processes of origin and development of various forms of Kazakh mausoleums are detailed [24]), E. G. Barsukova (the specifics of Sufism from the point of view of architectural and artistic solutions of religious buildings in medieval Central Asia are analyzed [25,26]), S. Chmelnizkij (the general characteristic of Timurid architecture as a historical and cultural phenomenon is given [27]), B. Glaudinov (the processes in the architecture of Kazakhstan at all stages of its ancient and medieval development are analyzed [28]), M. B. Kasymbekova, B. A. Glaudinov, and K. I. Samoilov (the specifics of Islamic memorials and religious buildings in medieval Kazakhstan are considered [29]), M. B. Kozha (a number of positions on the shaping of medieval monuments of Turkestan, obtained on the basis of new research, have been clarified [30]), A. Kh. Margulan, T. K. Basenov, and M. M. Mendikulov (the architecture of Kazakhstan before the beginning of the twentieth century is comprehensively considered [31]), G. A. Pugachenkova (the specifics of the architecture of the region in the fifteenth century are specified [32]), G. A. Pugachenkova and L. I. Rempel' (the specifics of the architecture in the context of the art of the region are shown [33]), D. J. Roxburgh (an overview of materials on architecture and art of the epoch is given [34]), K. I. Samoilov (the monuments of the memorial architecture of the Middle Ages are considered in the context of the centuries-old architecture of Kazakhstan [35]), A. Tondi (the architectural decor in the monuments of the epoch is investigated [36]), I. I. Umnyakov (the specifics of the research and restorative works of architectural monuments carried out in the 1920s in the region are shown [37]), V. L. Vyatkin (the ancient Samarkand monuments are described in detail [38]), B. V. Weimarn (a general overview of architecture in the context of the art of the region is given [39]), A. Ya. Yakubovsky (the construction activity of Emir Timur in Samarkand is described [40]), P. Zahidov (based on studies of Timurid architecture, conclusions are drawn about the presence of a peculiar system of proportionation [41–43]), B. N. Zasypkin (a general description of the architecture of the region is given, and the features of the most significant structures are specified [44]), among others (general information on the architecture of most of the medieval monuments preserved in the region is given [45,46]). In these works, an understanding of the processes of development of the architecture of the region is given from the perspective of the knowledge available during the period of the study. The body of knowledge is constantly expanding, and some of the provisions presented in the studies of the middle of the last century have now been supplemented and partially corrected. All these studies are important for understanding the process of formation of the modern appearance of the mausoleum-khanaka.

From the point of view of the assumption of the original design, which was not fully realized, it is important to study the identical buildings of that era: N. V. Gilmanova (features of the Ak-Sarai Palace [47]), S. N. Kabanov (description of the preserved ruins of the Ak-Sarai Palace [48]), Z. Khakimov (description of a number of buildings of the Timur era [49]), T. Kuziev (the specifics of the architectural composition of minarets in the region [50]), M. E. Masson (features of minarets in the context of the architecture of the Bibi-Khanim Mosque [51,52]), M. E. Masson, G. A. Pugachenkova and H. T. Sultanov (the development of the city of Shakhrisabz in its heyday [53,54]), L. Yu. Man'kovskaya and E. Paskaleva (some features of the Bibi-Khanim mosque [55–57]), among others (for example, the hidden meanings of shaping the Bibi-Khanim Mosque [58]). This is especially true of compositional analyses of the formation of various monuments of the region in general (M. S. Bulatov—the specifics of harmonization techniques in the architecture of Central Asia from the tenth–fifteenth centuries [59]), and individual examples in particular (Sh. E. Ratiya—features of the multidimensional study and restoration of the Bibi-Khanim mosque [60]).

Almost from the very beginning of the active study of the monument, there was an interest in the reconstruction of the original design of the structure, which is still in an unfinished state. However, apart from general indications of possible similarities to some modern buildings (Ak-Saray Palace in Shakhrisyabz, the main mosque of Bibi-Khanim in Samarkand), there were practically no attempts to, at least, graphically reconstruct the unfinished facade. The restoration works carried out since the middle of the twentieth century were focused on the restoration of partially destroyed cladding and the preservation of the monument. In this regard, it seems appropriate to conduct a kind of restoration experiment to recreate (at least graphically) the unrealized architectural and artistic design of the general appearance of the mausoleum-khanaka of Ahmed Yassawi in Turkestan. Additionally, there are not many questions about the decoration of the portal part of the mausoleum, since there are preserved fragments. Their interpretation, which is quite adequate from a historical and cultural point of view, is not particularly difficult. Determining the designed height of the portal and its flanking towers is very controversial. There are questions about the main dome. The definition of the form of the wedding of corner minarets is equally problematic. The geometric construction based on the compositional solution of similar structures in the region is methodically justified.

2. Materials and Methods

A common method of sequential data collection, analysis, and generalization was used to conduct this study. Initially, the differential method of studying the problem distinguishes examples with similar architectural forms from the aggregate. Next, the integral method groups selected examples according to the specifics of shaping. Continuing the study, the formal method examines the evolution of various architectural forms. Then, the iconographic method emphasizes the presence of certain features peculiar to the prototypes of the studied forms. The final analysis of the structural-semiotic method models the prospects for the evolution of architectural forms. The analysis methodology used is based on the principle of analogy. The principle of analogy is justified in this case since there is a triad—one historically short period, one region, and one customer. It is difficult to assume that in this combination of factors a cardinal difference in the principles of shaping is possible. The historical restoration of the beginning of the last century was based on the fantasy of the "ideal forms" of the epoch. In the situation under consideration, it is proposed to use standard shaping techniques that have been repeatedly implemented in buildings of similar size and function. At the beginning of the study, examples of construction completion in forms different from the original ones were mentioned. This "modernization" is seen by the authors as inadequate for the object under consideration. The proposed option suggests the possibility of reproducing the original design of the mausoleum-khanaka of Ahmed Yassawi. This approach is seen by the authors as the most attractive from a historical and educational point of view. It shows the greatness of the ancestors' plan. This is very important for the modern outlook of the country's citizens. This demonstrates the deep historical roots of the country's modern culture. The use of modern architectonic language in this case seems unacceptable to the authors. The goal is not just to complete the construction, the goal is to reproduce the original design.

A fairly common feature of a certain number of buildings is that their construction stretches over a considerable period of time. Most of them are still being completed preserving the original idea and function. For example, the Cologne Cathedral (50°56′28″ N, 6°57′24″ E) was built, almost completely preserving the original design, from 1248 to 1880. The Redemptive Church of the Holy Family (Temple Expiatori de la Sagrada Família) in Barcelona (41°24′13″ N, 2°10′28″ E) continues to be built according to the project of Antoni Plàcid Guillem Gaudí i Cornet since 1883. The main stages of this building construction are shown in Figure 1.

Figure 1. The Basílica i Temple Expiatori de la Sagrada Família, Barcelona (Francisco de Paula del Villar y Lozano, from 1877; Antoni Gaudí i Cornet, from 1883): 1—1953 [61]; 2—2021 [62]; 3—The model (for the ending in 2026 or late) [63].

Some buildings are being completed with changes that meet modern needs both in terms of function and imaginative solutions. St. Stephen's Cathedral in Vienna (48°12′30″ N, 16°22′23″ E) was started in 1137, went through several stages of rebuilding and expansion, and acquired a modern look by 1511. The building of the Cathedral of Santa Maria del Fiore in Florence (43°46′23″ N, 11°15′23″ E) began in 1296 according to the project of Arnolfo di Cambio. For several decades, the works were conducted under the direction of Giotto di Bondone, Francesco Talenti, Giovanni di Lapo Ghini, Andrea Pisano, Alberto Arnoldi, Giovanni d'Ambrogio, Neri di Fioravante, andAndrea Orcagna. These architects made some additions and changes to the original design. The construction was completed in 1436, when an innovative dome designed by Filippo Brunelleschi was erected for that era. The architectural and artistic solution of this dome, according to contemporaries, did not harmonize with the appearance of the main part of the cathedral. The later additions to the Cathedral of Our Lady of Chartres (48°26′50″ N, 1°29′16″ E), with towers of different shapes on the western facade (south–1140/1155, north 1513), look very contrasting. A similar contrast is seen in the Cathedral of Saint Etienne in Bourges (47°04′55″ N, 2°23′58″ E). Examples of construction completion in a modified form are shown in Figure 2.

Figure 2. Examples of construction completion in a modified form: 1—The Cathedral of Our Lady of Chartres, 1260 (Towers–1140/1155)—The North Tower, 1513 (Build. Jehan de Beauce), Chartres, France [64]; 2—The Cathédrale Saint-Étienne de Bourges, 1324—The North tower, 1542, Bourdes, France [65]; 3—The Cathédrale primatiale Notre-Dame de l'Assomption de Rouen, 1506—The Northern tower (the Saint-Romain tower), 1145—The Southern tower (the Butter tower), 1485–1769–1884–1956–1999, Rouen, France [66]; 4—The Cathédrale Notre-Dame d'Amiens, 1269 (Arch. Robert of Luzarches, Thomas de Cormont, Renaud de Cormont),—the South tower, 1372; The North tower, 1420, Amiens, France [67].

The mausoleum-mosque of Arystan-Baba in the Kogam village in the Turkestan region (42°51′05″ N, 68°16′05″ E) has repeatedly changed its appearance. The original construction dates back to the XII century; the first reconstruction was made in the XIV–XV centuries; the second in the XVIII century; the third in 1909; and, due to the threat of destruction, the building was demolished and rebuilt in 1971 [46]. Large towers with tall lanterns are located at the corners of the main facade. The wide frieze, highlighted by cornices with belts of figured masonry, is dissected by blades with panels between them. The main entrance is accentuated by a pointed arch with large detailed clinched archivolt masonry. The wedding is made in the form of a developed pediment of a keel shape and framed by turrets. The arch is flanked by buttresses topped with wall-mounted pinnacles with pointed domes. On the front face of the pylons there is a panel with a medallion in the middle. The large domes of the building have a pointed shape. One of them is crowned with a figured spire. The arched windows and door openings are framed by platbands with broken locks. The authors of this article suggest that the keeled shape of the pediment appeared only in the XVIII century, as a metaphor for the unfinished portal of the grand mausoleum-khanaka of Ahmed Yassawi in the XV century. The reason for it could be that the preacher Arystan Bab was the spiritual mentor of Khoja Ahmed Yassawi.

One of the earliest structures in Kazakhstan, which used curly brickwork in the form of fir trees, triangles, and rhombuses, is the mausoleum of Karakhan (42°54′02″ N, 71°23′14′E) in Taraz. The building dates back to the XI century. The mausoleum has a developed three-part portal with two turrets, not centered in the area of the walls of the portal proper. The portal is solved with figured brickwork. The masonry has several varieties with the use of both terracotta and glazed tiles. By the beginning of the twentieth century, the building began to need major repairs. After reconstruction in 1905—1906, the mausoleum of Karakhan has completely lost its original appearance. As A. S. Schenkov points out, restructuring in the modern forms at that time is explained by the peculiarity of the worldview, which put the sacred function of the structure first, which should continue to function while maintaining syncretic integrity. This determined the primacy of integrity during restoration "to the detriment of the authenticity of the monument" [68] (p. 178). In fact, the completely new building has a developed complexly dissected portal with blind corner towers topped with spheroconic domes on three-part cornices. A large lancet niche with a cantilever impost is placed in the shallow central two-part rectangular niche of the portal. Symmetrically from this niche, shallow two-part pointed and rectangular niches are arranged in three tiers. The side and rear facades have deep lancet niches along the axis, inscribed in rectangular shallow niches, which, in turn, are flanked by high lancet niches. A large spheroconic dome is placed on a low quad. From the construction that reached the beginning of the twentieth century, after the reconstruction, only the theme of the three-part division of the facades remained in appearance [35,69].

The mausoleum of Aisha-Bibi (42°50′01″ N, 71°12′37″ E) demonstrates an interesting interpretation of the architectural and artistic ideas of that period. The mausoleum is located in the village, which now has the name Aisha-Bibi and administratively belongs to the Dzhambul region. The building dates back approximately to the XI century. The corner columns of the original shape (high truncated cones articulated through the torus), are almost all continuous (with elements of figured masonry) facing with terracotta ornamented tiles (20 types, 4 types of themes), with a wide variety of compositional connections, figured capitals of columns under arches, ornamented tympanums, and rounded shaped tiles when moving from edges to smooth walls to form a unique a solution that later became widespread in the region [29]. Considering the Central Asian monuments of the XI–XII centuries, B. N. Zasypkin points out: "Facing with terracotta tiles attached to brickwork has opened a new page in architecture. (...) A sharp line is established between the structural brick body and decorative and ornamental cladding, which was especially pronounced in the architecture of the XIV century" [44] (p. 50). By the beginning of the twentieth century, a significant part of the wall of the western facade and fragments of other parts remained in place [68,70]. In 2002, under the leadership of N. Rameto, the building was restored. There are no questions about the shape and lining of the walls, since they are recreated on the basis of preserved fragments. The question of the smooth cone-shaped shape of the wedding and the polygonal drum under it remains debatable.

There are suggestions that this element could have large folds by analogy with the Babaji-Khatun mausoleum located next door (42°50′02″ N, 71°12′38″ E). This mausoleum, built a little later, is topped with a sixteen-rib cone. According to the well-preserved fragments, this element was restored in 1981. The cube-shaped building with a poorly developed portal and a ribbed conical dome on a sixteen-sided prismatic drum was plastically solved on a combination of differently recessed arched niches, simple round rosettes, frames, and toothed belts. The recessed planes of niches and frames are filled with ornamented tiles. A flat relief inscription on terracotta tiles is included in the low parapet. All facades, except for the smooth rear, have the same plastic solution. The building demonstrates the early stage of development of portal composition in the Region.

As an option for the reconstruction of the Aisha-Bibi mausoleum, a twelve–sided tent on a twelve-sided prismatic drum is possible—this is how the Mausoleum of Il-Arslan (Fahreddin Razi), built in the second half of the XII century in Kuneurgench (42°18′16″ N, 59°08′44″ E), was solved. In the analysis of the composition of the mausoleum, which was performed by M. S. Bulatov [59] (p. 111), the dome is depicted as spheroconic. The changes in the appearance of the Karakhan mausoleum and the Aisha-Bibi mausoleum are shown in Figure 3.

Figure 3. Examples of the reconstruction of medieval mausoleums in Kazakhstan: 1, 2—The Karakhan's Mausoleum, Taraz: the view in 1902 [35] (p. 25), the view after reconstruction in 1906 [71]; 3, 4—The Aisha-Bibi's Mausoleum, Aisha-Bibi village, Dzhambul region: the view in the middle of the twentieth century [35] (p. 27), the view after reconstruction at the beginning of the XXI century [72].

These examples show that the expediency of further use of a particular historical monument not only as a museum exhibit, but also as a part of modern life actively exploited for direct or indirect purposes, has deep historical roots. Completion, rebuilding, and adaptation are objective needs of vital activity. The degree of transformation of a historical monument depends on specific needs and ideological attitudes [35] (p. 849), [73]. These processes are present in all periods of the development of civilization. The current period of the evolution of the architectural and spatial environment is no exception.

Some buildings, now regarded as valuable monuments of the Central Asian Middle Ages, remained unfinished or were badly destroyed. This is especially true of the architecture of the Timur era. One of such grandiose structures is the mausoleum-khanaka of Ahmed Yassawi (43°17′52″ N, 68°16′15″ E). The mausoleum-khanaka of Khoja Ahmed Yasawi is the only construction undertaking of the emir, undertaken by him outside his hometown of Shakhrisyabz and the capital of his empire the city of Samarkand. The building is located in the fortress city of Yassy, which, from the middle of the XVI century, became known as Turkestan. The decree from the founder of the Empire defined the main parameters of the building and some details of its decoration. The construction of the building was carried out by Mawlyan Ubaidullah Said, Khoja Hussein al Shirazi, and Shelms Abd-al-Wahab al Shirazi. The construction took place in two stages in 1395—1399, and in 1591 partial completion and repair were carried out. In particular, the arch of the main facade was closed. However, the top of the arch was partially displaced relative to the central axis. In addition, the right part of the arch was partially deformed (arrow in Figures 4 and 6). The width of the portal (along the axes of the corner minarets) is 45.96 m. The pylons of the northern portal include the remains of a structure of the XII century, which had a glazed finish of predominantly green in combination with unglazed tiles [45]. The facades, almost devoid of fine plastic elements (except for the remaining unfinished portal part), have polychrome cladding made of glazed bricks. The remaining open square and rectangular niches in the lower part of the main portal and octagonal minarets were intended for the installation of mosaic panels. As pointed out by Sh. E. Ratiya [58] (p. 95), on the front side of this portal, the principle of dividing it into seven separate parts is clearly revealed; the equality between which, however, is violated here. The researcher admitted that with the timely completion of the construction and cladding of facades according to the plans of the architects who started the construction, this violation would not have been observed. It is possible that the inequality between the individual main divisions of the plan of the facade plane is caused only by the absence of cladding on its walls.

The question arises about the reasons for the cessation of work by the masters in 1399. The only work that remained was the completion of only the upper part of the portal and the cladding. Was it due to the need to start the construction of the Samarkand Mosque complex? Conversely, Emir Timur, according to the descriptions of his contemporaries, had great respect for the memory of Ahmed Yassawi. Accordingly, it is impossible to imagine that the building of the mausoleum-khanaka was halted only due to the transfer of all of the masters to another project. Moreover, the portal arch, the most important semantic element of the main facade, turned out to be incomplete. Additionally, judging by the composition and shape, the main dome is not completed. The termination of funding after the end of a successful military campaign in India also cannot be considered as an argument. The bulk of both the construction and finishing work was already completed. Given the large construction works that were carried out in Samarkand at the time, it is impossible to assume the complexities of financing an almost completely finished object in Turkestan. Maybe the Emir planned to finish the construction of the mausoleum-khanaka after the completion of the main works in Samarkand. This theory arises if you look at the intended appearance of the facades (Figures 15, 16, 17 and 18). The diagrams show the insufficient (from the point of view of composition) height of the main dome. This issue will be considered later. In general, the reason for the termination of the main construction work on the mausoleum-khanaka during the life of Emir Timur has not yet been clarified.

The portal, when completed in the XVI century, was crowned with teeth. These battlements served as a shelter for shooters in case of building defense or for sentries who notified residents of the enemy's approach (a large flag was hung out, and residents working in the fields hid inside the city walls). The arch was connected so as to form an unbroken protective line along the wall. The height of the portal with teeth has become about 30 meters. The basis of the decoration of the surfaces of the walls, drums, and domes is epigraphic [17,25]. The volumetric-plastic solution emphasizes the polyfunctionality of the structure. For example, the semantic importance of the mausoleum is emphasized by the allocation of its dome. This dome, unlike the main one, has a corrugated outer surface and a relatively high cylindrical drum. The main dome, being the largest in Central Asia (diameter 18.2 meters inside and 22.0 meters outside), is built on an octagon resting on a quad with beveled corners. The top of the dome is located at a height of 44 meters (inside the hall the height is 38.0 meters). According to the authors, the preserved shape of the main dome corresponds to the inner dome characteristics of the epoch. Accordingly, the authors assume that the construction of an external dome was originally envisaged (this assumption will be clarified further). Each individual volume of the structure has a clearly identified mosaic decoration with a three-part division. There is a functional differentiation of epigraphic decoration. In the appearance of the building, the sites erected in different periods are quite clearly traced.

Paradoxically, in the presence of an optimal design of walls and ceilings, the foundation of the building clearly does not correspond to the logic of construction; the depth of laying is only 0.3 m. The foundation itself is made of several rows of bricks. Clay layers were also made to a depth of 1.5 m. This became critical when, in 1846, the building was partially flooded with water for several months. As a result, the rate of destruction of the individual parts of the building has increased. In the early 1870s, the building was threatened with targeted destruction due to the fear that its cracked walls might collapse. The awareness of the cultural and historical value of the monument precluded its demolition and stimulated the activation of repair work.

In 1872, repairs were carried out, the main result of which was the cleaning of accumulated debris from the building. In 1884, the roof was repaired, spillways were arranged, collapsed sections of brickwork were laid in the lower part of the walls, and the lining of the outer surface of the main dome was started. As a result of the measurement work, the plan of the building was drawn. The ground movements that arose for various reasons led to the need to strengthen some sections of the southwestern and northwestern walls with buttresses (1886—1887). A detailed survey of the building began in 1905. Plans, sections, and facades of the building were drawn. In 1910, a partial re-laying of part of the walls and ceilings in the kitchen was carried out. Since 1922, regular inspections of the monument began to be carried out with a photo recording of the condition of the building as a whole and its individual parts. The inspection revealed multiple cracks, damage, destruction of arches in many parts of the building, and subsidence of the northwest corner. In 1925, major repairs of the building were also carried out: the lower parts of some rooms were shifted, the kudukhana dome was blocked, and the foundations were opened. The epigraphy of the mausoleum continued to be read.

In 1928, the lower parts of the walls inside the building and the main dome were repaired. In 1952, work began on the construction of concrete foundations under separate sections of the walls. The work was completed in the early 2000s. By the middle of the twentieth century, the building had lost some of the cladding on the walls and domes. This cladding was almost completely restored by the end of the century. To strengthen the lower part of the portal arch, steel cables were placed. In 1997, a well filled in, in the Middle Ages (diameter 0.9–1.0 m), was discovered in the geometric center of the square of the central room. Accordingly, a confirmed assumption was made that a twelfth-century complex existed at this place until the middle of the fourteenth century. The semantic center of this complex was likely a small mausoleum of Khoja Ahmed. In front of him was a vast courtyard with a well.

This expanded the understanding that appeared in the middle of the twentieth century that: "(1) a new building was not built entirely, before that there was another building with an inner courtyard (or a group of buildings forming a courtyard), later blocked by the dome of kazandyk; moreover, the mosque has older walls (the North-Western corner of the complex); (2) the main southern portal with minarets is attached to a previously existing building; for the installation of the dome the walls of the former courtyard were reinforced with linings, and the corridors-passages in these walls were narrowed (this explained the lack of foundations under the southern portal)" [74] (p. 146). That is, Emir Timur gave an order to build a portal and erect a dome over the courtyard of the existing complex.

The state of the monument at the beginning of the twentieth century and its modern appearance are shown in Figures 4–6.

Figure 4. The mausoleum-khanaka Ahmed Yassawi, Turkestan: 1—The view in the first half of the twentieth century [75]; 2—The cross section [76]; 3—The southeastern facade at the beginning of the twentieth century [77] (the arrow shows the archivolt deflection).

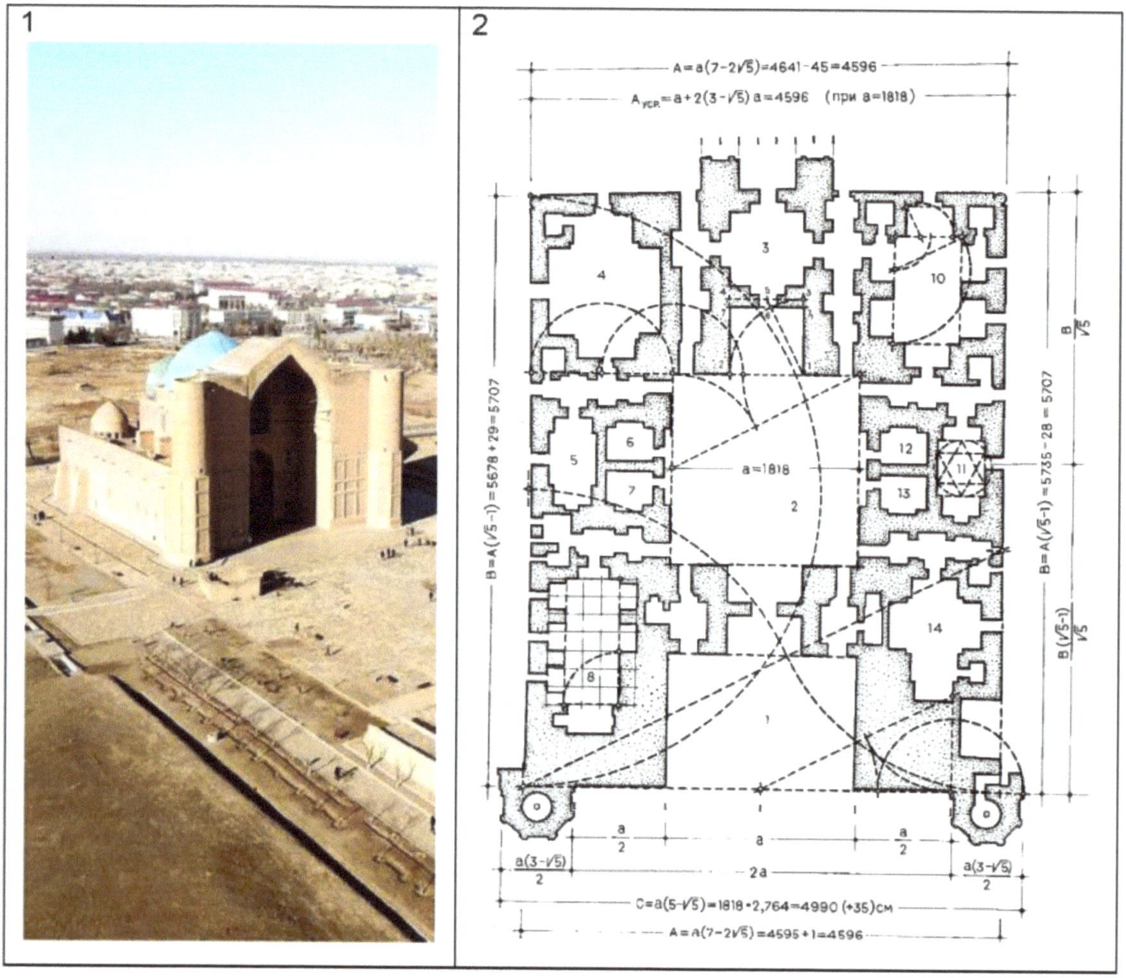

Figure 5. The mausoleum-khanaka of Ahmed Yassawi, Turkestan: 1—The general view in the middle of the twentieth century [78]; 2—The first level plan compositional analysis (M. S. Bulatov) [59] (p. 154).

For further research, it is advisable to consider the other buildings of the era. Their general composition and individual elements are important for determining the originally intended form of mausoleum-khanaka in Turkestan.

The Guri-Emir complex in Samarkand (39°38′54″ N, 66°58′08″ E) remained in an unfinished state. The main building was built in 1403—1404. It was included in the existing complex with a mosque and khanaka of Muhammad Sultan. Originally built octagonal in plan, the building is topped with a ribbed dome on a cylindrical drum (architect Muhammad ibn Mahmud Isfagani). The diameter of the dome is 15 m. In the future, various extensions will be made. The complex has a fenced yard from the north. The corners of the courtyard were fixed by four minarets. By the middle of the XIX century, only two minarets remained, adjacent directly to the main building. The northeast minaret collapsed in 1868, and the northwest minaret collapsed in 1903—they were restored in 1996. In the context of the problem considered in this article, the portal from the western side is of interest. This element remained unfinished; the pointed arch, brought to the height of the middle of the dome drum, rises above the unfinished walls on the sides. Part of the wall collapsed. The builders' plans clearly included the creation of a much larger portal than the one located on the north side. The northern portal forms the main entrance from the front yard. In the future, the restoration work that has been going on since the XVII century may include the completion of the western portal as well as the restoration of the madrasah and khanaka. Accordingly, the cladding with ceramic panels can be made using historical analogue techniques. However, the demonstrated options for the reconstruction of the complex, for some reason, do not imply the completion of the construction of the great western portal. However, without this, the complex cannot appear in all its glory, demonstrating the greatness of the plans of the architects of the Timur era. Figure 7 shows the plan of the Guri-Emir complex, an unfinished portal on the west side, as well as a reconstruction option in the form of a model.

Figure 6. The modern view of the mausoleum-khanaka Ahmed Yassawi, Turkestan: 1—The view from the south [79]; 2—The view from the northwest [80]; 3—The view from the north [81]; 4—The fragment of the main facade [82] (the arrow shows the archivolt deflection).

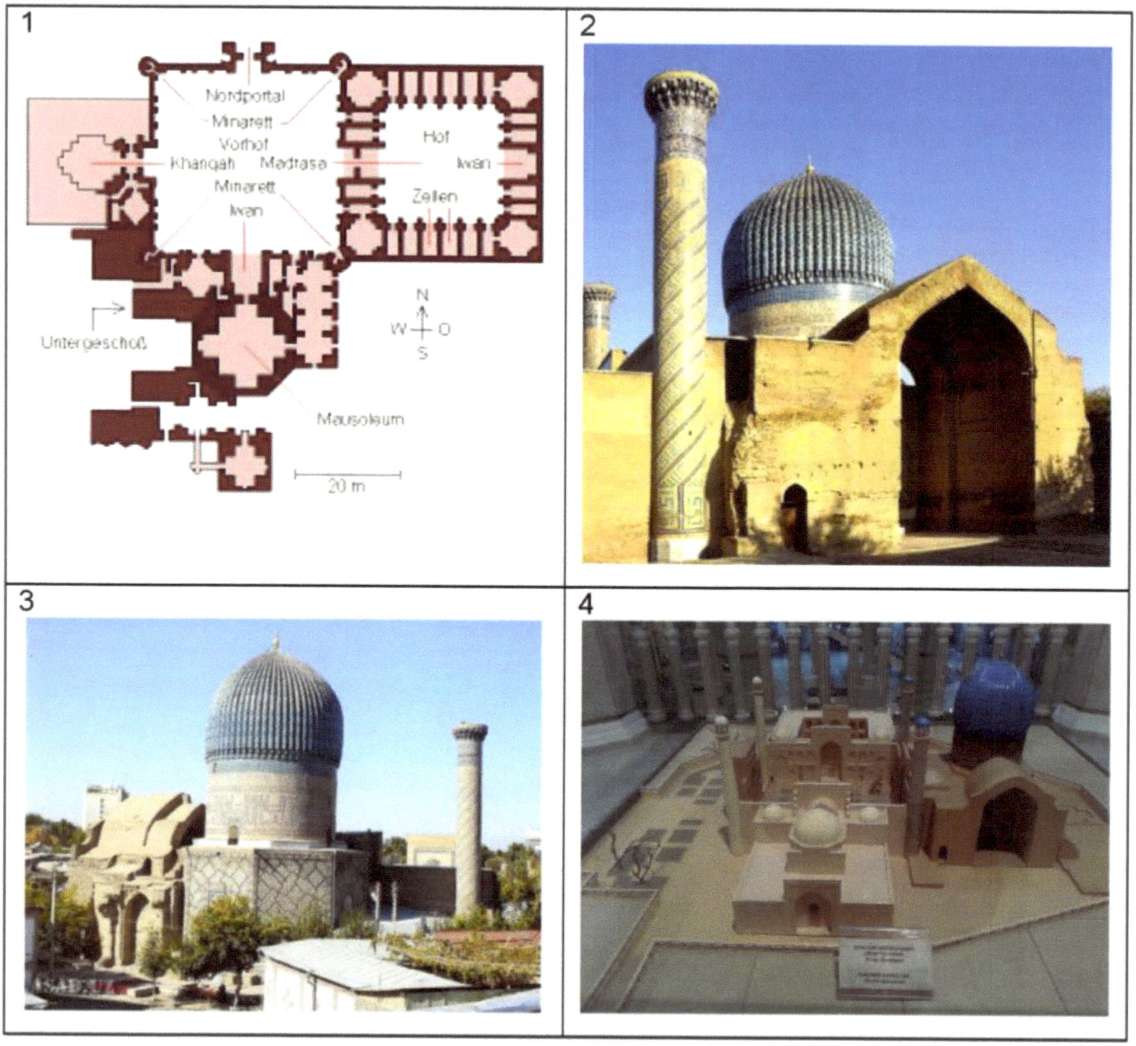

Figure 7. The Guri-Emir complex, Samarkand: 1—The plan (the reconstruction) [83]; 2, 3—The general view from the western side (beginning of the XXI century) [84,85]; 4—The reconstruction of the complex option (the model) [86].

The issue of the upper part of the minarets is debatable. In general, initially, minarets were independent structures. However, as noted by Sh. E. Ratiya [60] (p. 101), in the XIV century, for the first time in the architecture of Central Asia, there was a combination of two architectural forms—portal-domed buildings and minarets. Located symmetrically on both sides of the giant portal of the building, the minarets gave it the appearance of a complete, integral composition. Previously, minarets had only a functional religious purpose. Subsequently, during the period of the highest development of this type of portal structures, the minarets did not lose their functional-religious purpose, but acquired a new, architectural–tectonic significance. They became a characteristic part of the composition of structures. minarets are either directly adjacent to the portal or fixed to the corners of the extended facade.

Corner minarets, according to many researchers [4,10,50–52], had lanterns. This comes from the original function of minarets as observation towers. In addition to the comfort of staying on the observation deck of the guards, the lantern isolated the internal staircase from rain and snow. The confirmation of this hypothesis is in a comparison of detached medieval minarets and the corner minarets of one of the buildings of the late XVIII century. However, unlike Bukhara, where the lanterns have been preserved, in Samarkand, now almost all minarets are crowned only with a developed cornice. Additionally, this cornice, in fact, is the base of the lantern. There is a hypothesis about the destruction of lanterns in the process of historical development (earthquakes, deviation from the vertical due to subsidence of the soil). In the considered Guri-Emir complex, the crowning of minarets is difficult to determine—the painting by V. V. Vereshchagin (1870) shows a minaret with an almost completely destroyed top, on which only fragments of the "under-lantern" cornice are visible. The presented reconstruction option implies the presence of lanterns on both existing and restored minarets. The lanterns on the minarets in the Ulugbek Madrasah, which was built in 1415—1420 (39°39′17″ N, 66°58′28″ E), could have collapsed with a strong deviation of the minarets as a result of several earthquakes in the XIX century. Moreover, the most significant damage was created during the earthquakes of 1817/1818 and 1897. The minarets resisted, however, although they deviated. The largest deviation of the northeastern minaret was recorded in 1918. By the way, it is on this minaret that a low cylindrical superstructure above the "under-lantern" cornice has been preserved. This minaret was straightened in 1932 (engineers V. G. Shukhov, M. F. Mauer). The southeastern minaret was straightened in 1965 (engineers E. M. Handel, E. O. Nelly). Similar minarets in the Sher-Dor Madrasah opposite, which was built in 1619–1636 (39°39′17″ N, 66°58′34″ E), deviated slightly in the process of historical development. However, even in this building, the minarets do not have lanterns. The presence of an internal staircase implies the relatively active operation of these minarets. Accordingly, a lantern is needed to prevent rain from getting inside. Currently, the problem of insulation from rain is solved by a roof made of steel sheets. However, the question remains about the reasons for refusing to restore the lanterns. The complex restoration work carried out over many years included the straightening of existing minarets, the construction of destroyed minarets, and the restoration of cladding. However, the restoration of the lanterns crowning the minarets, which is a relatively simple event, was not carried out on any Samarkand monument of the Timurid era. Figure 8 shows modern roofs over Samarkand minarets.

The territory on which the city of Bukhara is located is much less susceptible to seismic impacts than Samarkand. This is one of the reasons for the significantly better preservation of medieval minarets in this region. Another factor is the significantly larger diameter of the trunk, which provides better stability for the building as well as powerful foundations with greater depth. The Kalyan minaret was built in 1127 in Bukhara (39°46′33″ N, 64°24′51″ E). The builder's name was Bako. The minaret has a height of 46.5 m. At the bottom, the trunk diameter is 9.7 m, and at the top—6.0 m. The rotunda lantern that completes the building has 16 through openings with a pointed arch. The upper part of the lantern has a multi-tiered cornice, above which there is a cone-like wedding, which is a continuation of the axial rod. The rotunda lantern was partially destroyed due to shelling in 1920. In 1925, the appearance of the minaret was restored. Earthquakes caused some damage to the stalactite cornices. These elements were restored in 1980. A complete restoration of the minaret was carried out in 1997. The Kalyan Minaret served as a prototype for similar buildings in the region. The earliest example of the imitation of this minaret is in the minaret in Vabkent (40°01′11″ N, 64°31′04″ E), which was built in 1198–1199. It has a height of 40.3 m. At the bottom, the trunk diameter is 6.2 m, and in the upper part—2.8 m. The rotunda lantern that completes the building has 8 through openings with a pointed arch. The diameter of the lantern is 3.66 m.

Figure 8. Medieval minarets of Samarkand: 1—The roof of the Ulugbek Madrasah minaret [87]; 2—The internal staircase in the minaret of the Ulugbek Madrasah [88]; 3—The roof of the Sherdor Madrasah minaret [89]; 4—The Bibi-Khanim mosque minaret top [90].

Corner minarets, designed in the medieval "Bukhara" style, are indicative in the Topchi-Bashi Madrasah in Bukhara. It was built in the second half of the XVIII century (its approximate coordinates are 39°46′14″ N, 64°25′73″ E). The building has corner minarets with lanterns, compositionally and plastically similar to medieval minarets in Bukhara and Vabkent. The building received significant damage as a result of shelling and was dismantled in the 1920s. The street named after Bahauddin Naqshbandi now runs through the place where the "Red" madrasah was located. The elegant plastic solution of the madrasah can be seen in the photo from S. M. Prokudin-Gorsky in 1911 [91,92]. This image, in some sources, mistakenly indicates the location of the object in Samarkand. In the context of the problem under consideration, the corner minarets are of interest. Their shape directly correlates with medieval minarets. The relatively thick trunk is crowned with a rotunda lantern with six arched openings. The supports of the pointed arches form paired columns. A belt with ceramic lining is placed under the lantern. Unlike the blue and purple cladding in medieval analogues, green cladding is used here. The rotunda lantern has a developed multi-tiered cornice. The main facade of the madrasah itself has a three-part structure with an arched portal in the center. The corners of the portal are accentuated by small turrets with sharp peaks. These kinds of "pinnacles" are the crowning of three-quarter round pilasters. These free-standing and corner minarets are shown in Figure 9.

Figure 9. minarets of the Central Asia: 1, 2—The Kalyan minaret, Bukhara [93,94]; 3, 4—The minaret, Vabkent [95,96]; 5—The Topchi-Bashi Madrasah, Bukhara [97,98].

The Ak-Saray Palace complex was built in 1380–1404 in Shakhrisyabz (39°03′39″ N, 66°49′45″ E). The architect's name is Muhammad Yusuf Tabrizi. It is assumed that the height of the portal of the main entrance was 56.0 m. The arch of this portal was the largest in Central Asia—22.3 m. It is believed that the targeted destruction of the complex began in 1570 by order of the then ruler of Bukhara, who controlled the region. The brick gradually began to be pulled apart for the construction of other buildings. However, in 1707, the Bukhara ruler Ubaydullah Khan lived in the palace for some time. According to some reports, the portal collapsed in the first half of the XVIII century. To date, two dilapidated pylons have been preserved. Their height is about 38.0 m. The general view of the preserved part of the portal of the main entrance to the Ak-Saray complex and the variant of its reconstruction in the form of a layout are shown in Figure 10.

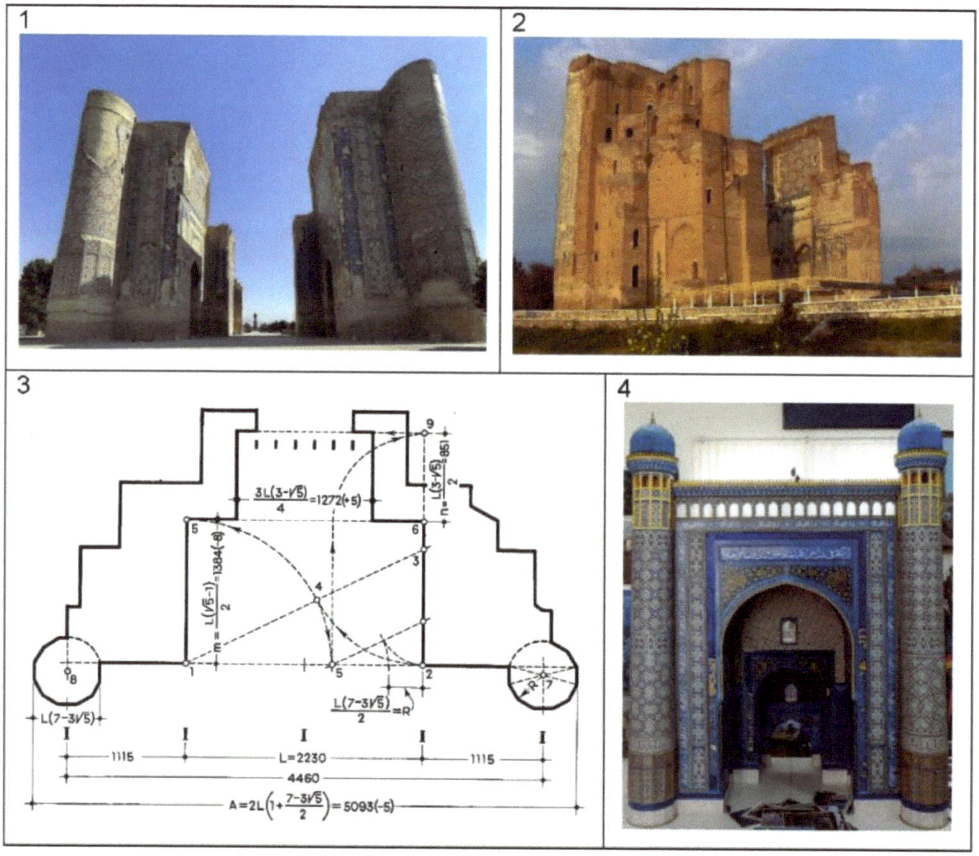

Figure 10. The Ak-Saray complex, Shakhrisyabz: 1, 2—The general view of the main entrance as of the beginning of the XXI century [99,100]; 3—The plan (the drawing—M. S. Bulatov) [59] (p. 166); 4—The reconstruction of the main entrance (the model) [101].

The Cathedral ("Jami") mosque Bibi-Khanim was built in 1399–1404 in Samarkand (39°39′38″ N, 66°58′45″ E). The complex was a rectangular courtyard 64 × 78 m enclosed by a colonnade of 480 columns. There was an entrance portal on the east side. Its height was 33.15 m. On the opposite side was the main mosque, and on the right and left (from the north and from the south) in the structure of the colonnade—small mosques. The corners of the complex were accentuated by minarets, which presumably had a height of 50.0 m. It is assumed that the minarets had a multi-tiered solution in the upper part. The corner minarets of the entrance portal had a circular cross-section, and the minarets of the main mosque had a multifaceted one. The main facade of this grandiose structure was a harmonious combination of a giant majestic portal with minarets, behind which a huge dome towered. It is significant that this dome was completely obscured by the portal and was not visible from the side of the main facade. The minarets narrowed upwards. The planes of the portal walls and minarets are divided into separate false arches and narrow rectangular panels decorated with various geometric ornaments and inscriptions from mosaics and majolica. The flat ornamented archivolt of the portal arch rests on the capitals of round three-quarter columns mounted at the corners of the portal opening. The walls of the portal niche are entirely lined with geometric stylized ornaments of brick mosaic, which are interspersed with tiled rosettes of carved mosaic and majolica. The basement was lined with marble. The destruction of the complex began due to natural reasons at the beginning of the XV century when the dome of the mosque began to crack. A few years later, the arch of the entrance portal collapsed. The most serious damage occurred as a result of the earthquake of 1897. After this earthquake, the upper part of the preserved north-western minaret was dismantled. By the beginning of the twentieth century, the colonnade had almost disappeared. The rest of the buildings have been preserved in varying degrees of destruction. The preserved minaret has a low parapet above the lantern cornice (Figure 8). Major restoration work began in 1968 and it continues to this day. Figure 11 shows: the complex preserved by the middle of the twentieth century; the plan of the complex; the geometric analysis of the composition of the plan of the main mosque; and the current state of the main mosque as a result of reconstruction. The sequence of construction of similar buildings of the Timur era: the Ak-Sarai Palace, Shakhrisyabz; the Mausoleum-Khanaka of Ahmed Yassawi, Turkestan; The main mosque of Bibi-Khanim, Samarkand—are shown in Figure 12. The reconstruction of the proposed view of the main mosque of Bibi Khanum in Samarkand is shown in Figure 13.

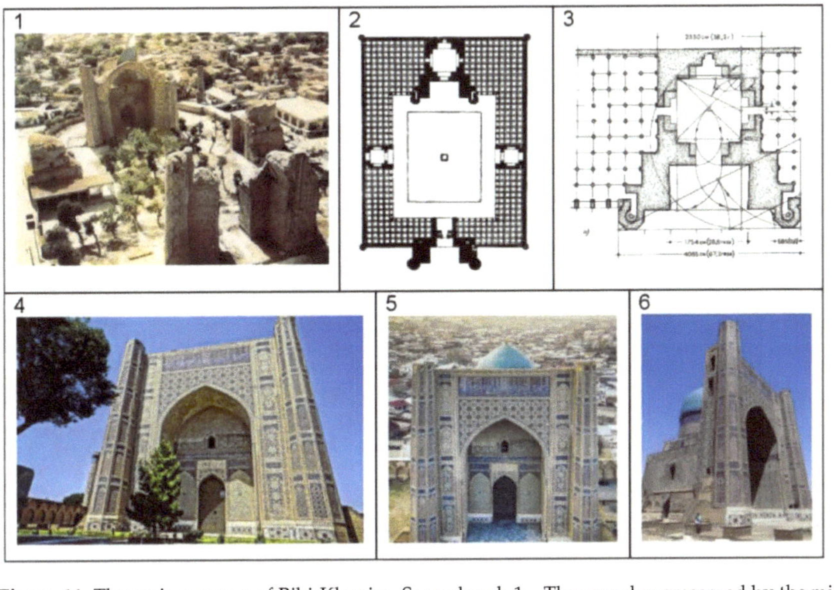

Figure 11. The main mosque of Bibi-Khanim, Samarkand: 1—The complex preserved by the middle of the twentieth century (photo from 1968) [102]; 2—The plan (the reconstruction) [103]; 3—The geometric analysis of the composition of the plan (M. S. Bulatov) [59] (p. 173); 4, 5, 6—The reconstruction of the late twentieth century [104–106].

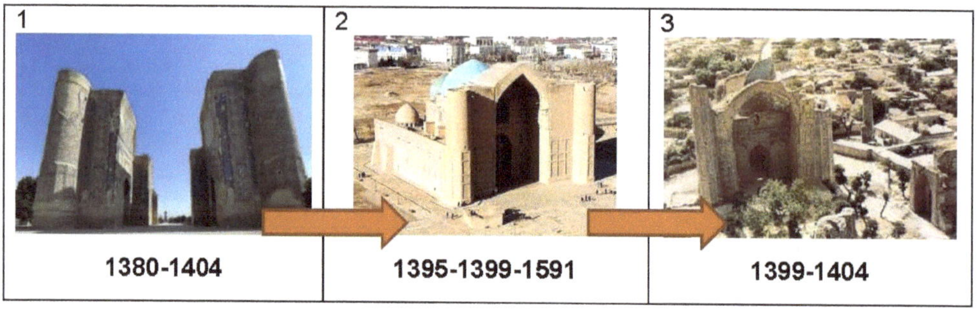

Figure 12. The sequence of construction of similar buildings of the Timur era (state at the beginning of the last quarter of the twentieth century): 1—Ak-Sarai Palace, Shakhrisyabz [99]; 2—Mausoleum-khanaka Ahmed Yassavi, Turkestan [80]; 3—The main mosque Bibi-Khanim, Samarkand [102].

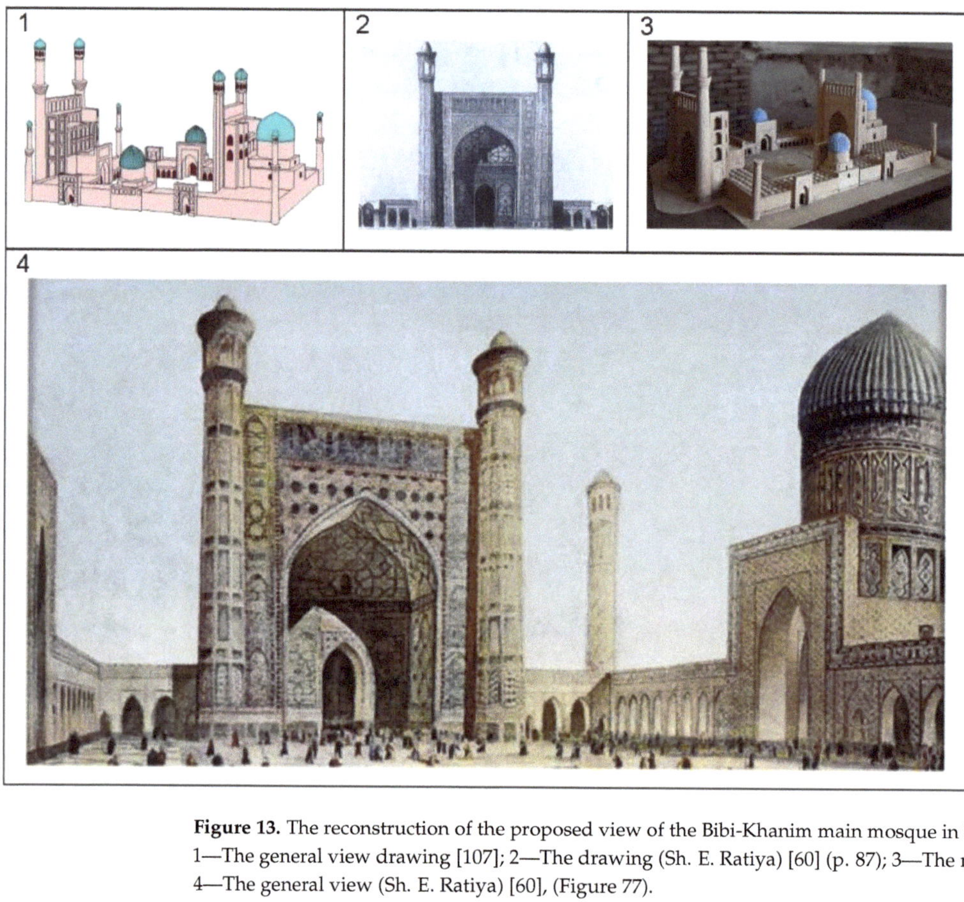

Figure 13. The reconstruction of the proposed view of the Bibi-Khanim main mosque in Samarkand: 1—The general view drawing [107]; 2—The drawing (Sh. E. Ratiya) [60] (p. 87); 3—The model [108]; 4—The general view (Sh. E. Ratiya) [60], (Figure 77).

The architectural composition of the buildings under consideration serves as the basis for the analysis carried out further in order to determine the unrealized original plan of the Turkestan mausoleum-khanaka.

3. Results and Discussion

In general, the idea of restoring the original condition of the building is methodologically based on reconstruction by analogy, the measure of quality of which is the criterion of reliability. Moreover, we are not talking here about restoration in kind, but only about conditional reconstruction in the drawing. In the case of the mausoleum-khanaka, this can serve as a starting point for discussion and stimulate further research on this issue. Moreover, despite the long-term study of the monument by many researchers, the issue of restoring the original design concept of the mausoleum-khanaka is being emphasized for the first time.

The following parameters are the initial data for the reconstruction of the proposed appearance of the south-eastern facade of the mausoleum-khanaka of Ahmed Yassawi:

(1) the width of the facade along the axes of the corner minarets—45.00 m;
(2) width of the north-western façade—45.96 m;
(3) the width of the facade in the dimensions of the corner minarets—49.9 m;
(4) the width of the base of the corner minarets is 6.9 m

The closest analogue of the portal, according to many researchers, is the portal of the main mosque of the Bibi-Khanim complex. It is even assumed that the construction of the mosque was led by craftsmen from the same corporation that previously built the mausoleum-khanaka [60] (p. 96). In the compositional analysis of Sh. E. Ratiya [60] (p. 92), the width of the facade of the main Bibi-Khanim mosque is 40.88 m. It consists of seven parts: two minarets (5.84 × 2), two piers (5.84 × 2), and an arched opening (5.84 × 3). Accordingly, it is believed that the height of the portal could be 40.88 m (based on the assumption of the same width and height). The height of the corner minarets to the cornice is 51.16 m (10.28 m above the portal). In the analysis of M. S. Bulatov [59] (p. 173), the width of the facade of the main Bibi-Khanim mosque is 40.85 m, and the corner minarets are 5.85 m wide. The assumption about the similarity of the width and height of the portal is based on the analysis of the existing building of the Ulugbek Madrasah mosque, built in 1415—1420 in Samarkand (39°39′17″ N, 66°58′28″ E). In this building, the width and height of the portal is 31.80 m (M.S.Bulatov [59] (p. 195)). These data are given in the drawings of Sh. E. Ratiya and M. S. Bulatov, shown in Figure 14. If we assume that this technique is suitable for calculating the height of the portal of the mausoleum-khanaka of Ahmed Yassawi, then it should be 49.9 m. The height of the minarets could be equal to $A2^{1/2} = 49.9 \times 1.41 = 70.36$ m. The drawing of this construction is shown in Figure 15. However, the assumption about the equality of the width and height of the portal comes from the analysis of the Ulugbek Madrasah. Additionally, it was built later. The portal of this building has no minarets (they are located on the edges of the facade). Accordingly, the width of the facade, which affects the height, can be considered the size of the northwest facade of the mausoleum-khanaka. The width of this facade is 45.9 m. Thus, the height of the portal could also be 45.9 m. The minarets towered above the portal to a height of triple thickness (6.9 × 3 = 20.7 m). The total height of the minarets could be 66.6 m. The drawing of this construction is shown in Figure 16.

G. A. Pugachenkova points out: "Timur's monumental buildings are interpreted as a volume closed from the external environment. The law of iron symmetry reigns in everything. The facade composition is based on a simple, clear idea of a huge portal arch, with powerful corner towers flanking the walls. The static equilibrium of the peshtak array is not disturbed by a certain vertical orientation of both arches and towers, which is ordered by a system of typical divisions, in the form of framing rectangular frames. At the same time, the architect uses the same compositional scheme both in the cathedral mosque of Samarkand, and in the Ak-Sarai Palace, and in the mausoleum of the cleric Hazret Ahmed Yasawi, and in the dynastic tomb of Dorus-Siadat" [74] (p. 70). Schematically, this is shown in Figure 17.

Figure 14. Geometric analysis of the composition of the facades of medieval buildings: 1—The Bibi-Khanim main mosque in Samarkand, the alleged view of the main facade (Sh. E. Ratiya) [60] (p. 92); 2—The Ulugbek Madrasah in Samarkand, geometric analysis of the composition of the facade (M. S. Bulatov) [59] (p. 195).

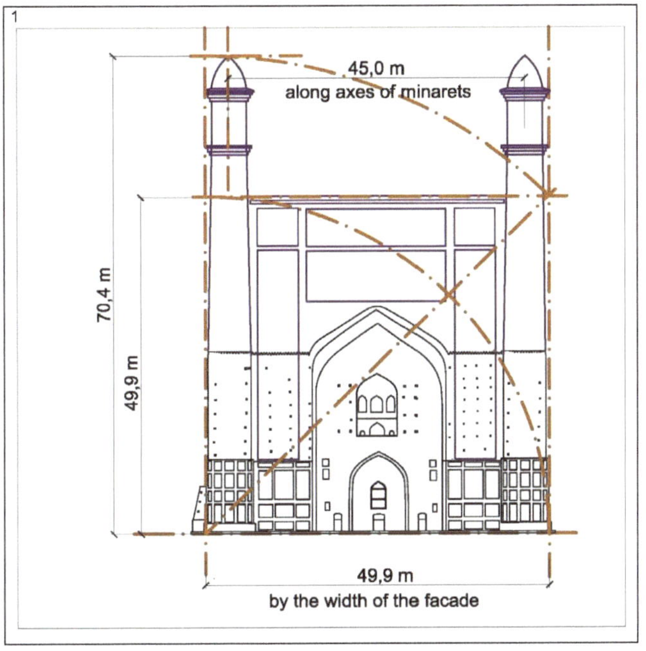

Figure 15. The portal of the mausoleum-khanaka of Ahmed Yassawi by analogy with the Bibi-Khanim main mosque in Samarkand (the drawing from the authors).

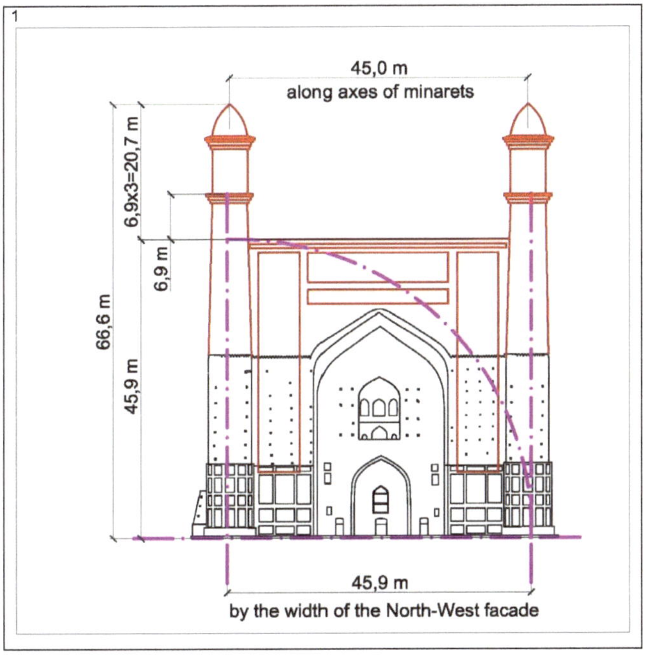

Figure 16. The portal of the mausoleum-khanaka of Ahmed Yassawi by analogy with the Ulugbek Madrasah in Samarkand (the drawing from the authors).

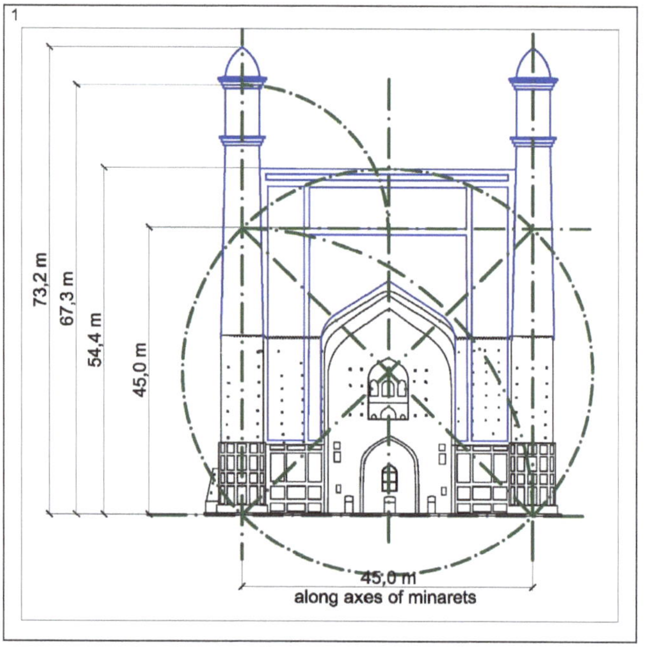

Figure 17. The portal of the mausoleum-khanaka of Ahmed Yassawi by analogy with the portal of the main entrance to the Ak-Sarai Palace in Shakhrisyabz (the drawing from the authors).

The appearance of the mausoleum-khanaka, as M. E. Masson pointed out [4] (p. 14), has similarities with the main entrance to the Ak-Saray complex in Shakhrisabz. The most similar are the angular minarets, which, having a polygonal lower part, turn into a cylinder above. "The wonderful scenery of the ruins of Aksaray and the surviving skeleton of the portal of the mausoleum of Khoja Ahmed in comparison complement each other in mental reconstruction. In order to restore the appearance that the peshtak of the Turkestan monument should have, it is necessary to imagine, instead of the ugly battlements of the portal walls, a rectangular frame embracing the castle of the arch brought down under Abdullakhan, which is now restlessly piercing into the air" [4] (p. 14).

In the compositional analysis of M. S. Bulatov [59] (p. 166), the portal of the Ak-Sarai Palace has a width of 50.93 m in the dimensions of the corner minarets, and the width along the axes of the minarets is 44.60 m. If we assume that the height of the portal was determined by its width, then the portal could be 50.93 m high.

However, as mentioned earlier, some studies indicate an estimated portal height of 56.0 m. This value is obtained if we use the well-known proportionation formula $A5^{1/2}:2 = 1.118A$, that is, $50.93 \times 1.118 = 56.94$ m. If we apply this formula to the mausoleum-khanaka of Ahmed Yassawi, then the height of its portal could be $49.9 \times 1.118 = 55.79$ m.

Approximately such a value can be obtained if the following geometric construction is applied. For the base size "A", take not the width of the facade, but the axes of the minarets. Since ancient times, and still yet in most cases, the marking of the dimensions of the building for construction was carried out along the axes of the main elements. It is fair to assume that in the era under consideration (XIV—XV centuries) this technologically simple and reliable technique could be used. The fact is that it is difficult to determine the actual dimensions of buildings (with an accuracy of centimeters, and sometimes decimeters) before the start of construction, since the thickness of the cladding is often unknown and errors are possible when performing foundational and masonry work (until now, the construction standards of some countries indicate the parameters of permissible deviations from the project up or down). Accordingly, it is technologically more rational to set the dimensions of the building on the construction site while focusing on the center axes. Based on this, it is possible to carry out a geometric construction of the facade based on the size of 45.0 m, corresponding to the axes of the minarets. The height of the portal is determined by the size of the upper part of the circle described around the square of the axes of the minarets. This value is approximately 54.4 m. If we assume the presence of crowning teeth (they are shown, among other things, on the reconstruction of the portal of the Ak-Sarai Palace—Figure 10), then an approximate value of 56.0 m is obtained. This is the size obtained on the basis of calculations by analogy with the estimated height of the portal of the Ak-Sarai Palace. minarets could have a height of 1.5 A to the cornice (67.3 m). Additionally, the total height of the minarets with domes could be 73.2 m (the domes have the size of the width of the base of the minaret). The drawing of this construction is shown in Figure 15.

Based on these geometric constructions, it can be suggested schematically how the facades of the mausoleum-khanaka of Ahmed Yassawi could look if the original plan had been fully implemented. This is shown in Figure 18. However, looking at these facades, the question arises about the height of the main dome. Unlike the other domes (above the mausoleum, above the mosque—Figure 4), which are double, this dome has a single solution. The profile of this dome is similar to the profiles of the inner domes. In addition, unlike most domes of the Timur era, the main dome of the mausoleum-khanaka has a low drum. It may have been planned later, upon completion of construction, to increase the height of the drum and make an additional external dome. At the same time, the top of the dome could be at the height of the portal (about 50 m—about 8–10 m higher than the preserved dome)—for example, the dome in the main mosque of the Bibi-Khanim complex was made. An approximate scheme of the possible solution of the main dome in comparison with analogues is shown in Figure 19. Its design is comparable to that of the dome of the Guri-Emir mausoleum (in the Turkestan mausoleum-khanaka, the dome diameter is about 18 m; in the Samarkand mausoleum, the dome diameter is about 15 m) and with the dome of the main mosque of the Bibi-Khanim complex (this dome is slightly smaller than in the mausoleum-khanaka). As indicated, likely due to the desire of the customer to complete the construction of the mausoleum-khanaka later, this partially explains the incompleteness of the arch and the entire portal. At the same time, the preserved main dome, hypothetically, was faced from the outside in order to preserve the building for a possibly long break in construction. Similar to the question of the shape of the portal, the question of the shape of the main dome of the mausoleum-khanaka requires additional study. Variants of reconstruction of similar buildings of the Timur era: Ak-Sarai Palace, Shakhrisabz (A. A. Akhmedov, 1980–2021); mausoleum-khanaka of Ahmed Yassawi, Turkestan (K. I. Samoilov, 2022); the main mosque Bibi-Khanim, Samarkand (S. E. Ratiya, 1950) are shown in Figure 20.

If, as a basis for constructing a possible scheme of decorative cladding, we take the examples discussed earlier (the entrance portal of the Ak-Saray Palace complex in Shakhrisyabz, the main mosque of the Bibi-Khanim complex in Samarkand, the Ulugbek Madrasah in Samarkand), then the portal of the Khanaka mausoleum could look as shown in Figure 21. As in the main volume of the building of the mausoleum-khanaka, the decoration of the portal could be as follows. The bottom of the walls is lined with smooth slabs of yellow sandstone. A wide strip of about 1.5 m with edges of blue samples could be located above it. The strip itself could have been laid out with a complex geometric ornament. The main element of the drawing could be an eight-ray star and alternating figured slabs of yellow sandstone and painted majolica. A narrow ribbon of a new row of sand tiles could serve as a border, above which all of the free space of the walls could be occupied by patterns of white and blue tiles on a yellow background of facing bricks. Some of the patterns could represent epigraphic decor.

Naturally, the study of recreating the original design of the general appearance of the mausoleum-khanaka is only part of the problem. It is necessary to conduct a significant amount of research in various fields. Determining the historically adequate general appearance of the building is the initial stage. Among the issues are determining the value and necessity of preserving historical layers; the condition of the main structures, walls, floors and the need to strengthen them (the foundations were reinforced with reinforced concrete during restoration); the use of historically adequate or new building materials and structures; the possibility of seasonal or year–round work; and much more.

Figure 18. The alleged schematic view of the mausoleum-khanaka of Ahmed Yassawi in Turkestan facades—the alleged view of the portal is shown in red, the alleged view of the main dome is shown in purple (the drawing from the authors).

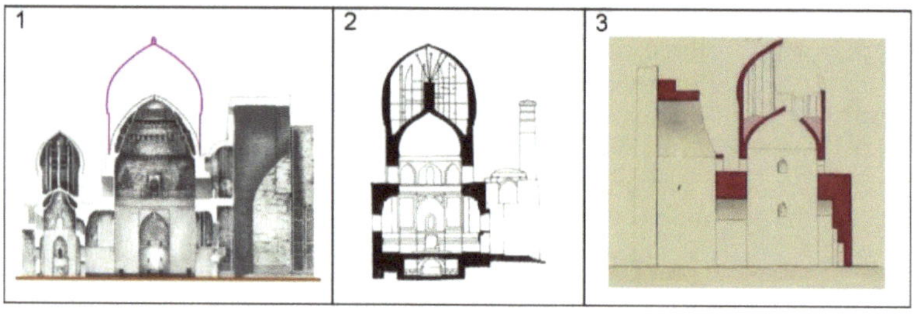

Figure 19. The assumed shape of the outer dome over the central part of the mausoleum-khanaka of Ahmed Yassawi in Turkestan (1 [109] + the drawing from the authors) in comparison with analogues—the Guri-Emir mausoleum (2 [110]) and the main mosque of Bibi-Khanim complex (3 [111]) in Samarkand.

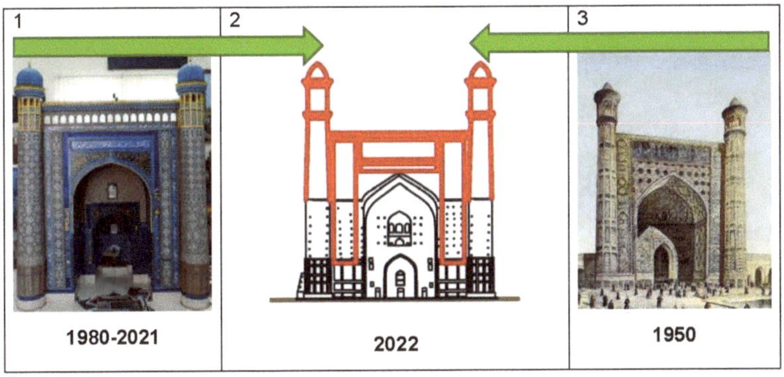

Figure 20. Reconstruction options for similar buildings of the Timur era: 1—Ak-Sarai Palace, Shakhrisyabz (A. A. Akhmedov, 1980–2021 [101]); 2—Mausoleum-khanaka Ahmed Yassavi, Turkestan (K. I. Samoilov, 2022—drawing from the authors); 3—The main mosque Bibi-Khanim, Samarkand (Sh. E. Ratiya, 1950 [58], (Figure 77)).

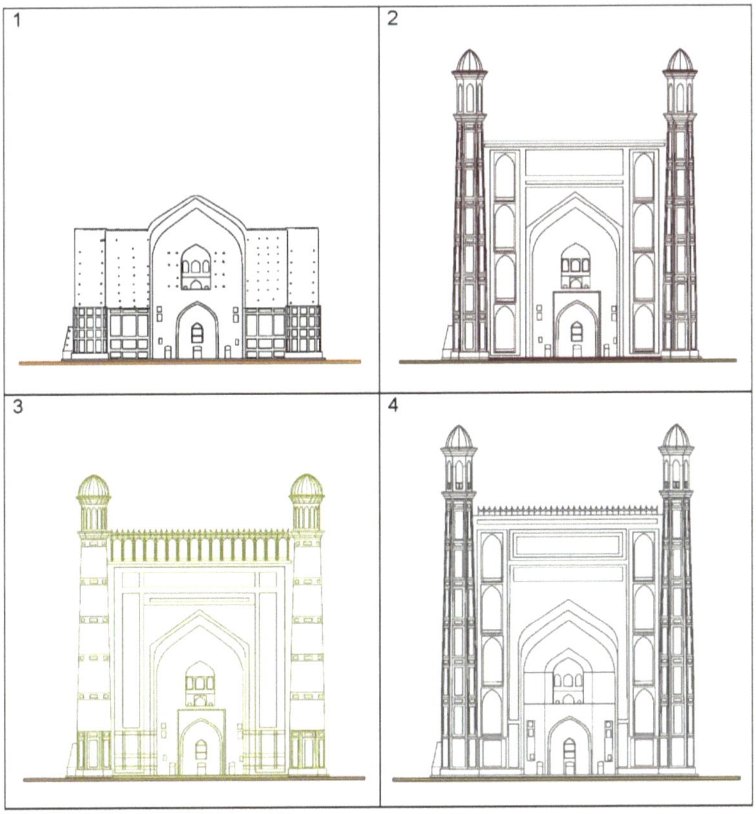

Figure 21. Views of the portal of the mausoleum-khanaka of Ahmed Yassawi in Turkestan (the drawing from the authors): 1—the current situation at the beginning of the XXI century; 2, 3, 4—the proposed types of the portal based on the interpretation of the techniques of decoration of the Ulugbek Madrasah in Samarkand, the main entrance to the Ak-Sarai Palace in Shakhrisyabz and the Bibi-Khanim main mosque in Samarkand.

In the Turkestan mausoleum-khan aka, it is necessary to determine the expediency of preserving "historical scars". These are the deflection of the right archivolt of the portal arch, and the reinforcing buttresses on the southwest and northwest facades. When the foundations are reinforced under the entire building, the buttresses cease to have a functional purpose. It is also advisable to redo the portal arch, since the archivolt deflection is a blatant deformation. It arose most likely due to a non-compliance with the masonry technology at its completion in the XVI century. The presence of modern steel cable ties shows the emergency condition of this structure. The construction of the portal to the design height will be one of the ways to strengthen this arch. Accordingly, the implementation of the original idea from a purely speculative problem passes into the category of a structurally necessary measure to preserve the strength of the architectural monument.

The next question is to determine the feasibility of using "historical" bricks produced in Sairam for possible completion. The bricks were laid on an alabaster mortar. Given the presence of cladding, the expediency of using historical materials for masonry walls is debatable. It is possible to use modern wall materials. In principle, it is possible to strengthen the walls with steel frames with steel grids. These elements will be hidden by the cladding. Accordingly, the appearance of the building, directly perceived by visitors, will not depend on the main structural materials.

When the building is completed, it will, in a sense, lose its historicity. However, since the mausoleum-khanaka of Ahmed Yassawi in Turkestan is now considered as the most important part of the historical and cultural heritage of Kazakhstan, the completion of the construction of this building will be an excellent illustration of the continuity of culture and the greatness of the deeds of Kazakhstan's ancestors. This is especially important if we recall the famous saying of Emir Timur, recorded on the gates of the Ak-Sarai Palace complex in Shakhrisyabz: "If you doubt our power, then look at our buildings".

The discussion about the prospect of the rapid completion of the mausoleum-khanaka is still unproductive, as extensive additional research is still needed. However, the potential of the history of the construction of this complex encourages the continuation of work:

(1) In the twelfth century, the first mausoleum was erected;
(2) By the middle of the thirteenth–early-fourteenth centuries, it became part of a complex of buildings with an open courtyard and a well in the middle;
(3) The end of the fourteenth century united all of the buildings and the courtyard into one building with a large dome and a majestic portal (the portal arch and the outer dome remained unfinished);
(4) At the end of the sixteenth century, the arch of the portal was completed, and the upper part of the unfinished portal and corner towers were strengthened;
(5) In the next two centuries, constant repairs protected the building from destruction;
(6) In the second half of the nineteenth century, repair and restoration work intensified, and the cladding of the main dome was started, which allowed it to be protected from destruction (the absence of an external dome became critical for the building);
(7) At the end of the nineteenth–mid-twentieth century, the building was completely cleared of later additions and almost all of the cladding was restored (with the exception of the portal);
(8) At the turn of the twentieth and twenty-first centuries, repair and restoration work was actively carried out.

Each of these stages was accompanied by a study of the complex (at a scientific and methodological level accessible for each period). Now, having conducted extensive research and public discussion, it is advisable to continue the work of the ancestors and complete the construction in memory of them, realizing their grandiose plan. Moreover, there are some examples of such solutions in the region.

Of course, we can leave the building in its "historical form" while continuing conservation and restoration work. However, research on how the khanaka mausoleum might look if it was completed at the beginning of the XV century is, in itself, of some specific scientific and general cultural interest. They are connected with the detailing of the forms of the unfinished portal, the features of the missing external dome on the high drum in the compositional center of the mausoleum-khanaka, and the specifics of the drawings of the cladding of these elements. At this new stage, detailed drawings and 3D models can be made.

4. Conclusions

Supporters of the complete restoration of monuments want to see them today in their former beauty and grace, making them available for tourist displays and the possibility of using them for museumification or adaptation for other purposes. Opponents of full restoration advocate for the minimization of restoration work, since, in this case, the monuments will lose their historicity, and then the historical and artistic value will be lost. On the one hand, there is the principle of holistic restoration and the idea of restoring the ruins in their original state for the sake of integrity of aesthetic perception. This principle is historically associated with the restorations of E. Viollet-le-Duc. On the other hand, there is the principle of conservation (in fact, anti-restoration). This is associated with the name of J. Ruskin. The "original state" is considered to be the ruin itself. The bearer of historical value is not the integral image lost and restored in the work of imagination, but the materiality of fragments preserved from antiquity. Both of these principles interact dialectically. There is no restoration without conservation intervention—without strengthening the structure and removing irreversible damage. There are no conservation projects that are completely free from the need to restore ruins.

Examples of unfinished constructions are a separate problem. In this case, it is sometimes advisable to recognize the need for the completion of construction. Numerous examples demonstrate the successful execution of these works. Two methods are usually used. One determines the completion of construction in other forms corresponding to a new view of architectural and artistic shaping. This method has historical value as it shows the dynamics of changing stylistic preferences. The second option seems interesting from a historical point of view. This is the completion of construction in forms presumably corresponding to the original plan.

In any case, for the remaining unfinished or dilapidated buildings, considered as important milestones in the development of the region, it is advisable to completely either "physically" or "virtually" restore them. Whether or not its continuity is considered by the State as fundamental is one of the features of the continuity of culture. The preservation of heritage in the form in which it has reached in the modern era is also very promising. It gives you the opportunity to feel the historical depth and the scale of times. The importance of this is confirmed by the romantic attention that periodically arises in the culture to the "Aesthetics of ruins".

Author Contributions: Conceptualization, K.S., B.K., G.S. and A.A.; methodology, K.S., B.K. and G.S.; software, K.S. and A.A.; validation, K.S., G.S. and A.A.; formal analysis, K.S., B.K., G.S.; investigation, K.S., B.K., G.S. and A.A.; resources, K.S., B.K. and A.A.; data curation, B.K. and K.S.; writing—original draft preparation, K.S., B.K., G.S. and A.A.; writing—review and editing, K.S., B.K. and G.S.; visualization, A.A. and K.S.; supervision, K.S. and G.S.; project administration, K.S. and B.K.; funding acquisition, K.S. and A.A. All authors have read and agreed to the published version of the manuscript.

Funding: This research received no external funding.

Data Availability Statement: Not applicable.

Acknowledgments: Authors express their gratitude to the Google search engine, the USA Library of Congress, and the British Library, whose resources are used for the selection of theoretical material.

Conflicts of Interest: The authors declare no conflict of interest.

References

1. Abdrassilova, G.S.; Murzagaliyeva, E.T.; Kuc, S. Mausoleum of Khoja Akhmet Yassawi as the element of regional identity formation in modern architecture of Kazakhstan. *Period. Eng. Nat. Sci.* **2021**, *9*, 127–138.
2. Gulama Kozha Akhmet Yasaui (The Wise Khoja Ahmed Yassawi). Al-Farabi Kazakh State University A. Bekenkyzy, Z. Kari 1-33. Available online: https://elibrary.kaznu.kz/wp-content/uploads/2020/05/kozha_ahmet_yasaui.pdf (accessed on 21 January 2023).
3. Bekchurin, M.S. Opisanie mecheti Azreta, nahodyashchejsya v gorode Turkestane (Description of the Azret mosque located in the city of Turkestan). Voen. Sb. 1866, 8; reprinted in Turkestanskaya Oblast' Zametki Statskogo Sov. Bekchurina. *Kazan' Univ. Tipogr.* **1872**, 52–63. Available online: https://www.vostlit.info/Texts/Dokumenty/M.Asien/XIX/1860-1880/Bekcurin_M_S/text1.htm (accessed on 21 January 2023).
4. Masson, M.E. Mavzolej Hodzhi Ahmeda Yassavi (Mausoleum of Khoja Ahmed Yassawi). *Tashkent Syr-Dar'inskoe Otd. Obs. Izucheniya Kaz.* **1930**, 1–24. Available online: https://www.academia.edu/39458202/%D0%9C%D0%B0%D1%81%D1%81%D0%BE%D0%BD_%D0%9C_%D0%95_%D0%9C%D0%B0%D0%B2%D0%B7%D0%BE%D0%BB%D0%B5%D0%B9_%D0%A5%D0%BE%D0%B4%D0%B6%D0%B0_%D0%90%D1%85%D0%BC%D0%B5%D0%B4%D0%B0_%D0%AF%D1%81%D0%B5%D0%B2%D0%B8_%D0%A2%D0%B0%D1%88%D0%BA%D0%B5%D0%BD%D1%82_1930 (accessed on 21 January 2023).
5. Masson, M.E. O postrojke mavzoleya Hodzha Ahmeda v gorode Turkestane (About the construction of the mausoleum of Khoja Ahmed in Turkestan). *Izv. Sredneaziat. Geogr. Obs. T. XIX* **1929**, 39–45.
6. Patsevich, G.I. Remont i restavraciya mavzoleya-mecheti Hodzha Ahmeda Yasavi v 1939–1941 gg. (Repair and restoration of the mausoleum-mosque of Khoja Ahmed Yassawi in 1939-1941). *Izv. Akad. Nauk Kaz. SSR. Seriya Arhit.* **1950**, *2*, 95–100.
7. Man'kovskaya, L.Y. Issledovanie arhitekturnogo kompleksa-mavzoleya Ahmeda Yassavi v gorode Turkestane i voprosy ego restavracii (The study of the architectural complex-the mausoleum of Ahmed Yassavi in Turkestan and the issues of its restoration): Avtoreferat dissertacii kandidata iskusstvovedeniya. *Tashkent Izd. AN UzSSR* **1963**, 1–18.
8. Man'kovskaya, L.Y. K izucheniyu priyomov sredneaziatskogo zodchestva konca XIV v.: (Mavzolej Hodzha Ahmeda Yassevi) (To study the techniques of Central Asian architecture of the late XIV century: (Mausoleum of Khoja Ahmed Yassavi)). *IZU* **1962**, *1*, 93–142.
9. Man'kovskaya, L.Y. Nekotorye arhitekturno-arheologicheskie nablyudeniya po restavracii kompleksa Hodzha Ahmeda Yassavi v g.Turkestane (Some architectural and archaeological observations on the restoration of the Khoja Ahmed Yassawi complex in Turkestan). *Izv. Akad. Nauk Kaz. SSR Seriya Istor. Arheol. I Etnogr.* **1960**, *3*, 52–69.
10. Man'kovskaya, L.Y. Tipologicheskie osnovy zodchestva Srednej Azii. IX–nachalo XX v. (Typological foundations of architecture of Central Asia. IX–early XX century). *Tashkent Fan* **1980**, 1–84.
11. Tuyakbayeva, B.T. Epigraficheskij dekor arhitekturnogo kompleksa Ahmeda Yassavi (Epigraphic decor of the architectural complex of Ahmed Yassavi). *Алма-Ата Өнер* **1989**, 1–176.
12. Alliyar, N.B.; Samoilov, K.I. Arhitekturno-hudozhestvennye osobennosti kompleksa mavzoleya-honako Ahmeda Yassavi v Turkestane (Architectural and artistic features of the mausoleum-khanaka complex of Ahmed Yasawi in Turkestan). *Nauka I Obraz. Segodnya* **2021**, *2*, 101–103.
13. Alliyar, N.B.; Samoilov, K.I. Issledovaniya arhitktury kompleksa mazoleya-honako Ahmeda Yassavi v g.Turkestan v XX veke (Studies of the architecture of the mausoleum-khanaka complex of Ahmed Yassavi in Turkestan in the XX century). *Nauka I Obraz. Segodnya* **2021**, *7*, 60–64.
14. Alliyar, N.B.; Samoilov, K.I. Razvitie arhitektury kompleksa mavzoleya-honako Ahmeda Yassavi v g. Turkestane v XVI–XIX vekah [The development of the architecture of the mausoleum complex-honako Ahmed Yassavi in Turkestan in the XVI–XIX centuries]. *Nauka I Obraz. Segodnya* **2021**, *6*, 68–71.
15. Baitanaev, B.A.; Yelgin, Y.A. Voprosy istoricheskoj onomastiki mavzoleya Hodzha Ahmeda YAsavi. Istoricheskaya onomastika i epigrafika (Questions of historical onomastics of the mausoleum of Khoja Ahmed Yasawi. Historical onomastics and epigraphy]. *Izv. Nac. Akad. Nauk. Resp. Kaz. Seriya Obs. I Gumanit. Nauk.* **2013**, *3*, 241–255. Available online: http://nblib.library.kz/elib/library.kz/journal/Baitanaev%20Elgin....3....13.pdf (accessed on 10 November 2022).
16. Basenov, T. Kompleks Mavzoleya Ahmeda Yasavi (The complex of the Mausoleum of Ahmed Yasawi). *Almaty Өner* **1982**, 1–71.
17. Mustapayeva, D. Issledovanie stroeniya mavzoleya Hodzhi Ahmeda YAsavi XII veka (Study of the structure of the mausoleum of Khoja Ahmed Yasawi of the XII century). *Yassaui Univ. Habar.* **2021**, *4*, 129–138. Available online: https://doi.org/10.47526/2021-4/2664-0686.12 (accessed on 10 November 2022).
18. Nurmukhammedov, N.-B. Mavzolej Hodzha Ahmed Yasavi (Mausoleum of Khoja Ahmed Yassawi). *Almaty Oner* **1988**, 1–166.
19. Ordabaev, A. Problemy izucheniya, ohrany i restavracii pamyatnikov arhitektury na primere Mavzoleya Hodzhi Ahmeda Yassavi [Problems of study, protection and restoration of architectural monuments on the example of the Mausoleum of Khoja Ahmed Yassawi]. *Videolekciya (Polnaya Rasshifrovka Lekc. S. Romashkina)* **2017**, 9. Available online: https://vlast.kz/gorod/22023-problemy-izucenia-ohrany-i-restavracii-pamatnikov-arhitektury-na-primere-mavzolea-hodzi-ahmeda-asavi.html (accessed on 10 November 2022).
20. Yelgin, Y.A. AArheologicheskie i istoriko-arhitekturnye issledovaniya mavzoleya Hodzhi Ahmeda Yasavi: Vtoraya polovina XIX v.—seredina 1950-h godov: (Ocherki i materialy) (Archaeological and historical-architectural studies of the mausoleum of Khoja Ahmed Yasawi: The second half of the XIX century—Mid-1950s: (Essays and materials)). *Almaty* **2013**, 1–225. Available online: http://tehne.com/library/yolgin-yu-arheologicheskie-i-istoriko-arhitekturnye-issledovaniya-mavzoleya-hodzhi-ahmeda-yasavi-vtoraya-polovina-xix-v-seredina-1950-h-godov-almaty-2013 (accessed on 10 November 2022).

21. Yelgin, Y.A. V.S.Gejncel'man i ego programma remonta mavzoleya Hodzhi Ahmeda YAsavi v 1906—1908 gg. (V.S.Heintzelman and his program of renovation of the mausoleum of Khoja Ahmed Yasawi in 1906-1908). *Arheol. Kaz.* **2018**, *1—2*, 201–214. [CrossRef]
22. Alliyar, N.B.; Samoilov, K.I. Features of the historical context of the development of architecture in Central Asia of the XIV-XV centuries. *Nauka I Obraz. Segodnya* **2020**, *12*, 96–98.
23. Baitenov, E.M. Memorial'noe zodchestvo Kazahstana: Evolyuciya i problemy formoobrazovaniya (Memorial architecture of Kazakhstan: Evolution and problems of shaping). *Almaty Kaz. Golovn. Arhit.-Stroit. Akad.* **2004**, 1–243.
24. Baitenov, E.; Tuyakayeva, A.; Abdrassilova, G. Medieval mausoleums of Kazakhstan: Genesis, architectural features, major centres. *Front. Archit. Res.* **2019**, *8*, 80–93. [CrossRef]
25. Barsukova, E.G. Features of Architectural Decoration of Religious Buildings of Uzbekistan in the Middle Ages (9–16 Centuries). *Int. J. Sci. Res.* **2019**, *8*, 1255–1257. Available online: https://www.ijsr.net/get_abstract.php?paper_id=ART20202799 (accessed on 10 November 2022).
26. Barsukova, E.G. Vliyanie sufizma na arhitekturu i dekor kul'tovyhsooruzhenij Srednej Azii IX-XVI vv. (The Influence of Sufism on the Architecture and Decor of Religious Buildings in Central Asia of the IX-XVI Centuries.): 18.00.01—Teoriya i istoriya arhitektury. Restavraciya pamyatnikov arhitektury, Avtoreferat dissertacii na soiskanie uchyonoj stepeni doktora filosofii (PhD) po arhitekture. *Tashkent* **2020**, 1–50. Available online: https://www.academia.edu/45619859/%D0%90%D0%B2%D1%82%D0%BE%D1%80%D0%B5%D1%84%D0%B5%D1%80%D0%B0%D1%82_%D0%91%D0%B0%D1%80%D1%81%D1%83%D0%BA%D0%BE%D0%B2%D0%B0_%D0%95_%D0%93 (accessed on 10 November 2022).
27. Chmelnizkij, S. Characteristics of Timurid Architecture. In *Islam: Art and Architecture*; Hattstein, M., Delius, P., Eds.; Cologne Könemann UK Ltd.: Köln, Germany, 2000.
28. Glaudinov, B. Istoriya arhitektury Kazahstana (s drevnih vremen do nachala XX veka) (History of architecture of Kazakhstan (from ancient times to the beginning of the XX century)). *Almaty KazGASA* **1999**, 1–295.
29. Kasymbekova, M.B.; Glaudinov, B.A.; Samoilov, K.I. Formoobrazovanie islamskih memorial'no-kul'tovyh sooruzhenij v Kazahstane X–XIX veka (Formation of Islamic memorial and religious buildings in Kazakhstan of the X–XIX century). *Almaty CISArchit.* **2020**, 1–143. Available online: https://ru.scribd.com/document/488132040/Kasymbekova-M-B-Glaudinov-B-A-Samoilov-K-I-THE-SHAPING-OF-ISLAMIC-MEMORIAL-CULT-STRUCTURES-IN-KAZAKHSTAN-OF-THE-X-XIX-CENTURY (accessed on 10 November 2022).
30. Kozha, M.B. Novye materialy o kul'tovyh sooruzheniyah g.Turkestana (XII-XVIII vv.) (New materials on religious buildings of Turkestan (XII-XVIII centuries)). *Vestn KazNPU Almaty* **2016**. Available online: https://articlekz.com/article/18266 (accessed on 10 November 2022).
31. Margulan, A.H.; Basenov, T.K.; Mendikulov, M.M. Arhitektura Kazahstana (Architecture of the Kazakhstan). *Alma-Ata Kaz. Gos. Izd.* **1959**, 1–170.
32. Pugachenkova, G.A. Zodchestvo Central'noj Azii XV v. Vedushchie tradicii i cherty (Architecture of Central Asia of the XV century. Leading traditions and traits). *Tashkent: Izd. Lit. I Iskusstv. Im. Gafura Gulyama* **1976**, 1–115.
33. Pugachenkova, G.A.; Rempel', L.I. Istoriya iskusstv Uzbekistana s drevnejshih vremyon do serediny XIX veka (Art history of Uzbekistan from ancient times to the middle of the XIX century). *Mosk. Iskusstv.* **1965**, 1–688.
34. Roxburgh, D.J. Timurid Art and Architecture. *Art Hist. Oxf. Bibliogr* **2014**, *1, 30*. Available online: https://www.oxfordbibliographies.com/view/document/obo-9780199920105/obo-9780199920105-0028.xml?rskey=JJsodK&result=142 (accessed on 10 November 2022).
35. Samoilov, K.I. Arhitektura Kazahstana XX veka (Razvitie arhitekturno-hudozhestvennyh form) (Architecture of Kazakhstan of the twentieth century (Development of architectural and artistic forms)). *Mosk.-Almaty M-ArIdizain* **2004**, 1–930. Available online: http://tehne.com/assets/i/upload/library/samoilov-arhitektura-kazakhstana-20-veka-2004.pdf (accessed on 10 November 2022).
36. Tondi, A. Arhitekturnyj dekor Srednej Azii epohi Timuridov (Architectural decor of Central Asia of the Timurid era). *Avtoreferat Diss. Kandidata Iskusstv. Spec. Mosk.* **2003**, 1–22.
37. Umnyakov, I.I. Arhitekturnye Pamyatniki Srednej Azii: Issledovanie. Remont. Restavraciya 1920-1928 gg. In *Architectural Monuments of Central Asia: A Study. Repair. Restoration of 1920–1928*; Glavnyj Sredne-Aziatskij Muzej: Tashkent, Uzbekistan, 1929; pp. 1–40.
38. Vyatkin, V.L. Pamyatniki drevnostej Samarkanda (Monuments of antiquities of Samarkand). *Samarkand Uzkomstaris* **1927**, 1–32.
39. Weimarn, B.V. Iskusstvo Srednej Azii (Art of Central Asia). *Mosk. Leningr. Gos. Izd. Iskusstv* **1940**, 1–205. Available online: https://www.academia.edu/79709099/%D0%92%D0%B5%D0%B9%D0%BC%D0%B0%D1%80%D0%BD_%D0%91_%D0%92_%D0%98%D1%81%D0%BA%D1%83%D1%81%D1%81%D1%82%D0%B2%D0%BE_%D0%A1%D1%80%D0%B5%D0%B4%D0%BD%D0%B5%D0%B9_%D0%90%D0%B7%D0%B8%D0%B8_%D0%9C_%D0%9B_1940 (accessed on 10 November 2022).
40. Yakubovsky, A.Y. Samarkand pri Timure i Timuridah (Samarkand under Timur and Timurids). *Leningr. Gos. Ermitazh* **1933**, 1–64.
41. Zahidov, P. Arhitekturnye pamyatniki Zarafshanskoj doliny (Architectural monuments of the Zarafshan valley). *Tashkent Uzb.* **1985**, 1–64.
42. Zahidov, P. Arhitekturnoe sozvezdie epohi Timura (Architectural constellation of the Timur era). *Tashkent Gl. Red. Izd. Poligr. Koncerna Shark* **1996**, 1–192.

43. Zakhidov, P.S. Kanon garmonii v arhitekture (The Canon of Harmony in architecture). *Teor. I Istor. Arhit. Restavraciya Pamyatnikov Arhit. Avtoreferat Diss. Na Soiskanie Uchyonoj Stepeni Dr. Arhit. Baku* **1990**, 1–37.
44. Zasypkin, B.N. Arhitektura Srednej Azii: Ocherki po istorii arhitektury narodov SSSR (Architecture of Central Asia: Essays on the history of architecture of the peoples of the USSR). *Mosk. Izd. Akad. Arhit. SSSR* **1948**, 1–160.
45. Arapov, A. The Artistic Culture of Central Asia and Azerbaijan in the 9th–15th Centuries. *Samarkand-Tashkent IICAS* **2013**, *4*, 1–284. Available online: https://www.academia.edu/6402071/The_Artistic_Culture_of_Central_Asia_and_Azerbaijan_in_the_9th_15th_centuries_Volume_IV_Architecture_Samarkand_Tashkent_IICAS_2013_280_p (accessed on 10 December 2022).
46. Svod pamyatnikov istorii i kul'tury Kazahstana. YUzhno-Kazahstanskaya oblast' (A set of historical and cultural monuments of Kazakhstan. South Kazakhstan region). *Almaty Gl. Red. Kazak Enciklopediyasy* **1994**, 1–368.
47. Gilmanova, N.V. The Ak-Saray palace. *Int. J. Adv. Res. Sci. Eng. Technol.* **2018**, *5*, 6080–6087. Available online: http://www.ijarset.com/upload/2018/june/12-IJARSET-_Gilmanova.pdf (accessed on 10 November 2022).
48. Kabanov, S.N. Ruiny dvorca Ak-Saraj v Shahrisabze (Ruins of the Ak-Saray Palace in Shakhrisabz). *TIIA AN UzSSR T. I. Tashkent* **1948**, *4*, 45–56.
49. Khakimov, Z. Arhitekturnye shedevry izumrudnogo goroda (Architectural masterpieces of the emerald city). *San'at J. Acad. Arts Uzb.* **2002**, 1. Available online: https://sanat.orexca.com/2002-rus/2002-1-2/history_art3-4/ (accessed on 10 November 2022).
50. Kuziev, T. Vzglyad na istoriyu minaretov na territorii Uzbekistana (A look at the history of minarets on the territory of Uzbekistan). *ASan'at J. Acad. Arts Uzb.* **2010**, 3. Available online: https://sanat.orexca.com/2010-rus/2010-3-2/tursunali_kuziev/ (accessed on 10 November 2022).
51. Masson, M.E. Sobornaya mechet' Timura, izvestnaya pod imenem mecheti Bibi-Hanym (The Cathedral Mosque of Timur, known as the Bibi-Khanim Mosque). *Sredneaziat. Kom. Po Delam Muzeev I Ohr. Pamyatnikov Stariny Iskusstv. I Prir. (Sredazkomstaris)* **1926**, 1–16.
52. Masson, M.E. Kratkaya istoricheskaya spravka o Sredne-Aziatskih minaretah (Brief historical information about Central Asian minarets) Materialy Uzkomstarisa; Vypusk 2-3. Uzb. Kom. Po Ohr. I Izucheniyu Pamyatnikov Mater. *Kul'tury Tashkent* **1933**, 1–18.
53. Masson, M.E.; Pugachenkova, G.A. Shahrisabz pri Timure i Ulugbeke (Shakhrisabz under Timur and Ulugbek). In *Trudy Sredneaziatskogo Gosudarstvennogo Universiteta*; Sredneaziatskogo Gosudarstvennogo Universiteta: Tashkent, Uzbekistan, 1953; pp. 17–99.
54. Sultanov, H.T. K istorii formirovaniya arhitekturnyh ansamblej Shahrisabza XIV-XV vv. (On the history of the formation of architectural ensembles of Shakhrisabz XIV-XV centuries). *Arheol. Avtoreferat Diss. Na Soiskanie Uchyonoj Stepeni Kandidata Istor. Nauk Samarkand* **1990**. Available online: https://cheloveknauka.com/k-istorii-formirovaniya-arhitekturnyh-ansambley-shahrisabza-xiv-xv-vv (accessed on 10 November 2022).
55. Man'kovskaya, L.Y. Bibi-Khanym. *Tashkent Uzbekistan* **1965**, 1–26.
56. Paskaleva, E. The Bibi Khanum Mosque in Samarqand: Its Mongol and Timurid Architecture. *Silk Road* **2012**, *10*, 81–98. Available online: http://www.silkroadfoundation.org/newsletter/vol10/SilkRoad_10_2012_paskaleva.pdf (accessed on 10 November 2022).
57. Paskaleva, E. A traveller's impressions, August 2013. *Samarqand Refashioned Silk Road* **2013**, *11*, 139–153. Available online: https://edspace.american.edu/silkroadjournal/wp-content/uploads/sites/984/2017/09/SilkRoad_11_2013_paskaleva.pdf (accessed on 10 November 2022).
58. Masjid-i Bibi Khanum, Hidden Architecture. 29 January 2020. Available online: https://hiddenarchitecture.net/masjid-i-bibi-khanum/ (accessed on 10 November 2022).
59. Bulatov, M.S. Geometricheskaya garmonizaciya v arhitekture Srednej Azii IX-XV vv. (Geometric harmonization in the architecture of Central Asia of the IX-XV centuries). *Mosk. Nauka* **1988**, 1–362. Available online: http://science.totalarch.com/book/0112.rar (accessed on 10 November 2022).
60. Ratiya, S.E. Mechet' Bibi-Hanym v Samarkande: Issledovaniya i restavracionnye opyty (Bibi-Khanim Mosque in Samarkand: Research and restoration experiments). *Mosk. Gos. Izd. Arhit. I Gradostroitel'stva* **1950**, 1–108.
61. The Basílica i Temple Expiatori de la Sagrada Família, Barselona. 1953. Available online: https://img1.liveinternet.ru/images/attach/b/3/13/839/13839931_1953.JPG (accessed on 30 December 2022).
62. The Basílica i Temple Expiatori de la Sagrada Família, Barselona. 2021. Available online: https://upload.wikimedia.org/wikipedia/commons/thumb/7/70/Sagrada_Familia_2021.jpg/800px-Sagrada_Familia_2021.jpg (accessed on 30 December 2022).
63. The Basílica i Temple Expiatori de la Sagrada Família, Barselona (the model). Available online: https://img0.liveinternet.ru/images/attach/b/3/13/840/13840229_Sagrada_Familiya.jpg (accessed on 30 December 2022).
64. The Cathedral of Our Lady of Chartres. Available online: https://upload.wikimedia.org/wikipedia/commons/thumb/1/13/Westfassade_Chartres.jpg/800px-Westfassade_Chartres.jpg (accessed on 30 December 2022).
65. The Cathédrale Saint-Étienne de Bourges. Available online: http://www.musiqueorguequebec.ca/orgues/france/bourgesse1.jpg (accessed on 30 December 2022).
66. The Cathédrale Primatiale Notre-Dame de l'Assomption de Rouen. Available online: https://upload.wikimedia.org/wikipedia/commons/thumb/1/1a/Rouen_Cathedral_as_seen_from_Gros_Horloge_140215_4.jpg/280px-Rouen_Cathedral_as_seen_from_Gros_Horloge_140215_4.jpg (accessed on 30 December 2022).

67. The Cathédrale Notre-Dame d'Amiens. Available online: https://upload.wikimedia.org/wikipedia/commons/6/60/Cath%C3%A9drale_Notre-Dame_d%27Amiens-3420.jpg (accessed on 30 December 2022).
68. Schenkov, A.S. «Vostochnye» i «zapadnye» tendencii v otnoshenii k pamyatnikam arhitektury ("Eastern" and "Western" trends in relation to architectural monuments). *Arhit. Mira Mater. Konf. Zapad-Vost. Vzaimodej. Tradic. V Arhit. Vyp. VNIITAG Mosk Archit.* **1993**, 177–179.
69. Samoilov, K.I. Mavzolej Karahana v Taraze–pervonachal'nyj vid i oblik nachala XX veka (Karakhan Mausoleum in Taraz—The original appearance and appearance of the early twentieth century). *Vestn. Kaz. Gos. Arhit.-Stroit. Akad.* **2001**, *4*, 15–18.
70. Turekulov, T.; Turekulova, N. Eshchyo raz ob Ajsha-Bibi (Once again about Aisha-BI bi). *Kumbez* **2001**, *3*, 74–80.
71. The Karakhan Mausoleum, Taraz (After the Reconstruction of the Early Twentieth Century). Available online: https://img.tourister.ru/files/2/2/2/2/4/8/1/8/original.jpg (accessed on 10 November 2022).
72. The Aisha-Bibi Mausoleum, Aisha-Bibi Village, Dzhambul Region (After Reconstruction of the 2000s). Available online: https://geocaching.su/photos/areas/104225.jpg (accessed on 10 November 2022).
73. Priemets, O.; Samoilov, K.; Kuspangaliyev, B.; Sadvokasova, G. The Regionalization of Forms in Architecture—A Use, An Adaptation, A Change, A Repulsion. *Civ. Eng. Archit.* **2023**, *11*, 450–472. [CrossRef]
74. Pugachenkova, G.A. On the reconstruction of the Dorus-Siadat ensemble, the Timurid tomb in Shakhrisabz. *Mater. Hist. Theory Archit. Uzbekistan. Issue 1. Publ. House Acad. Archit. USSR* **1950**, 58–72. Available online: https://archive.nyu.edu/bitstream/2451/39925/1/isawdca_000151.pdf (accessed on 11 December 2022).
75. The Mausoleum-Khanaka of Ahmed Yassawi, Turkestan (the General View of the Beginning of the Twentieth Century). Available online: https://s11.stc.yc.kpcdn.net/share/i/12/11443427/wr-960.webp (accessed on 29 January 2023).
76. The Mausoleum-Khanaka of Ahmed Yassawi, Turkestan (the Section). Available online: https://vlast.kz/media/upload/images/b6ae70b92b6b3eb89dcd787d2e478969.jpg (accessed on 10 November 2022).
77. The Mausoleum-Khanaka of Ahmed Yassawi, Turkestan (the Main Facade of the Early Twentieth Century). Available online: https://vlast.kz/media/upload/images/1bb9c716eda79f95f03e238a28bfe6ce.jpg (accessed on 10 November 2022).
78. The Mausoleum-Khanaka of Ahmed Yassawi, Turkestan (the General View). Available online: https://putidorogi-nn.ru/images/stories/aziya/kazahstan/mavzoley_hodzhi_ahmeda_yasavi_2.jpg (accessed on 10 November 2022).
79. The Mausoleum-Khanaka Ahmed Yassawi, Turkestan (the General View from the South—The Beginning of the XXI Century). Available online: https://mysl.kazgazeta.kz/wp-content/uploads/2018/07/photo_93789.jpg (accessed on 10 November 2022).
80. The Mausoleum-Khanaka of Ahmed Yassawi, Turkestan (the General View from the Northwest—The Beginning of the XXI Century). Available online: https://files.kazakhstan.travel/files/public/201812/11/f0cc6415933d4579 85ff7bc58fcc09b6/%D0%9C%D0%B0%D0%B2%D0%B7%D0%BE%D0%BB%D0%B5%D0%B9-%D0%9A%D0%90%D0%98-1.jpg (accessed on 10 November 2022).
81. The Mausoleum-Khanaka Ahmed Yassawi, Turkestan (the General View from the North—The Beginning of the XXI Century). Available online: https://qazaqstan3d.kz/upload/thumbed/1120x561/113635_1504017792_1120_561_0.jpg (accessed on 10 November 2022).
82. The Mausoleum-Khanaka Ahmed Yassawi, Turkestan (the Fragment of the Main Façade—The Beginning of the XXI Century). Available online: https://ic.pics.livejournal.com/denis_balin/14947129/4001769/4001769_original.jpg (accessed on 10 November 2022).
83. The Guri-Emir Complex, Samarkand (the Plan—The Reconstruction). Available online: https://encrypted-tbn0.gstatic.com/images?q=tbn:ANd9GcRsUNCRpGgP19MnlnfbT-tbJTcT8BdVBMmd0t3np13g7Qt7ouFL1mUAhT1NIxG67gBrUJ0&usqp=CAU (accessed on 10 November 2022).
84. The Guri-Emir Complex, Samarkand (the Beginning of the XXI Century). Available online: https://putidorogi-nn.ru/images/stories/aziya/uzbekistan/gur_emir_4.jpg (accessed on 10 November 2022).
85. The Guri-Emir Complex, Samarkand (the Beginning of the XXI Century). Available online: https://centralasia-adventures.com/image/data/info/uzbekistan/samarkand/gur_emir/03_big.jpg (accessed on 10 November 2022).
86. The Reconstruction of the Guri-Emir Complex Option (the Model). Available online: https://upload.wikimedia.org/wikipedia/commons/8/8e/%D0%9C%D1%83%D0%B7%D0%B5%D0%B9_%D0%B8%D1%81%D1%82%D0%BE%D1%80%D0%B8%D0%B8_%D0%A2%D0%B8%D0%BC%D1%83%D1%80%D0%B8%D0%B4%D0%BE%D0%B2_11.jpg (accessed on 10 November 2022).
87. The Roof of the Minaret of the Ulugbek Madrasah. Available online: https://pikabu.ru/4754725b-5837-4f60-b703-8e0a86de75a4 (accessed on 10 November 2022).
88. The Internal Staircase in the Minaret of the Ulugbek Madrasah. Available online: https://ic.pics.livejournal.com/uritsk/19436209/2779494/2779494_original.jpg (accessed on 16 October 2022).
89. The Roof of the Minaret of the Sherdor Madrasah. Available online: https://fotki.yandex.ru/users/uritsk/view/597560/ (accessed on 10 November 2022).
90. The Top of the Minaret in the Bibi-Khanim Mosque. Available online: http://photos.wikimapia.org/p/00/00/60/83/99_big.jpg (accessed on 10 November 2022).
91. Has the Mystery of the "Red" Madrasah in Bukhara Finally Been Solved?, "OldColor" for "Arch_Heritage" . 3 June 2018. Available online: https://arch-heritage.livejournal.com/2923864.html (accessed on 10 November 2022).
92. Behind the Scenes of Prokudin-Gorsky: Topchi-Bashi Madrasah in Bukhara, "Oldcolor" for "Arch_Heritage". 6 August 2018. Available online: https://oldcolor.livejournal.com/458746.html (accessed on 10 November 2022).

93. The Kalyan Minaret, Bukhara (the General View of the End of the Twentieth Century). Available online: https://img-fotki.yandex.ru/get/879536/50014243.12a/0_f8443_8ab70b8_orig (accessed on 10 November 2022).
94. The Kalyan Minaret, Bukhara (the rotunda lantern). Available online: http://www.maybeeabroad.com/wp-content/uploads/2020/04/maybeeabroad.com-buhara-15-1.jpg (accessed on 10 November 2022).
95. The Minaret, Vabkent (the General View of the end of the Twentieth Century). Available online: https://dolorestravel.com/wp-content/uploads/2016/12/vabkent1.jpg (accessed on 10 November 2022).
96. The Minaret, Vabkent (the rotunda lantern). Available online: https://sezamtravel.uz/thumb/2/NcTnVSb3PzGyKdYb5UIMmQ/r/d/v2.jpg (accessed on 10 November 2022).
97. The Topchi-Bashi Madrasah, Bukhara. Available online: https://upload.wikimedia.org/wikipedia/commons/2/20/Storks_samarkand.jpg (accessed on 10 November 2022).
98. Behind the Scenes of Prokudin-Gorsky: Topchi-Bashi Madrasah in Bukhara, "Oldcolor" for "Arch_Heritage". 6 August 2018. Available online: http://prokudin-gorskiy.ru/image.php?no=0001 (accessed on 10 November 2022).
99. The Ak-Saray Complex, Shakhrisabz (the General View at the Beginning of the XXI Century). Available online: https://fastly.4sqi.net/img/general/600x600/85645004_qMJc7XKMW4RgnbqRwBeLsosNVveu6Ddxflx2uMReICA.jpg (accessed on 10 November 2022).
100. The Ak-Saray Complex, Shakhrisabz (the General View at the Beginning of the XXI Century). Available online: https://silkadv.com/sites/default/files/Uzbekistan/Pamytniki/Shahrisabz/Ak_Sarai/0_3_%2824%29-min.JPG (accessed on 10 November 2022).
101. The Ak-Saray Complex, Shakhrisyabz (the Reconstruction of the Main Entrance—The Model). Available online: https://darsik.com/wp-content/uploads/2019/11/untitled-36.jpg (accessed on 10 November 2022).
102. The Main Mosque of Bibi-Khanim, Samarkand (a Complex Preserved by the Middle of the Twentieth Century). Available online: https://silkadv.com/sites/default/files/Uzbekistan/Pamytniki/Samarkand/Bibi_hanum_istoriya/0_4_-min.jpg (accessed on 10 November 2022).
103. The Main Mosque of Bibi-Khanim, Samarkand (the Plan—The Reconstruction). Available online: https://encrypted-tbn0.gstatic.com/images?q=tbn:ANd9GcSQQ_3gzGxraERfMQutpuhUwZwl8u1DS862PFNnwLlx_hBmKS_ZH7ylfmPOlLqVYKZI1NE&usqp=CAU (accessed on 10 November 2022).
104. The Main Mosque of Bibi-Khanim, Samarkand (the Restoration of the Late Twentieth Century). Available online: https://bravotours.uz/01BRAVOFOLDER/%D1%81%D0%B0%D0%BC%D0%B0%D1%80%D0%BA%D0%B0%D0%BD%D0%B4/%D0%B1%D0%B8%D0%B1%D0%B8%20%D1%85%D0%B0%D0%BD%D0%B8%D0%BC%20(2).jpg (accessed on 10 November 2022).
105. The Main Mosque of Bibi-Khanim, Samarkand (the Restoration of the Late Twentieth Century). Available online: https://lh3.googleusercontent.com/-dSaLNb7sbm4/Vwk6i_isJRI/AAAAAAABM1s/maoWxBL0_94/bibi-khanym-mosque-22.jpg?imgmax=800 (accessed on 10 November 2022).
106. The Main Mosque of Bibi-Khanim, Samarkand (the Restoration of the Late Twentieth Century). Available online: http://1.bp.blogspot.com/_vqXcmKuWsmg/SqneCPlZZnI/AAAAAAAAABM/C4AQYjuRFBo/s400/MosqueDome.JPG (accessed on 10 November 2022).
107. The Main Mosque of Bibi-Khanim, Samarkand (the Reconstruction Option—The General View, the Drawing). Available online: https://i.pinimg.com/736x/f6/d4/ce/f6d4ce9d2646e268d9e0d5d3392749f8.jpg (accessed on 10 November 2022).
108. The Main Mosque of Bibi-Khanim, Samarkand (Reconstruction Option—The Model). Available online: https://img.tourister.ru/files/3/2/0/4/3/3/9/original.jpg (accessed on 10 November 2022).
109. The Mausoleum-Khanaka Ahmed Yassawi, Turkestan (the Cross Section through the Dome). Available online: https://encrypted-tbn3.gstatic.com/images?q=tbn:ANd9GcTHzLGBzc690_LdvSvquOZ12azcb6FjI7vjZNzH3I9SI2YYD_eM (accessed on 10 November 2022).
110. The Guri-Emir Complex, Samarkand (the Cross Section through the Dome). Available online: http://www.artprojekt.ru/library/arthistory2.2/pic/000052.jpg (accessed on 10 November 2022).
111. The Main Mosque of Bibi-Khanim, Samarkand (the Cross Section through the Dome). Available online: https://hiddenarchitecture.net/wp-content/uploads/2020/01/1.jpg (accessed on 10 November 2022).

Disclaimer/Publisher's Note: The statements, opinions and data contained in all publications are solely those of the individual author(s) and contributor(s) and not of MDPI and/or the editor(s). MDPI and/or the editor(s) disclaim responsibility for any injury to people or property resulting from any ideas, methods, instructions or products referred to in the content.

heritage

MDPI

Article

Protection of Historical Mortars through Treatment with Suspensions of Nanoparticles

Efstathia I. Pavlakou [1,2], Christine Lemonia [1], Emily Zouvani [1], Christakis A. Paraskeva [1,2] and Petros G. Koutsoukos [2,3,*]

1 Laboratory of Transport Phenomena and Physicochemical Hydrodynamics, Department of Chemical Engineering, University of Patras, 26500 Patras, Greece
2 Institute of Chemical Engineering Sciences-Foundation of Research and Technology Hellas (FORTH-ICEHT), 26500 Patras, Greece
3 Laboratory of Inorganic and Analytical Chemistry, Department of Chemical Engineering, University of Patras, 26500 Patras, Greece
* Correspondence: pgk@chemeng.upatras.gr

Abstract: Mortars, which are very important elements for the integrity of historic monuments, consist mainly of calcium carbonate and silicates in different proportions. Chemical dissolution due to exposure in open air is very important for the degradation of mortars. Inorganic nanoparticles with chemical and crystallographic affinity with mortar components are expected to be effective structure stabilizers and agents offering resistance to chemical dissolution. In the present work, we have developed and applied suspensions of amorphous calcium carbonate (ACC), silicon oxide (am-SiO_2) and composite nanoparticles by the precipitation of ACC on am-SiO_2 and vice versa. The application of suspensions of the synthesized nanoparticles on three different historical mortars of Roman times (1st century AD), retarded their dissolution rate in solutions undersaturated with respect to calcite, in acid pH (6.50, 25 °C). All three test historic mortars, treated with suspensions of the nanoparticles prepared, showed high resistance towards dissolution at pH 6.50. The ability of the nanoparticles' suspension to consolidate the damaged mortar was the key factor in deciding the corresponding effectiveness in the retardation of the rate of dissolution. The combination of ACC with am-SiO_2 nanoparticles showed high efficiency for protection from the dissolution of calcite rich mortars.

Keywords: mortar treatment; amorphous calcium carbonate; amorphous silica; nanoparticles; dissolution rate

Citation: Pavlakou, E.I.; Lemonia, C.; Zouvani, E.; Paraskeva, C.A.; Koutsoukos, P.G. Protection of Historical Mortars through Treatment with Suspensions of Nanoparticles. *Heritage* **2023**, *6*, 1148–1168. https://doi.org/10.3390/heritage6020064

Academic Editors: Daniela Fico and Daniela Rizzo

Received: 29 December 2022
Revised: 20 January 2023
Accepted: 25 January 2023
Published: 27 January 2023

Copyright: © 2023 by the authors. Licensee MDPI, Basel, Switzerland. This article is an open access article distributed under the terms and conditions of the Creative Commons Attribution (CC BY) license (https://creativecommons.org/licenses/by/4.0/).

1. Introduction

The masonry of historic buildings is stabilized by the use of mortars, consisting of silica and calcite formed by the carbonation of $Ca(OH)_2$ [1]. The calcitic materials of the built cultural heritage deteriorate upon contact with water from wet precipitation [2,3]. Calcite is the basic component and model inorganic compound of calcitic building elements and is susceptible to the action of various physical, chemical and biological factors [4]. SO_2, NO_x and volatile organic carbon (VOC) gases in the atmosphere, temperature, light, the presence of microorganisms on the surface of the monuments and human interventions contribute to materials' erosion over time [5–7]. Given the value and importance of cultural heritage, the restoration of damage and conservation is urgent. Progress in nanotechnology provided novel methods and materials for the restoration of cultural heritage monuments and artefacts [7,8]. The use of $Ca(OH)_2$ and $Ba(OH)_2$ nanoparticles has been shown to be effective in the restoration of works of art [5,9]. Microemulsions, micellar solutions and cleaning gels have also been successfully used for the restoration of artefacts [5,10–12]. Polymer chemical coatings have been shown to protect limestone, but outdoor exposure is destructive due to the ageing of the polymeric materials [13,14]. As shown over the last two decades, nanoparticles of metal oxides and hydroxides (TiO_2, ZnO, $Ca(OH)_2$,

$Mg(OH)_2$, $Sr(OH)_2$, and metal nanoparticles (Au, Ag, Pt) have been successfully tested as consolidants and/or protective additives on different works of art [15–17]. Colloidal silica particles have been successfully used in the restoration and conservation of cultural heritage stone monuments, protecting them from further deterioration, and they contributed to their consolidation [18]. Silica nanoparticles protect calcareous materials penetrating inside pores in which their contact with calcium carbonate material results in rapid agglomeration and the development of a protective layer [19]. Using colloidal silica, suspended in aqueous systems or in water containing ethanol, hydroxyl groups on the surface of the particles condense upon drying, by the removal of water to form siloxane bonds (Si-O-Si), resulting in coalescence and inter-bonding [20]. The application of silica nanoparticles, average size 10–15 nm, has also been reported in mortars and it was concluded that they interact with calcium carbonate forming xerogels upon water loss [21]. The treatment of Tuff stone with suspensions of colloidal silica particles provided satisfactory protection, filling pores and penetrating the material in depth [22]. Although not well understood, there are significant interactions between silica and silicic species with calcium carbonate, which result to the consolidation of the latter [23]. The problem with silica colloidal particles formed by the hydrolysis of TEOS is that the layers formed inside the pores, or on surfaces, tend to crack [24].

Amorphous calcium carbonate (ACC) is a type of calcium carbonate that has a disordered, non-crystalline structure. ACC particles have been shown to be efficient consolidants for calcareous matrices, suggesting that ACC has a high potential for protection of stone cultural relics with calcareous matrix [25]. Nanolime particles have been applied for calcareous stone protection that involve the conversion of calcium hydroxide-CO_2 reaction products into ACC [26,27]. ACC and calcium oxalate particles in suspension applied on white marble, calcarenite limestone and gypsum plaster, showed enhanced hydrophobicity, less cracking, improved surface adhesion, and a marked improvement in acid resistance [28]. ACC is often used as a protective coating for stone surfaces to help protect against weathering. The use of ACC as a protective coating for stone surfaces is based on the fact that calcium carbonate is a relatively stable and durable material. When applied to a stone surface, ACC can help to form a protective layer, and is able to resist the effects of weathering, such as water and wind erosion, and UV radiation. ACC layers can help to preserve the appearance and structural integrity of the stone over time. ACC coating can be applied with a variety of methods, including spraying, brushing and dipping, to protect historic and cultural stone buildings, statues and other structures both indoors and outdoors. [29–31]. Several factors may affect the effectiveness of ACC as a protective coating, including the type of the treated material, the climate, and the type of weathering. In general, ACC is most effective at protecting against water and wind erosion but may not be as effective at protecting against other types of weathering, such as chemical attack or physical abrasion. Silica stabilization has been shown to improve the physical and chemical stability of ACC, making it less prone to dissolution and more resistant to changes in pH and temperature [32]. ACC may from directly on silica particles with methanol playing a stabilizing role for ACC [33]. Mortars are mixtures of inorganic or organic binders, predominantly fine aggregates, water and in some cases organic and/or inorganic additives (or mixture of binder only and water). They are mixed in appropriate proportions to provide mixture, in the fresh state, a suitable workability and, in the hardened state, adequate physical characteristics (porosity, water permeability etc.), mechanical characteristics (strength, deformability, adhesion, etc.) appearance, and durability. The evidence from the literature reports that the effectiveness of treatment of materials with nanoparticles depends on the extent of interactions with the matrix materials. Since mortars contain silica and calcium carbonate, both silica and calcium carbonate particles are expected to show compatibility. The preparation of nanoparticles consisting from ACC core and am-SiO_2 coverage and am-SiO_2 core with ACC over layering aimed at the treatment of composite materials, like the mortars, with chemically and crystallographically compatible components. In addition, with the information provided by literature reports on a broad range metal oxides nanoparticle, in the present study we have focused on nanoparticles compatible with the components of the treated materials using all possible combinations of ACC and am-SiO_2 nanoparticles. Historical mortars were used for treatment and they were tested with respect to their resistance in chemical weathering. The tests were based on the measurement of the rates of dissolution of the treated mortars, and of the nanoparticles used for the treatment, independently, in solutions undersaturated with respect to calcite at constant pH, which simulate better wet precipitation.

In the present work, the protective effect of am-SiO_2, ACC, (am-SiO_2)-ACC (composite 1, SY1) and (ACC)-am-SiO_2 (composite 2, SY2) nanoparticles was tested on mortars. Three historical mortars, consisting mainly of calcite and silica with various $CaCO_3/SiO_2$ proportions, were treated with suspensions of the nanoparticles. All specimens were treated in powdered form to maximize the exposed total surface area exposed in the weathering medium. The protection of the treated specimens from deterioration was evaluated by measurements of the rates of dissolution in undersaturated conditions of calcium carbonate (pH 6.50, 25.0 °C), and the effectiveness of the treatment materials and methods of application were evaluated. The composition of the undersaturated solutions containing calcium, carbonate, sodium and chloride ions simulated wet precipitation. [32]. The selected pH value was in the acid side and not very acid as in the case of heavily polluted atmosphere from acid gases (SO_x, NO_x).

2. Materials and Methods

Historical mortar samples were used for the evaluation of the nanoparticles tested with respect to their ability to protect them from chemical weathering in acid solutions. Three historical mortars obtained from Roman stadium in Patras (built between 80–90 AD, probably a gift from the Roman emperor Domitian, Patras, Greece) with different $CaCO_3/SiO_2$ proportions, were treated with suspensions of nanoparticles. The specific surface area and the porosity of the test materials were measured by nitrogen adsorption using the BET method (Micromeritics, Gemini 2370) (Table 1). The mortar specimens were treated with nanoparticles suspending 1 g of powdered material in nanoparticle suspensions formed as described in detail, next. The nanoparticles used in the treatment of mortars were ACC, am-SiO_2 and combinations of ACC and am-SiO_2, Composite 1 (SY1), Composite 2 (SY2)). All syntheses were done in a double walled borosilicate glass reactor, thermostated at 25.0 ± 0.5 °C, magnetically stirred to ensure homogeneity.

Table 1. Physico-chemical characteristics of test materials before treatment with suspensions of nanoparticles. (* the rest is silica and aluminosilicates).

Material Treated	% Ca *	Specific Surface Area (m^2/g)	Porosity (cm^3/g)
Mortar 1	61	10.0	0.021
Mortar 2	50	13.2	0.022
Mortar 3	43	14.0	0.064

2.1. Synthesis of Nanoparticles

ACC was prepared by the addition of 0.05 M dimethyl carbonate (DMC, Merck SA, For synthesis, Taufkirchen, Germany), sodium chloride (NaCl, Merck Greece SA, Athens, Greece) 0.1 M and calcium chloride ($CaCl_2$, Riedel de Haen, Seelze, Germany) 0.01 M in a 150 mL, magnetically stirred batch reactor (Figure 1) at 25.0 °C (concentrations correspond to the final solution prepared) [34]. Mixing of the solutions lasted 2.5 min before the separation of the precipitate from the solution. The solid was separated from the suspension by centrifugation at 2500 rpm for 4 min, then by filtration through membrane filters (0.20 μm, Sartorius, Goettingen, Germany) and it was rinsed with acetone and freeze dried. am-SiO_2 was prepared by the addition of tetraethyl orthosilicate (TEOS, Sigma Aldrich, Merck, Darmstadt, Germany) (0.29 M) solution, in a solution of 11 mL ethanol (C_2H_5OH, 95% w/v) and 5 mL ammonia (NH_3, 32% w/v). Before the addition of TEOS, the ethanol-ammonia solvent mixture was homogenized by magnetic stirring for 30 min TEOS hydrolysis started immediately upon addition in the ethanol-ammonia solution, resulting to the formation of am-SiO_2. The suspension was stirred for 1 h from the start of TEOS addition. [35]. Next, the solid was separated from the suspension by centrifugation at 40,000 rpm for 30 min. Composite nanoparticles SY1 were prepared by the addition of a 0.02 g of am-SiO2 in a batch reactor (150 mL) containing DMC, NaCl and $CaCl_2$ solutions. The pH was adjusted to 10 by the addition of NaOH as needed (20 mL of 0.5 M NaOH solution) (Figure 1). The suspension was homogenized by stirring with a magnetic stirrer for 2.5 min [35]. SY1 particles were separated by filtration was rinsed with acetone and freeze dried. SY2 nanoparticles, were prepared by the addition of ACC nanoparticles to a suspension of SiO_2 nanoparticles prepared in situ. Specifically, 0.07 g of ACC nanoparticles were introduced in a reactor, containing 115 mL of 0.67 M ammonia solution in 95% ethanol under stirring. Next, TEOS (0.29 M final concentration) was added and stirring continued for 10 min (Figure 1) [36].

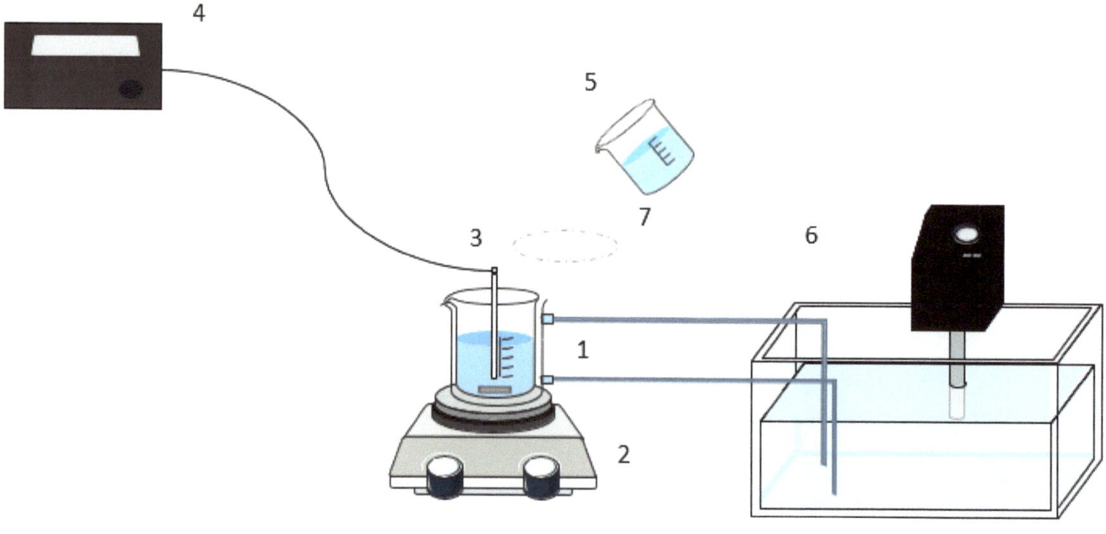

Figure 1. Experimental setup of the synthesis of nanoparticles (ACC, am-SiO_2, SY1 and SY2) and for the treatment of three historical mortars: 1: thermostated batch reactor; 2: magnetic stirrer; 3: glass/Ag/AgCl combination electrode; 4: pH meter; 5: added solutions; 6: bath; 7: polyamide reactor cap.

2.2. Treatment of the Test Specimens with the Nanoparticles

ACC, am-SiO_2, SY1, SY2 were used for the treatment of the test samples (Table 1). The treatment was carried out in a thermostated double walled borosilicate glass reactor, volume totaling 150 mL, at 25 °C. The suspensions in the reactor were homogenized by magnetic stirring using a Teflon® coated stirring bar (Figure 1). 1 g of the test sample specimens was suspended in a solution containing 0.05 M DMC and 0.01 M $CaCl_2$ solutions. The suspension of the test specimen particles was mixed thoroughly until the suspension was homogeneous. Next, 0.1 M standard sodium hydroxide solutions was added to adjust pH ≈ 10. At this alkaline solution DMC was hydrolyzed [37]. The release of carbonate ions, CO_3^{2-}, in combination with the presence of free calcium ions developed the supersaturation needed for the spontaneous formation of ACC:

$$(CH_3)_2CO_3 + 2NaOH \rightarrow 2CH_3OH + Na_2CO_3 \quad (1)$$

This treatment resulted in the formation of ACC, which was deposited, at least partly, on the suspended particles of the treated materials (three historic mortars). The methanol released during the hydrolysis of dimethyl carbonate has a stabilizing effect on the ACC [25]. In an analogous way, the specimens to be treated were suspended in a solution in which am-SiO_2 was precipitated. 1 g of the powdered test specimens (Table 1) was suspended in the 0.67 M ammonia solution in 95% ethanol. Next, TEOS was added to a final concentration of 0.29 M. Am-SiO_2 takes place in this solution by successive hydrolysis of TEOS and condensation reactions as shown in Figure 2:

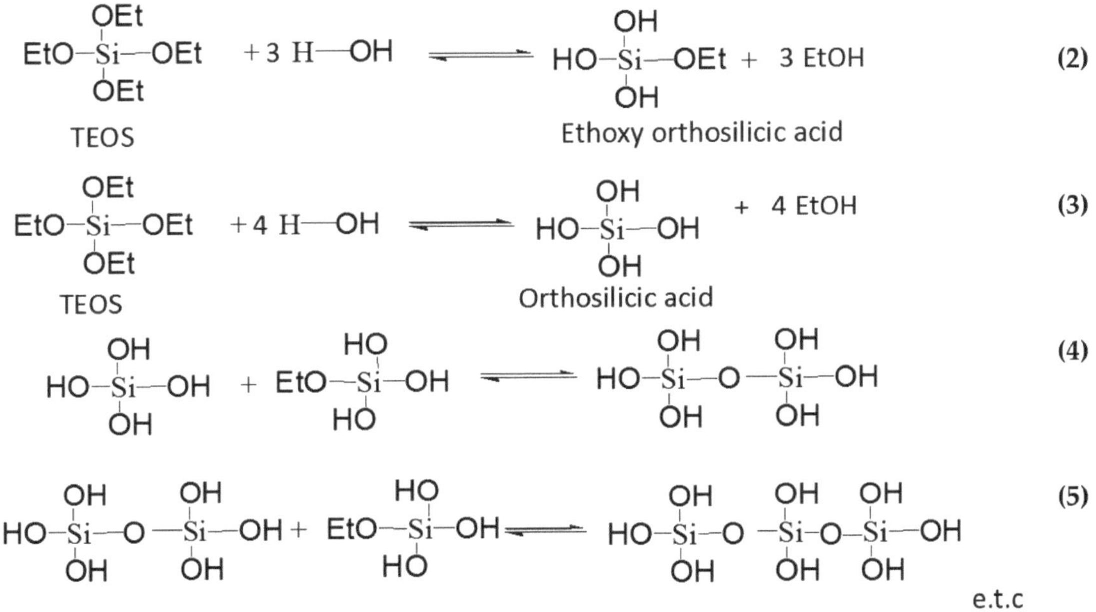

Figure 2. Synthesis of am-SiO_2 by hydrolysis of TEOS. Condensation reactions.

Ethoxy ortho silicic acd and orthosilicic acid are produced by the partial and full hydrolysis Equations (1) and (2) shown in Figure 2. Next, these two products are condensed according to Equation (3) forming disiloxane. At subsequent reaction (Equation (4)) trisiloxane and polysiloxanes are produced, until a critical size is reached characteristic of the am-SiO_2 precipitated.

Treatment with the composite particles SY1 and SY2 was carried out by the addition of 1.0 g of the specimens for treatment in suspensions of the corresponding particles under stirring. The contact time was limited to 10 min. Following suspension, SY1 and SY2 nanoparticles were separated from the liquid phase by filtration through membrane filters. The solid was freeze dried. The morphology of the nanoparticles and of the test materials were examined by scanning electron microscopy (SEM, Leo Supra 35 VP, Zeiss, Oberkohen, Germany). The treatment of the mortars with SY1 and SY2 did not change significantly the composition of the material tested by dissolution in the first two cases. The composition of mortar 1 was 61% w/w $CaCO_3$ and 39% w/w SiO_2. After treatment with SY1, the concentration for SiO_2 was 39% w/w and with SY2, the concentration of SiO_2 was 37% w/w. The composition of mortar 2 was 50% w/w $CaCO_3$ and 50% w/w SiO_2. After treatment with SY1, SiO_2 was 49% w/w and with SY2, SiO_2 was 47% w/w. The composition of mortar 3 was 43% w/w $CaCO_3$ and 57% w/w SiO_2. After treatment with SY1, the concentrations the concentration of SiO_2 was 42% w/w and with SY2, SiO_2 was 40% w/w.

2.3. Dissolution Tests

The dissolution of the test specimens was studied in undersaturated calcium carbonate solutions at constant pH. In all cases, the calcium concentration was monitored during dissolution, in undersaturated solutions with respect to calcite, the thermodynamically most stable polymorph of calcium carbonate. Monitoring was performed by sampling and analysis of the samples for their calcium content in the aqueous phase during dissolution. The undersaturated solutions were prepared in a thermostated batch reactor mixing equal volumes of standard $CaCl_2.2H_2O$ and $NaHCO_3$ solutions, 75 mL each. Both solutions were prepared from the respective stock solutions, prepared from the corresponding crystalline materials. The calcium chloride stock solutions were standardized with atomic absorption spectrometry (AAS, Perkin Elmer AAnalyst 300) and by titration with standard EDTA solutions with murexide indicator. The sodium bicarbonate stock solutions were prepared fresh in every experiment from the respective crystalline solid (Merck Puriss) dried overnight at 70 °C and used without any further standardization. The ionic strength of the solutions was adjusted with sodium chloride from the respective stock solution, made from the solid without any further standardization. In all measurements the total calcium concentration, (Ca_t), in the undersaturated solutions was equal to the total carbonate concentration (C_t) and equal to 1.25 mM. The ionic strength was 0.1 M in NaCl. The pH of the undersaturated solutions was acid (6.50) and it was adjusted by the addition of standard HCl solution (Merck, titrisol). Following the verification of the constancy of the solution pH, 10 mg of the test material was introduced in the undersaturated solutions and it was rapidly dispersed by magnetic stirring. Throughout the dissolution of the calcium carbonate component of the test specimens the pH of the undersaturated solutions was kept constant by the addition of standard hydrochloric acid to neutralize the protons released:

$$CaCO_3(s) + H_2O \leftrightarrows Ca^{2+}(aq) + HCO_3^-(aq) + OH^-(aq) \tag{5}$$

Increase of the solution pH by 0.005 pH units, caused by the dissolution of the carbonate component of the dissolving solid (Equation (5)), triggered the addition of standard hydrochloric acid solution from a motorized syringe controlled by a digital computer (Figure 3). During dissolution, samples were withdrawn, filtered through membrane filters and the filtrates were analyzed for (total) calcium by AAS. Analysis for total dissolved silicate was done spectrophotometrically at 350 nm (UV-VIS spectrophotometer Perkin Elmer Lambda) [38] by the construction of the appropriate reference curve of the absorbance of standard samples as a function of their concentration. The analytical method was based on the complexation of silicate ions with ammonium molybdate. The initial rates of dissolution were calculated from the calcium-time profiles and were expressed per unit surface area of the dissolving solid.

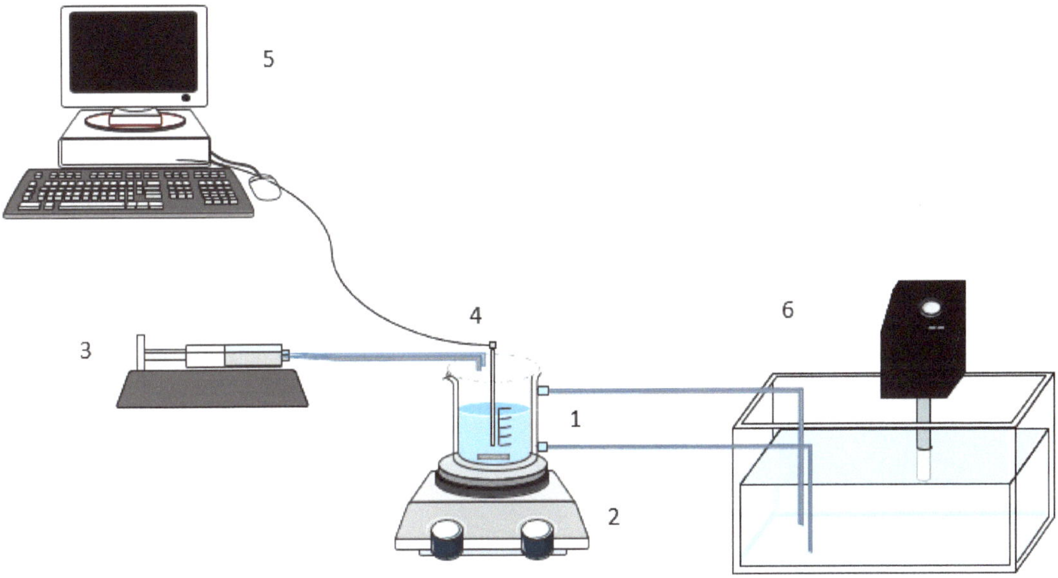

Figure 3. Experimental setup of the dissolution of the calcitic materials: 1: thermostated batch reactor; 2: magnetic stirrer; 3: motor driven syringes; 4: glass/Ag/AgCl combination electrode; 5: digital computer (data collection, syringe control); 6: bath.

3. Results

The deterioration of untreated and treated mortar samples was examined by measurements of the kinetics of dissolution, which was assessed by the rate of calcium increase in the solutions as a function of time and from the examination of the morphology of the dissolving material (before and after treatment with the nanoparticles). The morphology of the three mortars is shown in the SEM pictures in Figure 4a–c, respectively. In the first two mortars (60% and 50% calcite and the rest silica), the larger calcite crystals were partially covered by silica nanoparticles. In mortar 3 (Figure 4c) silica was in excess (60%) and the coverage of the calcitic particles was denser as may be seen from the texture of the mortar particles. The mortars were treated with amorphous calcium carbonate, (ACC), amorphous silica (am-SiO_2), SY1 and SY2 nanoparticles. Nanoparticles SY1 and SY2, consisted of ACC and am-SiO_2 in different proportions. The different calcium carbonate-silica content was due to the fact that in the case of SY1, the nanoparticles were prepared by a core of am-SiO_2 covered by ACC and SY2 by a core of ACC nanoparticles covered by the much smaller am-SiO_2. During the preparation of SY1, the silica nanoparticles, which were initially introduced into the synthesis solution, created a layer, on the surface of which a second layer, consisting of ACC nanoparticles was deposited (Figure 5a). Accordingly, SY2 was formed when a layer of silica nanoparticles was deposited on top of ACC nanoparticles (Figure 5b). As may be seen, the larger number of the am-SiO_2 nanoparticles resulted in the formation of an outer layer of ACC, but the presence of am-SiO_2 was also significant. The composition of SY1 was calculated as 69% ACC/31% am-SiO_2 while that of SY2 was 97% ACC/3% am-SiO_2.

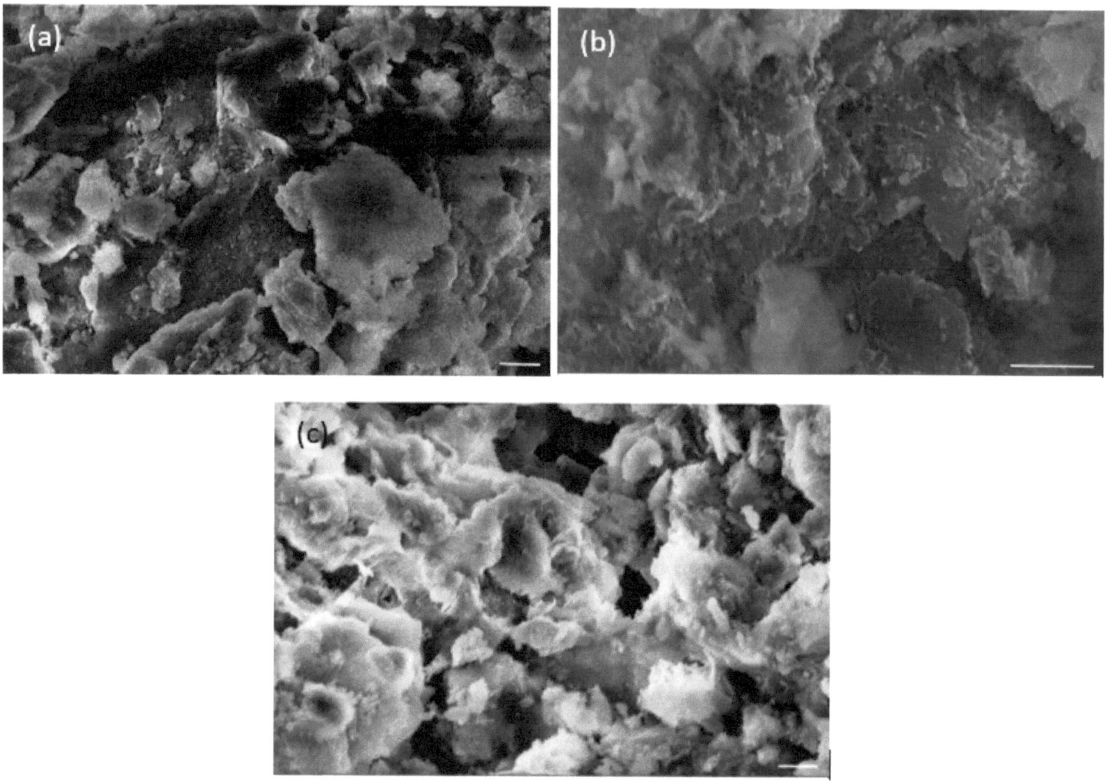

Figure 4. (**a**) Mortar 1 (bar = 2 μm), (**b**) mortar 2 (bar = 1 μm), (**c**) mortar 3 (bar = 2 μm).

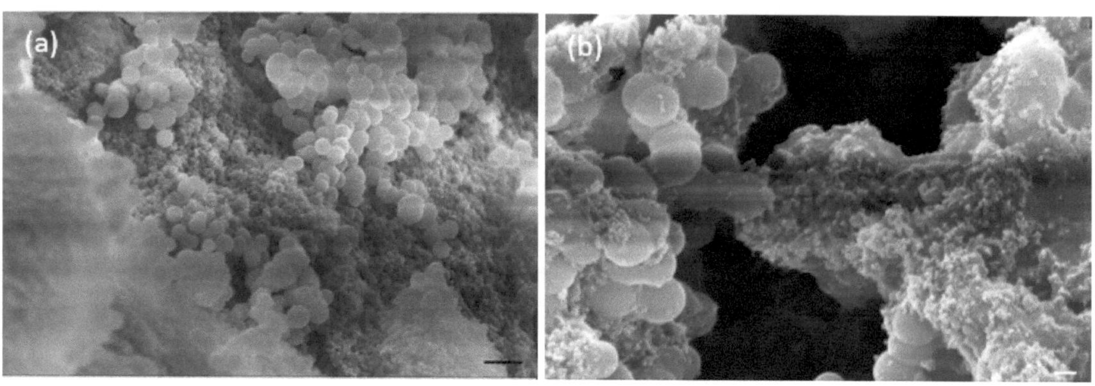

Figure 5. (**a**) SY1, bar = 1 μm; (**b**) SY2, bar = 200 nm.

The coating of mortar_1 with ACC nanoparticles is shown in Figure 6a. In some cases, the transformation of ACC to calcite is evident. The particle size of the deposits is in the nanometer range (ca. 450 nm), which is due to the initial formation of ACC. The coverage of the mortar grains appears to be complete. The chemical and crystallographic affinity of calcium carbonate and silica was the key factor for the coverage of the mortar surface [39]. The precipitated silica nanoparticles (≈70 nm) formed a layer covering completely the surface of the treated mortar as shown in Figure 6b. In Figure 6c the grains of mortar_1, treated with SY1 nanoparticles are shown. The amorphous silica nanoparticles, acting as substrate, penetrated the pores of the treated mortar. ACC nanoparticles precipitated on the template provided by the silica nanoparticles, reinforced the coating of the mortar grains. In Figure 6d the result of treatment of the mortar sample 2 with SY2 nanoparticles is shown. The coverage is excellent and as shown the silica nanoparticles covered the first layer of the underlying ACC nanoparticles, which are in direct contact with the mortar grains.

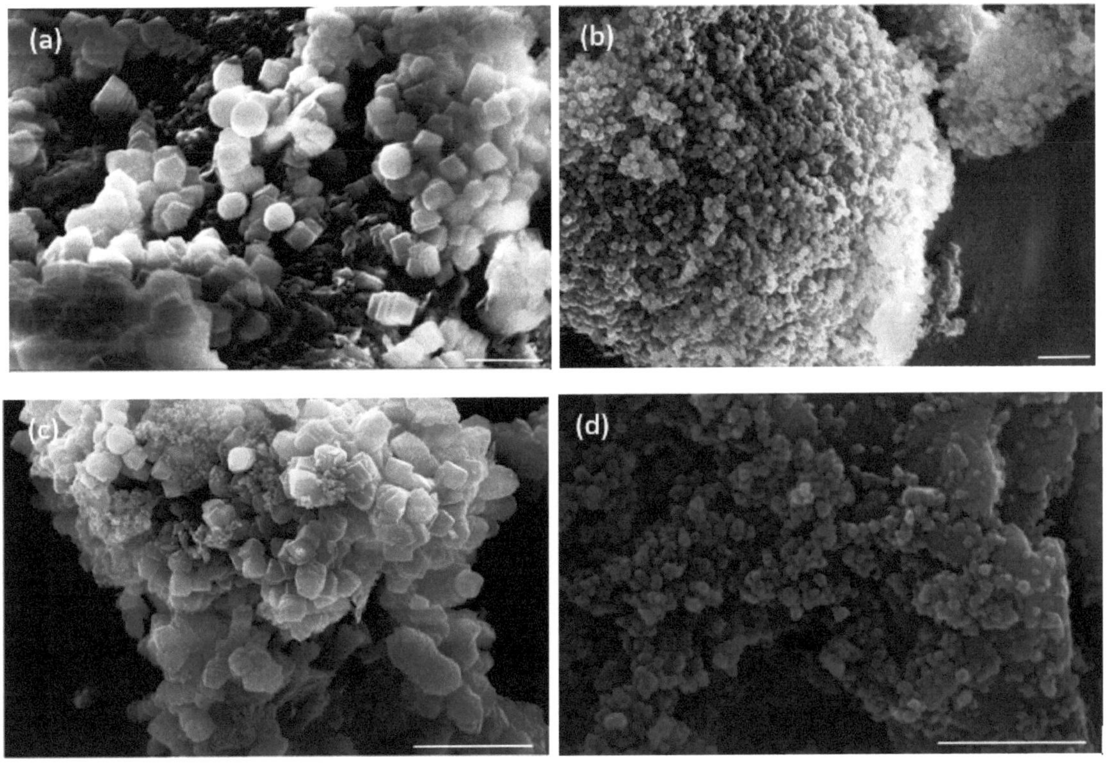

Figure 6. Mortar_1 with deposits (**a**) ACC deposits, bar = 1 μm; (**b**) SiO_2 deposits, bar = 1 μm; (**c**) SY1 deposits, bar = 5 μm; (**d**) SY2 deposits, bar = 1 μm.

In the SEM pictures shown in Figure 7, ACC crystals deposited on the surface of mortar sample 2, started transforming to the thermodynamically most stable calcium carbonate, rhombohedral calcite (Figure 7a). Upon coverage with am-SiO_2 (Figure 7b), it is very likely that the deposition of the respective nanoparticles took place not only on ACC but also inside the pores of the mortar sample. In the deposition of SY1 nanoparticles on the grains of mortar 2, ACC is still shown together with the calcite crystals from the transformation of ACC (Figure 7c). According to the treatment methodology with SY1 nanoparticles, am-SiO_2 covered the surface and the pores of mortar sample 2, followed by the deposition of ACC. In the treatment of the mortar samples with SY2 nanoparticles, ACC deposited first by spontaneous precipitation, followed by the deposition of am-SiO_2 by TEOS hydrolysis, which resulted in the rapid precipitation of the respective nanoparticles, as shown in the SEM picture presented in Figure 7d.

Figure 7. Mortar_2 treated with nanoparticles (**a**) ACC, bar = 200 nm; (**b**) am-SiO_2, bar = 1 μm; (**c**) SY1, bar = 3 μm; (**d**) SY2, bar = 2 μm.

In the case of treatment of the Si-rich mortar sample (mortar sample 3), as may be seen in the SEM picture shown in Figure 8a, ACC nanocrystals covered efficiently the mortar.

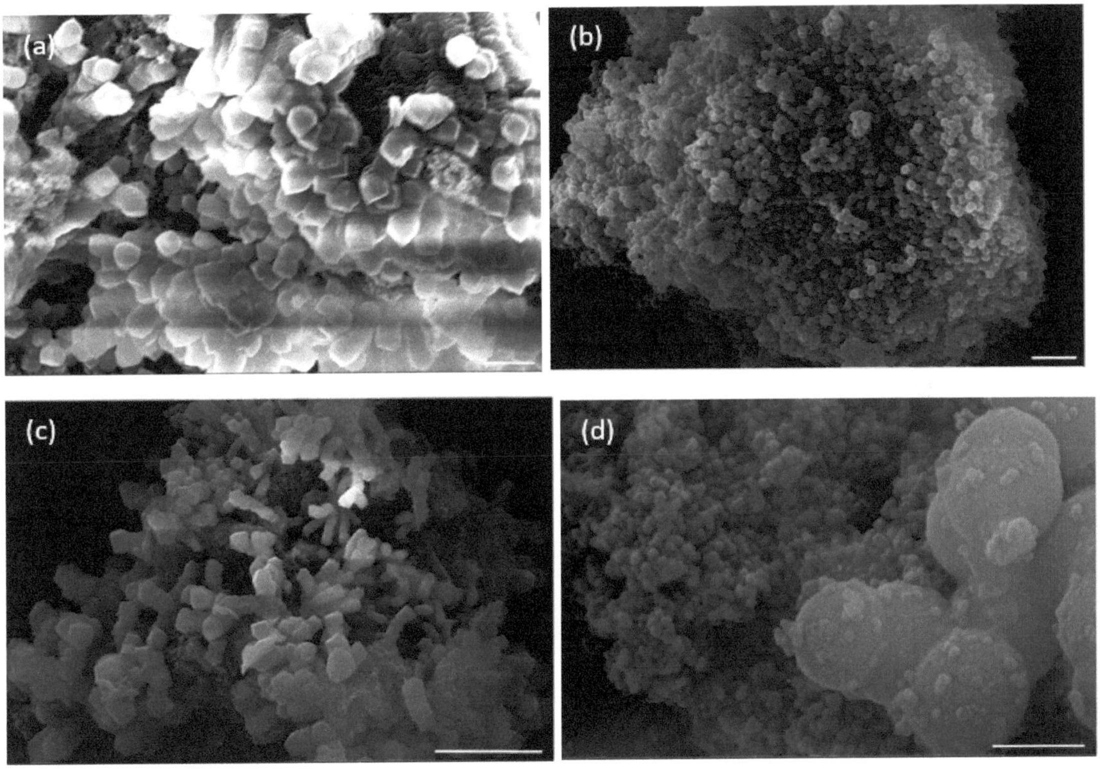

Figure 8. Mortar sample 3 treated with nanoparticles: (**a**) ACC, bar = 1 μm; (**b**) am-SiO_2, bar = 1 μm; (**c**) SY1, bar = 5 μm; (**d**) SY2, bar = 500 nm.

3.1. Composites of Silica Nanoparticles with ACC and Silica Particles

These methods involve the formation of a stable dispersion of ACC and silica particles, which can then be dried to form a solid composite material. Silica stabilization has been shown to improve the physical and chemical stability of ACC, making it less prone to dissolution and more resistant to changes in pH and temperature. ACC has been suggested to be stabilized by the presence of the large tetrahedral silicate ions destabilize calcite [40]. Moreover, silica particles in the presence of NaCl have been reported to enhance attachment to calcium carbonate particles [41]. Am-SiO_2 deposited on mortar sample 3 (Figure 8b), formed a uniform rather compact overlayer. In the SEM shown in Figure 8c, ACC nanoparticles layers on the mortar are visible. Some transformation to the thermodynamically more stable calcite is seen. In the case of treatment with SY1 nanoparticles, silica particles deposited served as a substrate for ACC over layering. The fact that in the mortar specimen 3, the silica content was in excess, favored the formation of the am-SiO_2 layer because of chemical affinity. In this case, the silica component acted as seed crystals, eliminating the nucleation barrier for am-SiO_2. In the treatment of the mortar sample 3 with SY2 nanoparticles, the silica nanoparticles that have settled on the larger ACC nanoparticles can be seen. The latter cover the mortar grains well, as shown in the SEM in Figure 8d.

3.2. Kinetics of Dissolution

Materials containing calcium carbonate dissolve in solutions undersaturated with respect to calcite, the thermodynamically most stable calcium carbonate polymorphs, because of the dissolution of the calcium carbonate component. The saturation of calcium carbonate solution with respect to calcite, $S_{calcite}$, is given by Equation (6):

$$S_{calcite} = \frac{\left(Ca^{2+}\right)\left(CO_3^{2-}\right)}{K^0_{s,calcite}} = \frac{(IAP)_{calcite}}{K^0_s} \quad (6)$$

In Equation (6), ( ) denote the activities of the ions inside parentheses, their product being the ion activity product, IAP, and $K^0{}_{s,calcite}$ is the thermodynamic solubility product of calcite. For $S_{calcite} = 1$ the solution is saturated with respect to calcite and for $S_{calcite} < 1$ it is undersaturated. The relative undersaturation with respect to calcite is defined as:

$$\sigma_{calcite} = 1 - S^{1/2}_{calcite} \quad (7)$$

The rate of dissolution, R_{diss}, of any calcitic material depends on the solution undersaturation:

$$R_{diss} = k_{diss}\sigma^n_{calcite} \quad (8)$$

In Equation (8) k_{diss} is the apparent rate constant of dissolution and n the apparent order of the process. In this work, the results reported refer to $\sigma_{calcite} = 0.89$. According to Equation (7), the relative undersaturation takes values in the range $0 < \sigma_{calcite} < 1$. The dissolution of mortar sample 1 (61% Ca/39% Si) in solutions undersaturated with respect to calcium carbonate, at constant pH 6.50, resulted in the increase of the calcium concentration in the undersaturated solutions as a function of time. At the initial stages the rate of calcium concentration change was rapid, slowing down with increasing calcium concentration in the solutions due to the dissolution of the carbonate content of the mortars. This results to decreasing relative undersaturation ($\sigma_{calcite} \to 0$) and to lower rates of dissolution. The concentration–time profiles of calcium and dissolved silicate, respectively, are shown in Figure 9a,b. The dissolution rates of the calcitic component of the mortars both untreated and treated with the methods described, were initial rates, calculated according to Equation (9):

$$R_{diss} = \left.\frac{d[Ca_t]}{dt}\right|_{t \to 0} \tag{9}$$

The initial rates of dissolution were calculated from the polynomial fit of the total calcium (Ca_t)–time profiles of each of the test samples [42]. The numerator in Equation (6) yielded the amount of calcium carbonate dissolved, as dictated by the stoichiometry of the dissolution Equation (5). The values calculated for the rate of dissolution for all mortar samples without and after treatment with the nanoparticles described are summarized in Table 2. The dissolution rate of mortar sample 1, treated with ACC nanoparticles, was lower in comparison to the respective rate of the untreated mortar, even though the rate of dissolution of ACC nanoparticles was higher in comparison to the respective untreated mortar sample. This reflects the intraparticle affinity between the mortar material and the nanoparticles, possibly with more contribution by the silicate component of it. The deposition of silica nanoparticles yielded dissolution rate for the treated mortar sample 1, one order of magnitude lower, in comparison with the respective rate obtained for the untreated sample. The treatment of mortar sample 1 with SY1 and SY2 nanoparticles yielded lower dissolution rates in comparison with the untreated mortar sample 1. Treatment with SY2 nanoparticles yielded lower dissolution rates in comparison with SY1, possibly because the effect of am-SiO_2 which sealed the pores and covered the surface of the mortar grains. ACC formed on the am-SiO_2 inlay. The ACC overlayer was stabilized by interactions with the am-SiO_2.

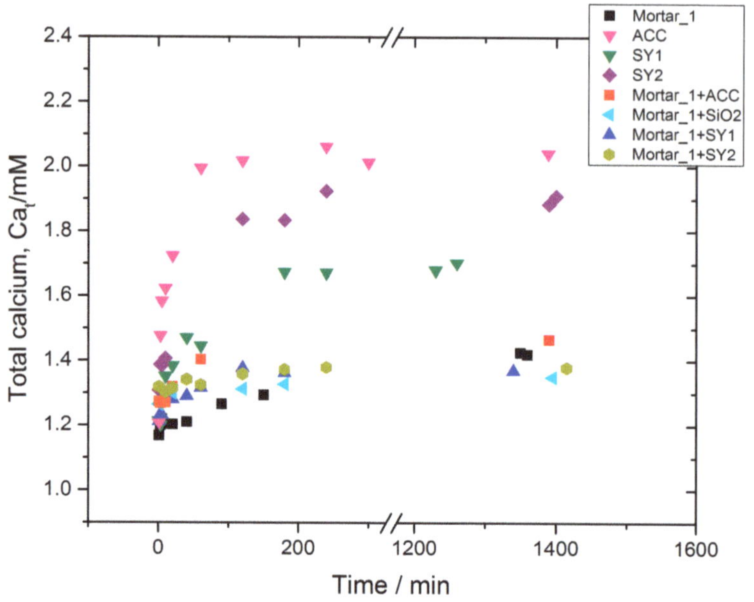

(a)

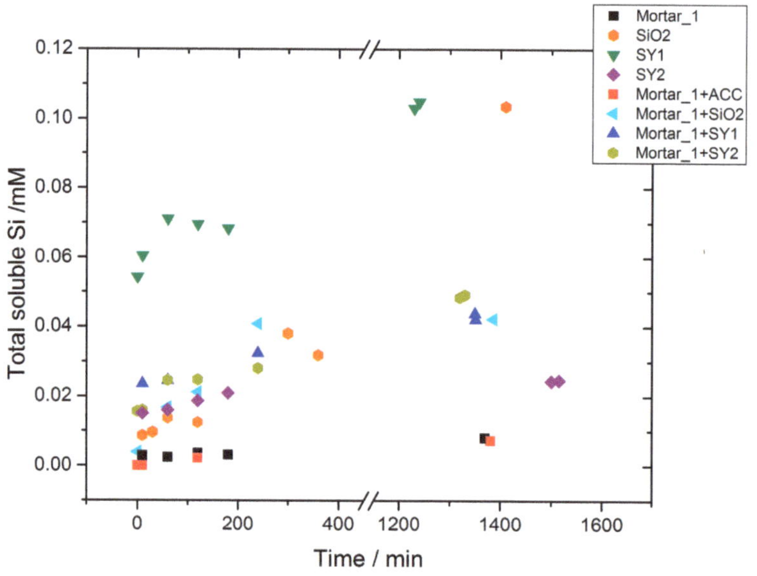

(b)

Figure 9. (**a**) Total calcium-timeprofile for the dissolution of mortar sample 1 in undersaturated calcium carbonate solutions (σ = 0.89) at constant pH; 25 °C; pH 6.50; 0.1 M NaCl; (■) Mortar 1; (■) treated with ACC nanoparticles; (◄) treated with SiO_2 nanoparticles; (▲) treated with SY1 nanoparticles; (•) treated with SY2nanoparticles; (**b**) total dissolved Si-time profiles. (■) Mortar 1; (⬢) am-SiO_2; (▼) SY1; (◆) SY2; (■) Mortar 1+ACC; (◄) Mortar 1+am-SiO_2; (▲) Mortar 1+SY1; (⬢) Mortar 1+SY2.

The total calcium and total silicate profiles during the dissolution of mortar sample 2 (50% Ca/50% Si) in solutions undersaturated with respect to calcium carbonate are presented in Figure 10, in which the total calcium and total silicate-time profiles are shown. The rate of dissolution of the mortar did not decrease significantly past treatment with ACC nanoparticles. It should be noted again that the rate of dissolution of ACC nanoparticles measured at the same conditions, was significantly higher than the corresponding rate of dissolution of the mortar. Treatment with am-SiO_2 nanoparticles reduced the dissolution rate of the mortar in comparison with the untreated mortar. The reduction in the rate was significant but not as high as in the case of mortar sample 1. It should be noted that in this case the dissolution of the coating contributed to the silicate concentrations in the undersaturated solutions, as shown in the silicate-time profile (Figure 10). The treatment of the mortar with SY1 and SY2 nanoparticles did not have a significant effect, but it is important to note that both calcium and silicate released in the solutions were lower in comparison with the corresponding free ACC and am-SiO_2 nanoparticles. This revealed the presence of **a** strong interaction between the treatment nanoparticles and the mortar treated. From the morphology examination it was clear that the relatively higher levels of calcium and silicate in the undersaturated solutions were due to partial dissolution of the coating nanoparticles, which provide protection to the mortar against dissolution through sacrificial dissolution, protecting the original material. This suggestion is due to the fact that the nanoparticles from the treatment cover the outer surface of the mortar and dissolution takes place at the outer surface of the treated solid.

The total calcium-time and total silicate-time profiles measured for the dissolution of mortar sample 3 (43% Ca/57% Si) in solutions undersaturated with respect to calcite at constant pH 6.50, are shown in the respective plots in Figure 11. The initial rates calculated for the dissolution of both untreated and treated with nanoparticles mortar sample 3 are summarized in Table 2. The treatment of the mortar with ACC nanoparticles, resulted in lower rates of dissolution in comparison with the corresponding to ACC nanoparticles, suggests that for this type of mortar (silica in excess) there is a strong interaction between the mortar and the ACC nanoparticles. Despite the fact that the measured rates were higher in comparison with the corresponding untreated mortar, in combination with the morphological examination of the mortar grains after dissolution, it was clear that the ACC nanoparticles functioned as sacrificial, dissolving at significantly lower rates in comparison with the free ACC nanoparticles, providing protection to the test historic mortar sample. Treatment with the silica particles resulted in rates of dissolution of the treated mortar, which were quite close to the corresponding am-SiO_2, suggesting that the treatment provides the mortar with sacrificial silica, thus protecting the mortar from dissolution. Apparently, in this case, the interaction between the deposited am-SiO_2 with the mortar rich in silica was not sufficiently strong to affect the performance of the treated mortar specimen with respect to dissolution in pH 6.50. The treatment of the mortar sample with SY1 and SY2 nanoparticles resulted in lower rates of dissolution of the mortar in comparison with the respective rates for each of the SY1, SY2 nanoparticles separately. Treatment with SY2 nanoparticles resulted in a significantly lower rate of dissolution at pH 6.50, in comparison with the rate of dissolution of SY2 nanoparticles. It may therefore be suggested that this reduction corresponded to stronger interactions of SY2 with the mortar. It is very likely that ACC is consolidated between the silica of the underlying mortar and the overlayer of am-SiO_2 of SY2. In the case of treatment with SY2 nanoparticles, it is possible that the higher silicate concentrations present in the undersaturated solutions reduce the rate of dissolution of the ACC.

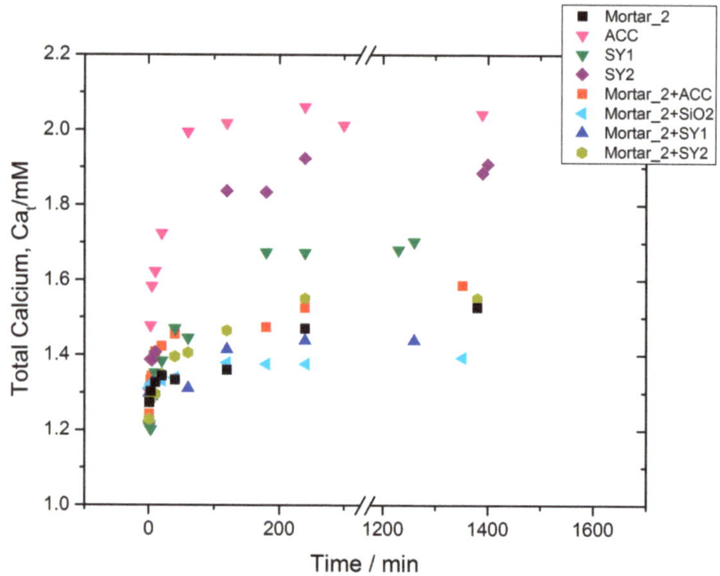

(a)

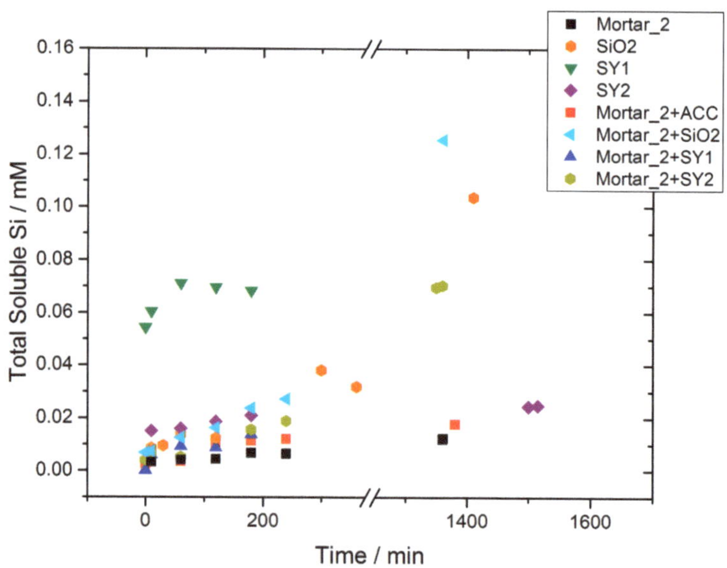

(b)

Figure 10. (**a**) Total calcium-time profile during the dissolution of mortar sample 2 in solutions undersaturated with respect to calcite at constant pH; ($\sigma_{calcite}$ = 0.89); 25 °C; pH 6.50; 0.1 M NaCl; (■) Mortar 2; (▼) ACC nanoparticles; (▼) SY1; (◆) SY2 nanoparticles; (▲) mortar 2 treated with SY1 nanoparticles; (⬢) Mortar 2 treated with SY2 nanoparticles. (**b**) Total silicate-time profile: (■) Mortar 2; (⬢) am-SiO2; (▼) SY1; (◆) SY2; (■) Mortar 2+ACC; (◄) Mortar 2+am-SiO_2; (▲) Mortar 2 + SY1; (⬢) Mortar 2 + SY2.

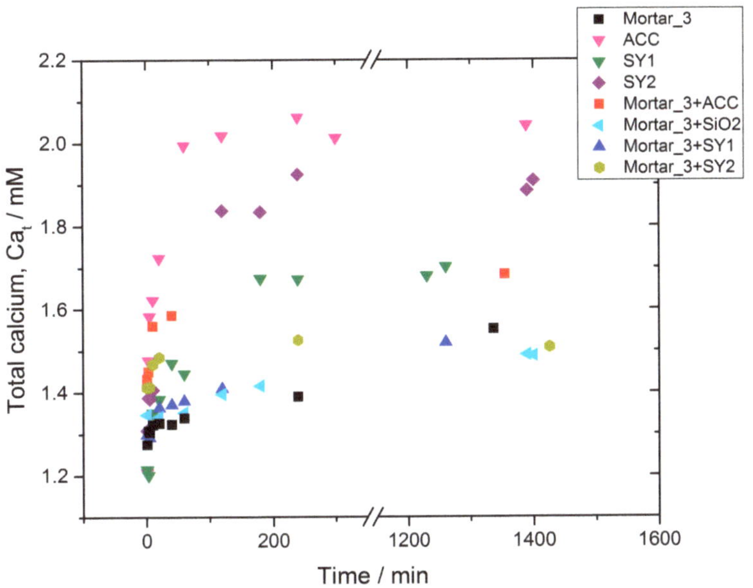

(a)

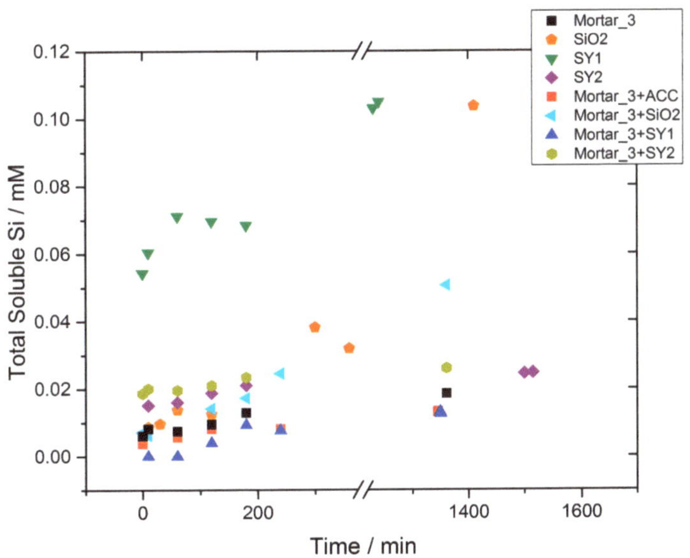

(b)

Figure 11. Total calcium-time profile measured for the dissolution of mortar sample 3, in solutions undersaturated with respect to calcite ($\sigma_{calcite}$ = 0.89) at constant pH; 25 °C; pH 6.50; 0.1 M NaCl; (■) Mortar 3; (■) Mortar 3 treated with ACC nanoparticles; (◄) Mortar 3 treated with am-SiO_2 nanoparticles; (▲) Mortar 3 treated with SY1 nanoparticles; (⬣) Mortar 3 treated with SY2 nanoparticles. (**b**) Total soluble silicate- time profile. Symbols as in (**a**).

Table 2. Initial rates of dissolution of the calcium carbonate component of nanoparticles of: ACC, SY1, SY2 and mortar samples 1,2,3 untreated and treated with ACC, am-SiO_2, SY1 and SY2 in solutions undersaturated with respect to calcite, $\sigma_{calcite}$ = 0.89, at constant pH 6.50; 25 °C, 0.1 M NaCl. Rates calculated per unit surface area of the dissolving sample.

Material	Rate of Dissolution, $R_{diss}/\times 10^{-8}$ mol·m^{-2}·s^{-1}		
Mortar	1	2	3
	3.8	9.9	10.0
ACC	170.0		
SY1	10.0		
SY2	43.0		
ACC Treated	3.0	2.8	5.9
am-SiO2 Treated	0.5	1.3	0.2
SY1 Treated	3.0	5.6	6.9
SY2 Treated	0.9	3.2	6.2

The results of the measured rates of dissolution of the three mortar samples without (blank) and with treatment with nanoparticles, summarized in Table 2, are presented in Figure 12.

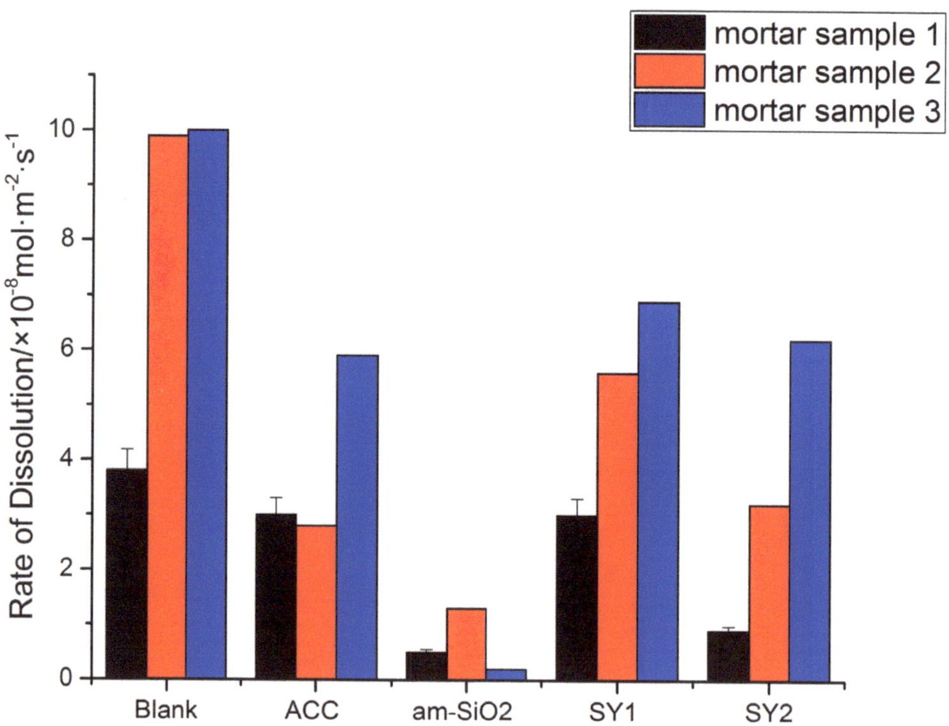

Figure 12. Rates of dissolution of historic mortar samples 1, 2 and 3 without (blank) and with treatment with ACC, am-SiO_2, SY1 and SY2 nanoparticles, in solutions undersaturated with respect to calcite ($\sigma_{calcite}$ = 0.89) at constant pH 6.50; 0.10 M NaCl, 25 °C.

It is interesting to note that the historic mortars used in this study showed different behavior with respect to their dissolution in calcium carbonate solutions undersaturated with respect to calcite. Mortar sample 1, with the highest content in calcium carbonate, dissolved faster, while samples 2 and 3 with ca. 50% and 60% silica content dissolved faster. Since mortar sample 1 had the least porosity (Table 1) of all three samples, the higher calcium carbonate content in this sample corresponded to more complete consolidation of the respective mortar, which showed up in the slower rate of dissolution. The treatment with ACC resulted in the reduction of the rates of dissolution, most probably because of the interaction of ACC with the silica content which consolidated the mortar. The decrease in the dissolution rate was more impressive for mortar samples 1 and 2. This suggestion was corroborated by the effect of the treatment of all mortar samples with am-SiO_2 nanoparticles, which because of their very small size are able to penetrate into the nanopores and consolidate the mortar samples. All mortar samples, upon treatment with am-SiO_2 particles, showed a drastic reduction of their rates of dissolution in the undersaturated acidic solutions. The effect was pronounced to the more porous material, mortar sample 3 (Table 1). The treatment of mortar samples with SY1 and SY2 nanoparticles did reduce significantly the dissolution rates of all three mortar samples for the treatment of which they were used. Clearly, SY2 treatment was more efficient. These types of nanoparticles were prepared by precipitation am-SiO_2 on ACC. The formation of layers of am-SiO_2 provided better consolidation. The highest effect (reduction of the rate of dissolution) on the calcite rich mortar sample 1 is due, on one hand, to the higher affinity of ACC for the calcitic component (61%), and to the consolidation effect of am-SiO_2 on the other.

4. Conclusions

Historic mortars consist mainly of calcite and silica. Deterioration due to weathering and more specifically to chemical dissolution due to wet precipitation can be greatly reduced by treatment with nanoparticles, which have both chemical and/or crystallographic affinity for the mortars. Strong bonding between the nanoparticles and the treatment matrix consolidated the deteriorated material, as shown by the morphological examination of the treated mortar samples. ACC, am-SiO_2 and composite nanoparticles prepared by the formation of ACC on am-SiO_2 nanoparticles, or the formation of am-SiO_2 on ACC, were shown to offer protection towards chemical dissolution in acid solutions (pH 6.50), undersaturated with respect to calcium carbonate. As shown from measurements of the rates of dissolution at constant pH 6.50, consolidation achieved through the interaction between silica and ACC is a key factor for the achievement of effective protection of historic materials from weathering. The nanoparticles deposited on the test historic mortars acted not only as consolidants, increasing the resistance of the material towards dissolution, but they also acted as sacrificial materials because of their apparent solubility. The novelty of the methodology proposed is in the use of ACC and am-SiO_2, and composites of these two components, at conditions in which they are stabilized for sufficiently long time periods, allowing for interaction with the mortar matrix treated. The treatment may be applied in situ (brushing or spraying) for the treatment of building elements and mortars.

Author Contributions: Conceptualization, C.A.P.; Methodology, E.I.P., C.A.P. and P.G.K.; Validation, E.I.P. and C.L.; Formal analysis, E.Z.; Investigation, E.I.P. and C.L.; Data curation, E. I P. and E.Z.; Writing—original draft, E.I.P.; Writing—review & editing, C.A.P. and P.G.K.; Supervision, C.A.P. and P.G.K. All authors have read and agreed to the published version of the manuscript.

Funding: This research received no external funding.

Institutional Review Board Statement: Not applicable.

Informed Consent Statement: Not applicable.

Data Availability Statement: Not applicable.

Conflicts of Interest: The authors declare no conflict of interest.

References

1. Syrmakezis, C.A. Seismic protection of historical structures and monuments. *Struct. Control Health Monit.* **2006**, *13*, 958–979. [CrossRef]
2. Howard, A.J.; Hancox, E.; Hanson, J.; Jackson, R. Protecting the Historic Environment from Inland Flooding in the UK: Some Thoughts on Current Approaches to Asset Management in the Light of Planning Policy, Changing Catchment Hydrology and Climate Change. *Hist. Environ. Policy Pract.* **2017**, *8*, 125–142. [CrossRef]
3. Brimblecombe, P.; Richards, J. Moisture as a Driver of Long-Term Threats to Timber Heritage—Part II: Risks Imposed on Structures at Local Sites. *Heritage* **2022**, *5*, 154. [CrossRef]
4. Spathis, P.; Triantafyllidis, K.; Prochaska, C.; Karapanagiotis, I.; Pavlidou, E.; Stefanidou, M. Characterization and Properties of Silicate and Nanocomposite Coatings for the Protection of Dolomite Marble against Weathering: Natural and Anthropogenic Hazards and Sustainable Preservation. In *10th International Symposium on the Conservation of Monuments in the Mediterranean Basin*; Springer International Publishing AG: Cham, Switzerland, 2018; pp. 287–294.
5. Baglioni, P.; Chelazzi, D.; Giorgi, R.; Poggi, G. Colloid and Materials Science for the Conservation of Cultural Heritage: Cleaning, Consolidation, and Deacidification. *Langmuir* **2013**, *29*, 5110–5122. [CrossRef]
6. Sadat-Shojai, M.; Ershad-Langroudi, A. Polymeric coatings for protection of historic monuments: Opportunities and challenges. *J. Appl. Polym. Sci.* **2009**, *112*, 2535–2551. [CrossRef]
7. Brunelli, A.; Calgaro, L.; Semenzin, E.; Cazzagon, V.; Giubilato, E.; Marcomini, A.; Badetti, E. Leaching of nanoparticles from nano-enabled products for the protection of cultural heritage surfaces: A review. *ESEU* **2021**, *33*, 48. [CrossRef]
8. Becerra, J.; Zaderenko, A.P.; Gómez-Morón, M.A.; Ortiz, P. Nanoparticles Applied to Stone Buildings. *Int. J. Archit. Herit.* **2021**, *15*, 1320–1335. [CrossRef]
9. Giorgi, R.; Ambrosi, M.; Toccafondi, N.; Baglioni, P. Nanoparticles for Cultural Heritage Conservation: Calcium and Barium Hydroxide Nanoparticles for Wall Painting Consolidation. *Chem. Eur. J.* **2010**, *16*, 9374–9382. [CrossRef]
10. Giorgi, R.; Baglioni, M.; Berti, D.; Baglioni, P. New Methodologies for the Conservation of Cultural Heritage: Micellar Solutions, Microemulsions, and Hydroxide Nanoparticles. *Acc. Chem. Res.* **2010**, *43*, 695–704. [CrossRef]
11. Baglioni, P.; Carretti, E.; Chelazzi, D. Nanomaterials in art conservation. *Nat. Nanotechnol.* **2015**, *10*, 287–290. [CrossRef]
12. Serafini, I.; Ciccola, A. Chapter 14—Nanotechnologies and Nanomaterials: An Overview for Cultural Heritage. In *Nanotechnologies and Nanomaterials for Diagnostic, Conservation and Restoration of Cultural Heritage*; Lazzara, G., Fakhrullin, R., Eds.; Elsevier: Amsterdam, The Netherlands, 2019; pp. 325–380.
13. Abdellah, M.Y.; Gelany, A.F.; Mohamed, A.F.; Khoshaim, A.B. Protection of Limestone Coated with Different Polymeric Materials. *Am. J. Mech. Eng.* **2017**, *5*, 51–57. [CrossRef]
14. Lettieri, M.; Masieri, M.; Frigione, M. Durability to simulated bird guano of nano-filled oleo/hydrophobic coatings for the protection of stone materials. *Prog. Org. Coat.* **2020**, *148*, 105900. [CrossRef]
15. Ion, R.-M.; Nyokong, T.; Nwahara, N. Wood preservation with gold hydroxyapatite system. *Herit. Sci.* **2018**, *6*, 37. [CrossRef]
16. Ion, R.-M.; Doncea, S.-M.; Turcanu-Carutiu, D. Nanotechnologies in Cultural Heritage-Materials and Instruments for Diagnosis and Treatment. In *Novel Nanomaterials*; IntechOpen: London, UK, 2018; pp. 173–190.
17. Blee, A.; Matisons, J. Nanoparticles and the conservation of cultural heritage. *Mater. Forum* **2008**, *32*, 121–128.
18. Chandra, S.; Wu, L. Kolloidale Kieselsäure für das Festigen von Naturstein/Colloidal Silica in Stone Conservation. *RBM* **1999**, *5*, 15–28. [CrossRef]
19. Stepien, P.; Kozlowski, R.; Tokarz, M. Gypstop—Colloidal silica for protective coating of porous building materials: Practical experience at the Wawel Castle, Cracow, Poland. *WIT Trans. Built Environ.* **1993**, *4*, 303–310.
20. Storemyr, P.; Wendler, E.; Zehnder, K. Weathering and Conservation of Soapstone and Greenschist Used at Nidaros Cathedral (Norway). In *Report Raphael II Nidaros Cathedral Restoration Trondheim Norway 2000. EC Raphael Programme—European Heritage Laboratory*; Report no. 2/2001; Lunde, Ø., Gunnarsjaa, A., Eds.; The Restoration Workshop of Nidaros Cathedral: Trondheim, Norway, 2001.
21. Sierra-Fernandez, A.; Gomez-Villalba, L.S.; Rabanalb, M.E.; Fort, R. New nanomaterials for applications in conservation and restoration of stony materials: A review. *Mater. Construc.* **2017**, *67*, e107. [CrossRef]
22. Stuff, M.; Rübner, K.; Prinz, C.; Rische, N.; Chronz, M.; Kühne, H.-C. Weathering and conservation of tuff stone. In Proceedings of the EGU General Assembly 2020, Online, 4–8 May 2020.
23. Zendri, E.; Biscontin, G.; Nardini, I.; Riato, S. Characterization and reactivity of silicatic consolidants. *Constr. Build. Mater.* **2007**, *21*, 1098–1106. [CrossRef]
24. Illescas, J.F.; Mosquera, M. Surfactant-Synthesized PDMS/Silica Nanomaterials Improve Robustness and Stain Resistance of Carbonate Stone. *J. Phys. Chem. C* **2011**, *115*, 14624–14634. [CrossRef]
25. Wang, W.; Wang, S.; Liu, Q.; Wang, X.; Zhu, J.; Luo, H.; Ji, S. Study on the efficacy of amorphous calcium carbonate as a consolidant for calcareous matrix. *Herit. Sci.* **2022**, *10*, 165. [CrossRef]
26. Rodriguez-Navarro, C.; Elert, K.; Ševčík, R. Amorphous and crystalline calcium carbonate phases during carbonation of nanolimes: Implications in heritage conservation. *CrystEngComm* **2016**, *18*, 6594–6607. [CrossRef]
27. Baglioni, M.; Poggi, G.; Chelazzi, D.; Baglioni, P. Advanced Materials in Cultural Heritage Conservation. *Molecules* **2021**, *26*, 3967. [CrossRef] [PubMed]

28. Burgos-Cara, A.; Rodríguez-Navarro, C.; Ortega-Huertas, M.; Ruiz-Agudo, E. Bioinspired Alkoxysilane Conservation Treatments for Building Materials Based on Amorphous Calcium Carbonate and Oxalate Nanoparticles. *ACS Appl. Nano Mater.* **2019**, *2*, 4954–4967. [CrossRef]
29. Karthikeyan, A.; Pillai, S.G.; Unnikrishnan, K.K. Amorphous calcium carbonate coated sandstone: Preparation, characterization and weathering resistance. *Surf. Coat. Technol.* **2009**, *203*, 3138–3145.
30. Unnikrishnan, K.K.; Pillai, S.G.; Karthikeyan, A. Amorphous calcium carbonate coatings on sandstone: Preparation, characterization and weathering resistance. *Mater. Lett.* **2009**, *63*, 2386–2389.
31. Zhang, Z.; Gao, Y.; Gao, Y.; Li, Y. Amorphous calcium carbonate coatings on marble surfaces: Preparation, characterization, and weathering resistance. *Surf. Coat. Technol.* **2015**, *268*, 89–98.
32. Kellermeier, M.; Melero-García, E.; Glaab, F.; Klein, R.; Drechsler, M.; Rachel, R.; García-Ruiz, J.M.; Kunz, W. Stabilization of amorphous calcium carbonate in inorganic silica-rich environments. *J. Am. Chem. Soc.* **2010**, *132*, 17859–17866. [CrossRef] [PubMed]
33. Nakashima, Y.; Takai, C.; Razavi-Khosroshahi, H.; Suthabanditpong, W.; Fuji, M. Synthesis of ultra-small hollow silica nanoparticles using the prepared amorphous calcium carbonate in one-pot process. *Adv. Powder Technol.* **2018**, *29*, 904–908. [CrossRef]
34. Salve, P.R.; Maurya, A.; Wate, S.R. Chemical Composition of Major Ions in Rainwater. *Bull. Environ. Contam. Toxicol.* **2008**, *80*, 242–246. [CrossRef]
35. Faatz, M.; Gröhn, F.; Wegner, G. Amorphous Calcium Carbonate: Synthesis and Potential Intermediate in Biomineralization. *Adv. Mater.* **2004**, *16*, 996–1000. [CrossRef]
36. Singh, T.K.; Jain, C.L.; Sharma, S.K.; Singh, S.S. Preparation of dispersed silica by hydrolysis of tetraethyl orthosilicate. *Indian J. Eng. Mater. Sci.* **1999**, *6*, 349–351.
37. Voinescu, A.E.; Touraud, D.; Lecker, A.; Pfitzner, A.; Kunz, W.; Ninham, B.W. Mineralization of $CaCO_3$ in the presence of egg white lysozyme. *Langmuir* **2007**, *23*, 12269–12274. [CrossRef] [PubMed]
38. Babko, A.K.; Pilipenko, A.T. *Photometric Analysis, Methods of Determining Non-Metals*; Mir Publishers: Moscow, Russia, 1976.
39. Lioliou, M.G.; Paraskeva, C.A.; Koutsoukos, P.G.; Payatakes, A.C. Heterogeneous nucleation and growth of calcium carbonate on calcite and quartz. *J. Colloid Interface Sci.* **2007**, *308*, 421–428. [CrossRef]
40. Gal, A.; Weiner, S.; Addadi, L. The Stabilizing Effect of Silicate on Biogenic and Synthetic Amorphous Calcium Carbonate. *J. Am. Chem. Soc.* **2010**, *132*, 13208–13211. [CrossRef] [PubMed]
41. Li, Y.V.; Cathles, L.M. Retention of silica nanoparticles on calcium carbonate sands immersed in electrolyte solutions. *J. Colloid Interface Sci.* **2014**, *436*, 1–8. [CrossRef]
42. Tavare, N.S. *Characterization of Crystallization Kinetics from Batch Experiments, Industrial Crystallization*; The Springer Chemical Engineering Series; Springer: Boston, MA, USA, 1995; pp. 141–196.

Disclaimer/Publisher's Note: The statements, opinions and data contained in all publications are solely those of the individual author(s) and contributor(s) and not of MDPI and/or the editor(s). MDPI and/or the editor(s) disclaim responsibility for any injury to people or property resulting from any ideas, methods, instructions or products referred to in the content.

heritage

Article

Mortar Characterization and Radiocarbon Dating as Support for the Restoration Work of the Abbey of *Santa Maria di Cerrate* (Lecce, South Italy)

Giovanna Vasco, Antonio Serra, Giovanni Buccolieri, Daniela Manno, Lucio Calcagnile, Gianluca Quarta and Alessandro Buccolieri *

Center of Applied Physics, Dating and Diagnostics (CEDAD), Dipartimento di Matematica e Fisica "Ennio De Giorgi", University of Salento, 73100 Lecce, Italy
* Correspondence: alessandro.buccolieri@unisalento.it

Abstract: During the restoration work promoted by the FAI foundation (*Fondo Ambiente Italiano*) of the abbey of *Santa Maria di Cerrate* (Lecce, Apulia, South Italy), multidisciplinary analyses have been realized to support conservators and art historians for the safeguard and valorization of the most important byzantine evidence (12th–13th century) in Apulia. In this paper, mortar samples have been investigated using scanning electron microscopy (SEM-EDX) and X-ray diffraction (XRD) to characterize the observed materials, directing conservators for the integration interventions with compatible realizations. Moreover, the samples were compared with specimens taken from local quarries and the nearest coastline area, and vegetal fibers, embedded into the mortars were analyzed by applying radiocarbon dating by accelerator mass spectrometry (AMS). It was thus possible to give a contribution to the historical-artistic research related to the building techniques, the ratio aggregates/binder and the employed materials of the mortars and the dating of the mural paintings between the end of the 12th century and the first half of the 13th century.

Keywords: medieval mortars; *Santa Maria di Cerrate*; SEM-EDX; XRD; radiocarbon dating

Citation: Vasco, G.; Serra, A.; Buccolieri, G.; Manno, D.; Calcagnile, L.; Quarta, G.; Buccolieri, A. Mortar Characterization and Radiocarbon Dating as Support for the Restoration Work of the Abbey of *Santa Maria di Cerrate* (Lecce, South Italy). *Heritage* **2022**, *5*, 4161–4173. https://doi.org/10.3390/heritage5040215

Academic Editors: Daniela Fico and Daniela Rizzo

Received: 24 November 2022
Accepted: 12 December 2022
Published: 15 December 2022

Publisher's Note: MDPI stays neutral with regard to jurisdictional claims in published maps and institutional affiliations.

Copyright: © 2022 by the authors. Licensee MDPI, Basel, Switzerland. This article is an open access article distributed under the terms and conditions of the Creative Commons Attribution (CC BY) license (https://creativecommons.org/licenses/by/4.0/).

1. Introduction

The monastery of *Santa Maria di Cerrate* (Lecce, Apulia, South Italy) is a complex of structures (12th–20th century) due to several uses of the area that modified it over time (Figure 1). In fact, the Italian-Greek community of monks that founded the monastery was gradually replaced by the organization of a typical Southern Italian farm.

The conservation activities for its redevelopment offered an interesting meeting point between diagnostic and historical studies, together with the stylistic knowledge correlated to the context. In fact, scientific methodologies of investigation have been applied to characterize the frescos of the abbey and their degradation state to support the conservation treatments. This opportunity was also seized to nail their chronology down and to relate the materials with the employed artistic techniques.

In agreement with the Superintendence of Cultural Heritage and the conservation director of the FAI (Fondo Ambiente Italiano) foundation, through multidisciplinary collaboration with conservators, some specimens were sampled, trying to limit as much as possible the invasiveness of the sampling and considering the required instrumental quantities.

For conservators, it was particularly necessary to identify the original pigments used for the frescos to properly clean their surface and to characterize the degradation of the bleached areas. Moreover, the composition of the mortars inside the church and their binder/aggregates ratio were required to replenish the gaps of renders and plaster layers using materials with suitable physicochemical properties.

(a) (b)

Figure 1. External (**a**) and internal (**b**) photos of the abbey of *Santa Maria di Cerrate*.

The characterization of the frescoes regarding their conservation state and the results of the pigment analysis was reported in previous work [1]. In particular, the bleaching alterations were related to the photocatalytic effect of titanium dioxide, which was found together with chromium, attributable to previous documented restorations (1960s). About the pigments, in addition to the common presence of hematite and goethite for the landscapes and the characters, the rare use of the expensive lapis lazuli blue with carbon black was identified in the background of the wall paintings. Finally, for historical-artistic research, it was possible to define the *mezzo fresco* technique through morphological and qualitative analyses on a cross-section.

Regarding the building techniques, petrographic, mineralogical and chemical characteristics of the mortars were investigated during previous conservation work, concerning the external walls of the church, the sculptured portico and the buildings around the courtyard [2]. This study highlighted the differences between medieval mortars, based on carbonate aggregate, and post-medieval mortars with mostly siliciclastic aggregate.

Moreover, to complete the restoration work, it was still necessary to characterize the mortars inside the church in the points required by conservators to finish the integrations. Therefore, in this study, we report the granulometric and mineralogical analyses performed on the samples from the mortars inside the church, which also gave the opportunity to contribute to the dating of the frescoes with scientific methods.

1.1. *The Abbey of Santa Maria di Cerrate*

Between the 9th and the 11th century, Apulia was a byzantine region, culturally influenced by the presence of political and ecclesiastical officials coming from different areas of the Empire, followed by Greek artists. The artistic production was thus the distinctive result of a combination of Greek conservative features and updated byzantine art trends of monumental painting, grafted into local specificities thanks to the cultural absorption by local inhabitants. Therefore, after the end of the byzantine domination because of the Norman conquest, the byzantine language partially survived, above all in the area known as Terra d'Otranto where entrenched connections were established with Greece, Cappadocia, Syria, Egypt, Cyprus and Corfu, thanks to its geographical position that made it a crossroad for pilgrims and trades.

During the 12th century, where the Apulian Romanesque architecture and sculpture reached their apex, the Apulian-byzantine wall paintings have gradually gone and disappeared. However, byzantine paintings survived until the 14th century in the Salento peninsula, where Italian-Greek communities existed until the 17th century, creating a continuous dialectic between western and eastern cultures in a kaleidoscopic society [3].

Therefore, the church of *Santa Maria di Cerrate* (14 km north of Lecce, South Italy), stands out as one of the few and most important evidences of monumental byzantine painting of the 12th–13th century in a unique framework of the architectural and sculptural Apulian Romanesque style [4]. In addition to the church (12th century), the complex consists of a carved portico (13th century), Baroque additions (16th–17th century) realized during the restoration work by the Roman curia after the Basilian monks left in the 14th century, and buildings (18th–20th century) dedicated to stalls, granaries and underground mills for cereal production and lamp oil-making developed after Turkish pirate raids [5–10]. These different phases led to several modifications of the wall paintings, which were updated in between the 14th and the 15th century and whitewashed when the complex was completely abandoned by the Church, becoming exclusively a farm.

From 1965 AD, the Province of Lecce opened the Museum of Arts and Folklore of Salento, also promoting the architectural restoration (up to the 1970s) of the complex by the architect Franco Minissi [10,11]. In the same period (1975 AD), the Central Institute for Restoration of Rome (ICR) restored the whitewashed frescoes and detached those of the 14th–15th century, discovering the original wall paintings (12th–13th century) in the underlying layer [10,11].

Finally, in 2012 AD, the abbey was entrusted to the FAI foundation for thirty years, becoming its first Apulian beneficiary, with the purpose of making it a cultural center.

1.2. The Wall Paintings of Santa Maria di Cerrate

Regarding the different phases of the frescoes, those in the apses are the oldest, stylistically dated at the end of the 12th century [3,4,12]. In the central apse, the *Ascension* is represented according to pure byzantine canons, without any contamination of the *Maiestas Domini* [3,4,12]. In fact, the whole arrangement of the apse shows a common imprint with specific references to Macedonia and in particular to Kurbinovo, Kastoria, Nerezi, Skopje and Thessalonica [3,4,12]. The images are framed by vegetal motifs and a decoration with white pseudo-Kufic letters on a blue background, as a reference to the Egyptian Coptic churches, built under the Fatimids [3,4,12]. In the lower part of the central apse, five bishop saints were painted, including Saint John Chrysostom.

The sequence of bishops continues in the right apse, under the representation of Saint John the Baptist, while the wall paintings of the left apse have been lost. Monks and deacons are represented in the arches with old repaints probably realized during the conservation treatments of the 17th century [3,4,12].

In the aisles, a series of monks, martyrs, bishops, hymnographer saints and knights were painted [3,4,12]. In particular, in the right aisle, the presence of the Virgin child between Saint Anne and Saint Joachim is recognizable, although part of the masonry collapsed. This event led to a casual reconstruction with the same stone bricks, without relocating them in the exact position according to the fresco representation.

On the counter-façade, the *koimesis* was realized together with holy figures.

All these frescoes have been considered contemporary with the realization of the ciborium which has a Greek inscription dated 1269 AD [3,4,12]. The only exception is the portion framed by the baroque altar of Saint Oronzo (right aisle), dated around the 14th century [3,4,12].

Regarding the detached frescoes located by ICR in the Museum of Arts and Folklore, both the *Death of the Virgin* (14th century), the *Annunciation*, the *Archangel Gabriel, Saint George rescuing the princess from the dragon* and the *Conversion of Saint Eustace* (15th century) are marked by a complex scenario with architectural structures and decorated clothes of the characters [12]. The authors are not known, but in the *Death of the Virgin*, a praying character defined by an inscription as "*The Pilgrim of Morciano*" was considered the commissioner of this wall painting [12].

2. Materials and Methods

2.1. Research Aims

In this study, the composition of the mortars inside the church and their binder/aggregates ratio were required by conservators to replenish the gaps at the level of renders and plaster layers using materials with suitable physicochemical properties. Therefore, some samples from the required parts were analyzed using scanning electron microscopy (SEM-EDX) and X-ray diffraction (XRD), also testing a new method of image elaboration of macro and micro images for statistical analysis of particles [13].

In fact, historic mortars are composite materials with different compositions, depending on their typology (i.e., areal, hydraulic or mixed mortar), the ratio between their components (sand, lime, water, organic and inorganic additives), the production processes (with different kinetics of the carbonation mechanism during setting, hardening and ageing) and the material sources, also having differences in the mortar technologies in relation to the historical period [14–16]. Consequently, in conservation, materials with compatible physicochemical characteristics are necessary to prevent internal disconnections and tensions given by a different material reaction to external stresses and to environmental factors [17,18].

Moreover, the presence of different quarries (located next to the towns of Monteroni and Surbo) and the mouth of the Idume (Frigole) river close to the area of the abbey were considered as possible sources of the raw materials. Therefore, some reference samples were taken to compare the results of the mortars.

While sampling in the abbey, it was also possible to take vegetal remains embedded into the mortars and the plasters coming from the gaps to date the frescos on the walls of the aisles with radiocarbon dating using accelerator mass spectrometry (AMS) [19–21], thus giving art historians new information about the development of the building.

2.2. Sample Selection

For this study, five specimens were taken from the abbey. The sampling areas have been set out in Figures 2 and 3. The sampling areas were chosen by conservators together with the Superintendence, in the most representative parts of the aisles. In fact, differently from the apses, these areas needed strong conservative interventions and more information for their reconstruction. The most extended gaps were selected to avoid invasive sampling and new possible losses.

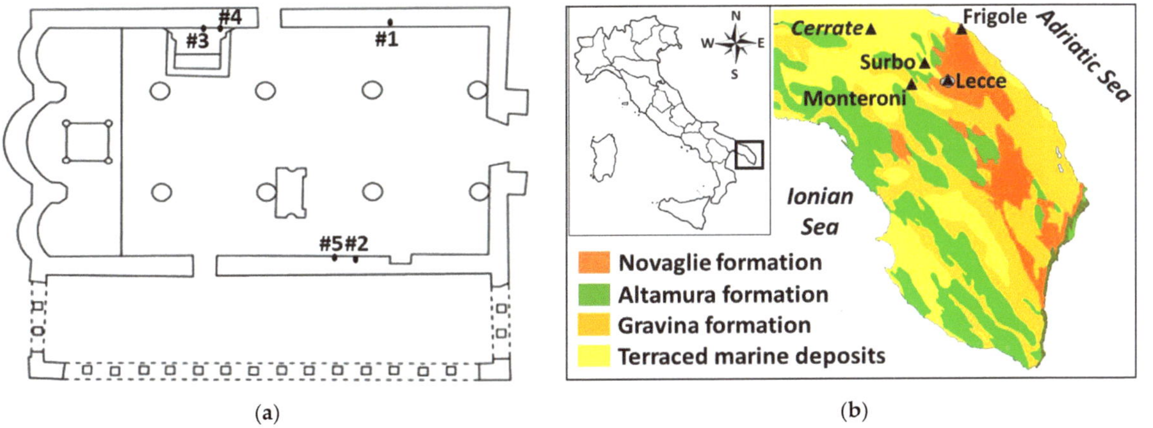

(a) (b)

Figure 2. Planimetry and measuring points in the abbey of *Santa Maria di Cerrate*: (**a**) Sampling area and geological characteristics of the area near Cerrate (**b**) showing the formations of Novaglie (bioconstructed calcarenites and brecce and limestones—Late Miocene), Altamura (laminated bioclastic micritic and calcarenitic limestone, with banks of dolomicritic parts with biostromal and rudist shells—Late Cretaceous), Gravina (biolimestone and biocalcirudites in intercalated banks with rare calcilutites—Lower Pleistocene) and Terrace marine deposits (Terraced Marine Deposits: sands, conglomerates, loam—Middle-Upper Pleistocene).

Figure 3. Sampling areas (**a**–**f**) inside the abbey of *Santa Maria di Cerrate*.

Three samples of mortars were taken from the aisles (Samples #1, #2 and #3), together with two samples of vegetal remains (Samples #4 and #5). Among the mortar samples, two of them were collected from the render layer and the last one from the plaster layer belonging to the frescoes. In particular, Sample #1 was collected as a compact material, while Samples #2 and #3 crushed during the sampling phase. Therefore, image elaboration of macro and micro images were repeated on Sample #1, completely crushing it to compare the method of particles' statistical analysis. In this second step, all material of the crushed Sample #1 was used, thus leading to a large area of analysis.

Regarding the vegetal remains, they were found coming out from the *arriccio* and render layers of the frescoes in the gaps.

Moreover, two rock samples were taken from the nearby quarries of Monteroni (Sample #6) and Surbo (Sample #7), while one sand sample (Sample #8) was collected in the coastline area of Frigole, from the mouth of the river Idume, just a few kilometers from the abbey. The quarries and the mouth were chosen because they were considered the most probable source of the employed raw materials due to their geographical proximity to the abbey. All the examined samples are listed in Table 1.

Table 1. Description of the analyzed samples.

Sample	Typology	Description
#1 #2 #3	Mortars	Render layer near the collapsed masonry of the right aisle Render layer of the left aisle Plaster layer next to the 14th-century fresco of the right aisle
#4 #5	Vegetal remains	Vegetal remains from the *arriccio* next to the Sample #3 Vegetal remains and one wooden fragment from the left aisle
#6 #7	Rocks	Quarry of Monteroni—limestone Quarry of Surbo—limestone
#8	Sand	Coastline area of Frigole (river Idume)—sandstone

2.3. Scanning Electron Microscopy (SEM-EDX)

All micro images were acquired by scanning electron microscope JEOL model JSM-5410 LV with an emission source of tungsten filament, a resolution of 3.5 nm, a magnification of 100–200,000× and an acceleration voltage of 0.5–30 kV. The opportunity to perform investigations in low vacuum allowed to avoid charge accumulation for samples with insulating properties without the need for a conductive coating. Analysis was processed using Link Isis software by Oxford Instruments and Image Pro-Plus by Media Cybernetics. For each sample, more than fifty images were taken, together with the characteristic X-ray spectra. The topographic distribution of the detected elements was acquired through elemental maps.

The digital processing of macro and micro images, taken respectively with the camera Nikon D80 and the scanning electron microscope, was carried out with ImageJ software, based on Sun-Java and developed by the National Institutes of Health [22]. The images, converted into 8-bit grayscale, were segmented into a new black-and-white figure using the threshold function, highlighting the components by assigning a defined threshold value to the pixel depending on their brightness. Using the analysis functions, images were processed and the components were converted into elliptical shapes. By associating the number of pixels with the dimensional reference value, it was possible to obtain the object dimensions. The area of the aggregates was thus estimated, identifying their volume ratio with the binder.

2.4. Structural Analysis

A diffractometer Rigaku model Miniflex was used for X-ray diffraction (XRD) analysis, employing Cu Kα radiation at 30 kV and 100 mA operating in step-scan mode from 10° to 80°, with a 0.020° step and a scan speed of 0.25 $^{\circ}\cdot\text{min}^{-1}$.

2.5. Radiocarbon Dating with AMS

After a careful selection carried out at the microscope Nikon SMZ 1500 in order to select the organic remains less affected by the degradation due to carbonation, the weight of the specimens was re-measured and a new lab code was assigned to each sample. In fact, the carbonation process involved in the hardening and setting phases of the frescoes consumes and degrades the vegetal materials, which could not survive the chemical treatments for radiocarbon dating.

Samples were treated by following the standard acid-alkali-acid procedure used at CEDAD [23]. For the first step, used to remove possible microscopic traces of carbonates and fulvic acids, 10 mL of 1% HCl solution were added. After reaching a pH value below 1, samples were washed with ultrapure deionized water until a pH higher than 5 was obtained. Humic acids were then removed with the second step, with 10 mL of 1% NaOH solution, followed by warm washings with ultrapure deionized water until reaching a pH below 9. The third step with 10 mL of 1% HCl solution was performed to remove the possible CO_2 dissolved during the previous treatment, until reaching a pH higher than 5. The extracted material was then combusted to carbon dioxide together with silver wool and copper oxide in quartz tubes flame-sealed under vacuum. The CO_2 was then cryogenically purified and transferred to the graphitization lines to be converted at 600 °C to graphite using hydrogen as the reducing agent and iron powder as the catalyst. The measurements of the radiocarbon content were carried out at the ^{14}C AMS beamline at CEDAD using a 3 MV HVEE Tandetron accelerator. The measured $^{14}C/^{12}C$ isotopic ratios were corrected for machine and processing background and isotopic fractionation by using the $\delta^{13}C$ term measured online with the AMS system. The measurement uncertainty was conservatively estimated as the larger between the ^{14}C counting statistic error and the data scattering of the ten measurements repeated for each sample. Saccharose C6 provided by the IAEA and oxalic acid II supplied by NIST (National Institute of Standards and Technology) were used as standard reference materials.

3. Results and Discussion

3.1. Granulometric Studies

All macro images and the images acquired through the SEM-EDX showed the presence of ground aggregates with elliptical and subspherical shapes immersed in a particularly fine particle size component. The presence of some micro cracks was probably related to the shrinkage phenomenon during the carbonation reaction in the setting and hardening phase of the mortar. Moreover, a few lime lumps were observed only in Sample #2, with the rare presence of small granules, thus denoting a generally remarkable making process, with a good setting, hardening and ageing [14,24]. Furthermore, in all samples, the presence of vegetal remains was observed, used as reinforcing materials in agreement with the typical byzantine building technique [24–26], while no traces of fossils were found, despite being expected due to the known characteristics of the local limestone and sand [27].

Along with the abovementioned common features, important granulometric differences were identified together with the chemical characterization of the mortar components.

The digital processing of macro photographs allowed the investigation of the dimensional characteristics of the macro aggregates, while images acquired by SEM-EDX were used to investigate the micro components. Through the data obtained for macro photography, it was possible to estimate the percentage of aggregate and its volume ratio with respect to the binder.

Considering that among the three samples just Sample #1 remained fully compact, two processes were followed to test an explorative strategy for the granulometric test of imaging analysis for historical mortars: Sample #2 and Sample #3 were completely arranged on two different glass slides with proper tape in order to have all the forming particles laid on the same plane, while Sample #1 was first analyzed as a block.

Then the analysis was repeated unravelling Sample #1 it on a glass slide to check both the ratio between aggregates and binders was compatible, considering the aggregates are casually located in the matrix and their disposition was statistically representative even applying imaging analysis.

The steps of the digital imaging process of elaboration using ImageJ, given in 2.3 scanning electron microscopy (SEM-EDX), are illustrated in Figure 4 with the application on Sample #1. In particular, since the aggregates had mainly a subspherical shape, their dimensions were calculated using the diameter of Feret [13], defined as the mean value of a series of measures of the distance between two parallel tangents to the perimeter of the projected area from the particle, assumed to be hypothetical spherical.

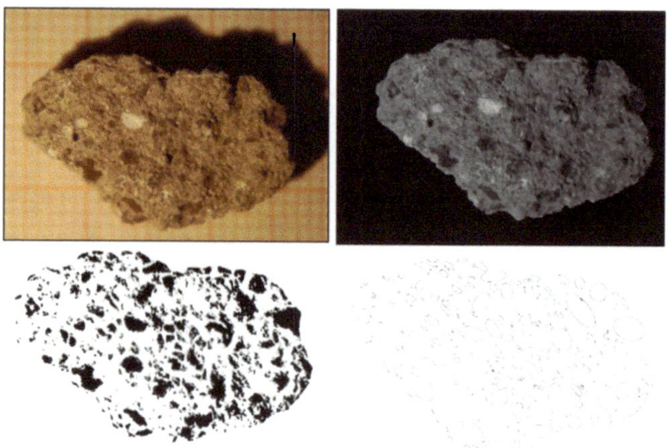

Figure 4. Digital elaboration of a macro photograph of Sample #1.

Throughout the macro images analysis, it was possible to estimate the areas covered by the aggregates and to verify the differences in the ratio between aggregates and binders (Table 2). In particular, the values found for Sample #1 (1:3) and Sample #2 (1:4) were in agreement with the historical byzantine practices [15,28–30], used both to reinforce the mortars and to slow down the carbonation process, thus giving more time to the artists to use the pigments for the fresco technique. Moreover, the considerable difference in the aggregates/binder ratio for Sample #3 (1:13) depended on the sampling realized on the outer layer within the plaster, which commonly have a low content of aggregates with respect to the internal layers of the mortars. The elaboration on the crushed Sample #1 gave comparable results, confirming the volume aggregates/binder ratio (1:3).

Table 2. Percentage of aggregates in the binder and the volume aggregates to binder ratio.

Sample	Total Area of the Sample (mm^2)	Total Area of the Aggregates (mm^2)	Aggregates (%)	Aggregates/Binder Ratio
#1 (compact)	235	80.94	34.4	1:3
#1 (crushed)	1967	640.7	32.6	1:3
#2	204	52.75	25.8	1:4
#3	198	15.02	7.6	1:13

Micro images using SEM-EDX were used to investigate the differences among their micro components. In particular, the aggregates were identified with help of the X-ray microanalysis. Then, the morphological information obtained from SE images was obtained by applying the previously described imaging method of statistical analysis for particles. The results were given according to the Wentworth scale. Regarding their granulometric distribution, as shown in Table 3, Sample #1 was characterized by the higher granulometric dispersion, with a considerable presence of clay particles (80.20%) and silt (19.70%) with the smallest amount of sand whose diameter ranges from medium to very fine grains (0.10% in total). Sample #2 was quite homogeneous, but coarser than the previous one, with the presence of the smallest components classifiable in the silt category (95.69%) and with a higher amount of sand (4.31% in total) than Sample #1 also including a few coarse grains (0.01 %). Sample #3 resulted to have a rather thin granulometry mostly consisting of silt components (98.00%) and a small amount of aggregate ranging from medium to very fine sand (2% in total).

Table 3. Percentage of micro components of aggregates in the binders.

Sample	Granulometry (μm)	Number of Particles Based on the Diameter of Feret (μm)	Total Percentage (%)
#1	Clay particle (<3.9)	936.67	80.20
	Silt (3.9–62.5)	230.13	19.70
	Very fine sand grain (62.5–125)	0.93	0.07
	Fine sand grain (125–250)	0.2	0.02
	Medium sand (250–500)	0.07	0.01
	Coarse sand (500–1000)	0	0
#2	Clay particle (<3.9)	0	0
	Silt (3.9–62.5)	1079.67	95.69
	Very fine sand grain (62.5–125)	37.6	3.33
	Fine sand grain (125–250)	9.3	0.83
	Medium sand (250–500)	1.53	0.14
	Coarse sand (500–1000)	0.13	0.01
#3	Clay particle (<3.9)	0	0
	Silt (3.9–62.5)	829.87	98.00
	Very fine sand grain (62.5–125)	10.37	1.23
	Fine sand grain (125–250)	6.25	0.74
	Medium sand (250–500)	0.25	0.03
	Coarse sand (500–1000)	0	0

Regarding the X-ray microanalysis, the maps showed the widespread aggregates of different granulometry consisting of silica immersed in a matrix with very high calcium content and a general, but not so relevant, presence of iron for Sample #1 and Sample #2. Sample #3 was characterized by higher content of calcium and a considerable low amount of aggregates and iron in accordance with its plaster function. The recorded X-ray spectra allowed the revelation of a very similar qualitative composition for all samples, which were further investigated realizing a relative percentage analysis of the main elements of the three samples. The analysis was performed with the ZAF correction for the matrix effects due to the atomic number of the elements, the phenomenon of absorption and the secondary fluorescence. The normalized concentrations of the major elements, conventionally expressed as %wt, are reported in Table 4.

Table 4. Normalized concentrations of the major elements obtained from X-ray microanalysis.

Elements	#1 (%wt)	#2 (%wt)	#3 (%wt)
Mg	2.6	2.9	1.8
Al	6.7	11.5	2.8
Si	21.8	38.5	12.1
S	0.4	0.4	3.7
Cl	1.2	<0.1	0.4
Ca	64.4	41.3	78.0
Fe	2.8	5.3	1.3

Hence, it was possible to determine a greater presence of silica, iron, sulfur, chlorine, potassium and sodium in Sample #2 than in Sample #1 and a meaningful lower measurement of the same elements in Sample #3, due to the use of a smaller amount of aggregate.

The high silica-based component for the aggregates of all samples was indicative of the excellent quality of mortars and a careful material selection, while the presence of aluminum and iron within the binder and not being related to the principal siliceous aggregates is a clue for the probable presence of clay, used as an additive. Therefore, despite appearing, at first, like aerial mortars due to the absence of crushed bricks or any kind of ceramic material and the abundance of calcium, the analyzed mortars could be ascribed to a typology having mixed aerial and hydraulic characteristics. In fact, hydraulic mortars can be distinguished for the development of silicates and aluminates coming from ceramics to form conglomerates or the use of marlstones, rich in carbonate minerals, clays and silt [14,31–36]. Moreover, in this case, the content of aluminum and iron was below the usual amount in the so-called hydraulic mortars. Moreover, the traces of sulfur and chlorine could be attributable respectively to gypsum and sodium chloride due to degradation for the sulfation process of the lime binder fraction and marine environment. Furthermore, the quality of the mortars regarding the silica and lime content was verified by means of structural analysis by using X-ray diffraction.

3.2. Structural Analysis

Through the X-ray diffractometric analysis, the mineral crystalline components of the mortars were investigated, highlighting the presence of quartz (SiO_2) and calcite ($CaCO_3$). A biphasic analysis of the crystalline components was carried out using the Rietveld method [37], allowing the detection of quartz and calcite for the three samples of mortars with some small impurities, ascribable to hematite (Fe_2O_3). The obtained percentages are reported in Table 5, which also shows the related percentages of silica and calcium resulting from microanalysis by using SEM-EDX. Regarding the render layer of the mortars, the results exhibit the careful selection of high-quality materials. In particular, the ratio between calcite and quartz in Sample #2 (0.90) corresponded with the one between Ca and Si (0.93), meaning nearly the totality of the silica and calcium amount is in the crystalline form. The different ratio between the silica and calcium content (2.9) and one of their mineral components (3.69) in Sample #1 is due to a lower quantity of quartz with respect to the %wt silica content, so as for Sample #3, which has completely different ratios (6.44 for the crystalline content against 1.63 for the total amount of calcium and silica). The undetected presence of gypsum ($CaSO_4 \cdot 2(H_2O)$) and halite (NaCl) confirmed well-manufactured mortars made good preservation possible so that the small contaminants of sulfur and chlorine can be related to the ion absorption from the soil.

Table 5. Crystalline components in the mortars related to the percentages of silica and calcium.

Sample	Quartz (%wt)	Calcite (%wt)	Si (%wt)	Ca (%wt)
#1	21.3	78.7	21.8	64.4
#2	52.4	47.6	38.5	41.3
#3	38.0	62.0	12.1	78.0

3.3. Comparison with Local Materials

Different kinds of limestones from nearby quarries (Monteroni and Surbo) were analyzed by using XRD to verify a possible extraction from local sources. As shown in Table 6, Sample #6 and Sample #7, coming respectively from Monteroni and Surbo, had a similar percentage of calcite (67.4% for Sample #6 and 65.9% for Sample #7), but a different content of quartz (4.8% for Sample #6 and 1.1% for Sample #7). Moreover, Sample #6 had a considerable amount of periclase (MgO) (27.8%), while Sample #7 was characterized by wollastonite ($CaSiO_3$) (33%). Sample #8, taken from the Idume mouth (Frigole), consisted of a considerable quantity of seashell fragments, as proved by the presence of aragonite (41.7%), together with a small percentage of calcite (25.3%) and quartz (33%). Regarding the limestone source, the small content of magnesium revealed in the samples by means of SEM-EDX excluded the provenience from Monteroni. Similarly, the absence of wollastonite from the mortar samples did not make a correlation with Surbo possible. In fact, quicklime is produced by baking the raw materials based on calcium carbonate at a temperature of 850–900 °C, which is not sufficient for the degradation of wollastonite, considering it is stable up to 1150 °C [38]. Finally, the characteristic calcareous sand of Frigole was not compatible with the aggregates used in the mortars.

Table 6. Crystalline components in the Samples #6, #7 and #8.

Sample	Crystalline Components (%)
#6	Calcite (67.4%), Periclase (27.8%), Quartz (4.8%)
#7	Calcite (65.9%), Wollastonite (33%), Quartz (1.1%)
#8	Aragonite (41.7%), Quartz (33%), Calcite (25.3%)

Therefore, raw materials from distant sources were used for the lime, the aggregates and the presumed clay content. The effort of importing different materials, increasing the manufacturing costs, highlighted the important role of the abbey for which specific material selection was realized. In fact, the import of expensive materials has already been documented for this abbey by the extended use of the blue of lapis lazuli in the background of the wall paintings [1].

3.4. Radiocarbon Dating

Straw remains and vegetal fibers inside the mortars were exploited to determine the absolute dating of the frescoes. In fact, the dating of the fibers does not allow a direct chronology of the wall painting, as it does not refer to the plaster itself but the underlying layers. Anyway, it gives a useful timing threshold reference for the realization of the frescoes. While selecting through the stereomicroscope the fibers not strongly affected by the carbonation process of the mortars and the plasters, it was possible to identify among the fibers some glumes and culms belonging to the *genus triticum* of the family of graminaceous plants, compatible with the rural production that the area has had over the centuries.

For the two samples, conventional radiocarbon ages expressed in years in BP years were calibrated to calendar age using the Oxcal 4.4 software and the INTCAL20 curve based on atmospheric data (Figure 5).

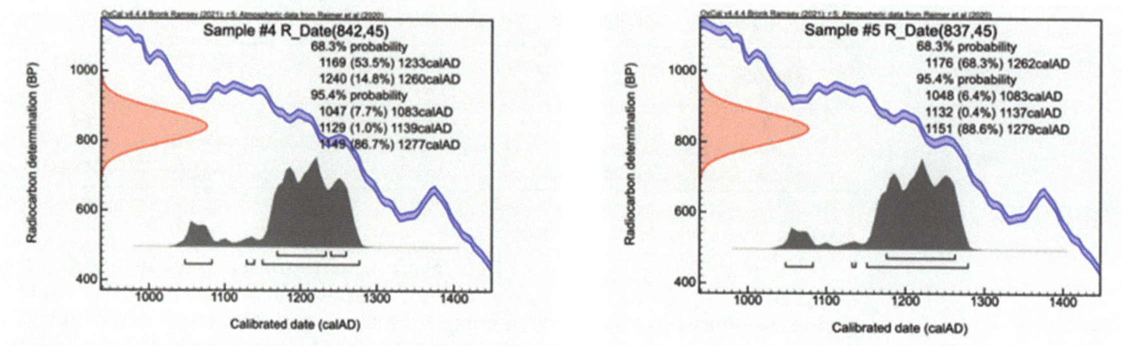

(a) (b)

Figure 5. Radiocarbon dating of Sample #4 (**a**) and of Sample #5 (**b**).

The Sample #4 and Sample #5 gave overlapping ages and they are dated between the end of the 12th century and the first half of the 13th century, scientifically confirming the stylistic dating previously carried out through careful historical-artistic research [3,4,12]. In Sample #4, although the fibers were taken next to the portion of the 14th-century frescoes, their dating is almost identical to the older frescoes on the left aisle. Moreover, this result shows that the portion of the wall taken into consideration was not affected by the collapse of the right masonry and that the 14th-century frescoes were realized starting from the plaster layer, using the previous *arriccio*.

4. Conclusions

Interdisciplinary analyses were performed to support the conservation and historical-artistic studies in the abbey of *Santa Maria di Cerrate* by means of characterization of materials for the cleaning, consolidation and integration actions. In particular, the three mortar samples, taken from significantly different points, had the purpose of suggesting the granulometry, and the most suitable materials fill the gaps in the masonry walls. Moreover, digital imaging processing allowed the identification of the aggregates-binder ratio from macro images and the presence of clay from micro images, expressing the perspective of an easy and non-destructive method.

SEM-EDX and XRD analyses applied to all the mortar samples showed a low percentage of impurities and a consistent siliceous component, denoting strong accuracy in the selection of the sand used for several building phases, while the small clay content highlighted the use of mixed aerial-hydraulic lime. The results were compared with local raw materials coming from the nearby quarries of Monteroni and Surbo and the sand of the Idume river mouth from Frigole, finding different and non-compatible characteristics.

Some vegetal remains, identified as a common byzantine practice, were sampled and dated, with radiocarbon dating by particle accelerator with AMS, between the end of the 12th century and the first half of the 13th century, also showing that one sample from the right aisle was not affected by the collapse of part of the same masonry. Therefore, as found in our previous studies regarding the pigments of the frescoes, mortars also highlighted the valuable role of the abbey through the careful selection of far and, consequently, expensive raw materials.

Author Contributions: Conceptualization, G.V., A.S., G.B., D.M., L.C., G.Q. and A.B.; methodology, G.V., A.S., G.B., D.M., L.C., G.Q. and A.B.; validation, G.V., A.S., G.B., D.M., L.C., G.Q. and A.B.; formal analysis, G.V., A.S., G.B., D.M., L.C., G.Q. and A.B.; investigation, G.V., A.S., G.B., D.M. and A.B.; resources, G.V., A.S., G.B., D.M., L.C., G.Q. and A.B.; data curation, G.V., A.S., G.B., D.M., L.C., G.Q. and A.B.; writing—original draft preparation, G.V., A.S., G.B., D.M., L.C., G.Q. and A.B.; writing—review and editing, G.V., A.S., G.B., D.M., L.C., G.Q. and A.B. All authors have read and agreed to the published version of the manuscript.

Funding: This research received no external funding.

Data Availability Statement: Data sharing not applicable.

Acknowledgments: The authors thank the FAI foundation, conservators and the Superintendence of Cultural Heritage for their collaboration.

Conflicts of Interest: The authors declare no conflict of interest.

References

1. Vasco, G.; Serra, A.; Manno, D.; Buccolieri, G.; Calcagnile, L.; Buccolieri, A. Investigations of Byzantine Wall Paintings in the Abbey of Santa Maria Di Cerrate (Italy) in View of Their Restoration. *Spectrochim. Acta Part A Mol. Biomol. Spectrosc.* **2020**, *239*, 118557. [CrossRef] [PubMed]
2. Conte, A.M.; Corda, L.; Esposito, D.; Giorgi, E. Characterization of Mortars from the Medieval Abbey of Cerrate (Southern Italy). *J. Archaeol. Sci. Rep.* **2017**, *12*, 463–479. [CrossRef]
3. Falla Castelfranchi, M. La Pittura Bizantina in Puglia (Secoli IX–XIV). In *Percorsi Bizantini nel Salento*; De Gruyter: Berlin, Germany, 2008; pp. 25–36.
4. Falla Castelfranchi, M. *Pittura Monumentale Bizantina in Puglia*, 1st ed.; Electa: Milano, Italy, 1991; ISBN 9788843537051.
5. Trono, A.; Fiore, N. (Eds.) *Travelling to the East on the Via Francigena: Through Art, Cultures and History*; Edizioni Esperidi: San Cesario di Lecce, Italy, 2012; ISBN 9788897895008.
6. Houben, H.; Vetere, B. (Eds.) *Pellegrinaggio e "Kulturtransfer" nel Medioevo Europeo: Atti del 1. Seminario di Studio dei Dottorati di Ricerca di Ambito Medievistico delle Università di Lecce e di Erlangen, Lecce, 2–3 Maggio 2003*; Congedo: Galatina, Italy, 2006; ISBN 9788880866671.
7. De Giorgi, C.T. *La Chiesa di S. Maria di Cerrate in Territorio di Lecce: Note Storiche Ed Archeologiche*; Tip. Lit. Luigi Lazzeretti e Figli: Lecce, Italy, 1889.
8. Barletta, R. *Santa Maria di Cerrate: La Storia Nascosta*; Edizioni del Grifo: Lecce, Italy, 2003.
9. Delli Ponti, G. *Cerrate e Il Suo Museo*; Amministrazione Provinciale di Lecce: Lecce, Italy, 1976.
10. Castromediano, S. *La Chiesa di S. Maria di Cerrate Nel Contado di Lecce: Ricerche*; Forgotten Books: London, UK, 2018; ISBN 9780483143425.
11. Pellegrino, T.; Luca, V.D. *Santa Maria a Cerrate: Antica Abbadia*; Capone Editore: Lecce, Italy, 2004; ISBN 9788883490934.
12. De Giorgi, M. *Il Transito della Vergine: Testi e Immagini dall'Oriente al Mezzogiorno Medievale*; Fondazione CISAM: Spoleto, Italy, 2016.
13. Miriello, D.; Crisci, G.M. Image Analysis and Flatbed Scanners. A Visual Procedure in Order to Study the Macro-Porosity of the Archaeological and Historical Mortars. *J. Cult. Herit.* **2006**, *7*, 186–192. [CrossRef]
14. Moropoulou, A.; Bakolas, A.; Bisbikou, K. Investigation of the Technology of Historic Mortars. *J. Cult. Herit.* **2000**, *1*, 45–58. [CrossRef]
15. Moropoulou, A.; Bakolas, A.; Bisbikou, K. Physico-Chemical Adhesion and Cohesion Bonds in Joint Mortars Imparting Durability to the Historic Structures. *Constr. Build. Mater.* **2000**, *14*, 35–46. [CrossRef]
16. Rampazzi, L.; Colombini, M.P.; Conti, C.; Corti, C.; Lluveras-Tenorio, A.; Sansonetti, A.; Zanaboni, M. Technology of Medieval Mortars: An Investigation into the Use of Organic Additives. *Archaeometry* **2016**, *58*, 115–130. [CrossRef]
17. Akcay, C.; Sayin, B.; Yildizlar, B. The Conservation and Repair of Historical Masonry Ruins Based on Laboratory Analyses. *Constr. Build. Mater.* **2017**, *132*, 383–394. [CrossRef]
18. Miriello, D.; Lezzerini, M.; Chiaravalloti, F.; Bloise, A.; Apollaro, C.; Crisci, G.M. Replicating the Chemical Composition of the Binder for Restoration of Historic Mortars as an Optimization Problem. *Comput. Concr.* **2013**, *12*, 553–563. [CrossRef]
19. Sanjurjo-Sánchez, J. An Overview of the Use of Absolute Dating Techniques in Ancient Construction Materials. *Geosciences* **2016**, *6*, 22. [CrossRef]
20. Hajdas, I.; Ascough, P.; Garnett, M.H.; Fallon, S.J.; Pearson, C.L.; Quarta, G.; Spalding, K.L.; Yamaguchi, H.; Yoneda, M. Radiocarbon Dating. *Nat. Rev. Methods Prim.* **2021**, *1*, 62. [CrossRef]
21. Sanjurjo-Sánchez, J. Dating Historical Buildings: An Update on the Possibilities of Absolute Dating Methods. *Int. J. Archit. Herit.* **2016**, *10*, 620–635. [CrossRef]
22. Abramoff, M.D.; Magalhães, P.J.; Ram, S.J. Image Processing with ImageJ. *Biophotonics Int.* **2004**, *11*, 36–42.
23. D'Elia, M.; Calcagnile, L.; Quarta, G.; Sanapo, C.; Laudisa, M.; Toma, U.; Rizzo, A. Sample Preparation and Blank Values at the AMS Radiocarbon Facility of the University of Lecce. *Nucl. Instrum. Methods Phys. Res. Sect. B Beam Interact. Mater. At.* **2004**, *223–224*, 278–283. [CrossRef]

24. Taglieri, G.; Rigaglia, D.; Arrizza, L.; Daniele, V.; Macera, L.; Rosatelli, G.; Romè, V.; Musolino, G. Microanalytical Investigations on a Byzantine Fresco of the Dormitio Virginis from Sicily. *J. Cult. Herit.* **2019**, *40*, 155–162. [CrossRef]
25. Miriello, D.; Antonelli, F.; Apollaro, C.; Bloise, A.; Bruno, N.; Catalano, M.; Columbu, S.; Crisci, G.M.; De Luca, R.; Lezzerini, M.; et al. A Petro-Chemical Study of Ancient Mortars from the Archaeological Site of Kyme (Turkey). *Period. Mineral.* **2015**, *84*, 497–517. [CrossRef]
26. Singh, M.; Vinodh Kumar, S.; Waghmare, S.A. Characterization of 6–11th Century A.D Decorative Lime Plasters of Rock Cut Caves of Ellora. *Constr. Build. Mater.* **2015**, *98*, 156–170. [CrossRef]
27. Sansò, P.; Margiotta, S.; Mastronuzzi, G.; Vitale, A. The Geological Heritage of Salento Leccese Area (Apulia, Southern Italy). *Geoheritage* **2015**, *7*, 85–101. [CrossRef]
28. Bakolas, A.; Biscontin, G.; Moropoulou, A.; Zendri, E. Characterization of Structural Byzantine Mortars by Thermogravimetric Analysis. *Thermochim. Acta* **1998**, *321*, 151–160. [CrossRef]
29. Moropoulou, A.; Labropoulos, K.; Moundoulas, P.; Bakolas, A. The Contribution of Historic Mortars on the Earthquake Resistance of Byzantine Monuments. In *Measuring, Monitoring and Modeling Concrete Properties*; Konsta-Gdoutos, M.S., Ed.; Springer: Dordrecht, The Netherlands, 2006; pp. 643–652.
30. Moropoulou, A.; Cakmak, A.S.; Lohvyn, N. Earthquake Resistant Construction Techniques and Materials on Byzantine Monuments in Kiev. *Soil Dyn. Earthq. Eng.* **2000**, *19*, 603–615. [CrossRef]
31. Pintér, F.; Weber, J.; Bajnóczi, B.; Tóth, T. Brick-Lime Mortars and Plasters of a Sixteenth Century Ottoman Bath from Budapest, Hungary. In Proceedings of the 37th International Symposium on Archaeometry, Siena, Italy, 13–16 May 2008; Turbanti-Memmi, I., Ed.; Springer: Berlin/Heidelberg, Germany, 2008. [CrossRef]
32. Pintér, F.; Vidovszky, I.; Weber, J.; Bayer, K. Mineralogical and microstructural characteristics of historic Roman cement renders from Budapest, Hungary. *J. Cult. Herit.* **2014**, *15*, 219–226. [CrossRef]
33. Franquelo, M.; Robador, M.; Ramirez-Valle, V.; Durán, A.; Jimenéz de Haro, M.; Pérez-Rodruigez, J. Roman ceramics of hydraulic mortars used to build the Mithraeum house of Mérida (Spain). *J. Therm. Anal. Calorim.* **2008**, *92*, 331–335. [CrossRef]
34. Keppert, K.; Kobera, L.; Scheinherrová, L.; Doleželová, M.; Brus, J.; Černý, R. Kinetics of pozzolanic reaction and carbonation in ceramic-lime system: Thermogravimetry and solid-state NMR spectroscopy study. *J. Build. Eng.* **2020**, *32*, 101729. [CrossRef]
35. Arizzi, A.; Cultrone, G. Mortars and plasters—How to characterse hydraulic mortars. *Archaeol. Anthropol. Sci.* **2021**, *13*, 144. [CrossRef]
36. Michalska, D.; Czernik, J. Carbonates in leaching reactions in context of 14C dating. *Nucl. Instrum. Methods Phys. Res. Sect. B Beam Interact. Mater. At.* **2015**, *361*, 431–439. [CrossRef]
37. Rietveld, H.M. The Rietveld Method. *Phys. Scr.* **2014**, *89*, 098002. [CrossRef]
38. Sokolova, T.S.; Dorogokupets, P.I. Equations of State of Ca-Silicates and Phase Diagram of the $CaSiO_3$ System under Upper Mantle Conditions. *Minerals* **2021**, *11*, 322. [CrossRef]

heritage

MDPI

Article

The Comparative Analysis for the New Approach to Three Tourism-Oriented Heritage Districts in the United Arab Emirates

Jihad Awad [1], Mohammad Arar [1], Chuloh Jung [1,*] and Bouzid Boudiaf [2]

1 Department of Architecture, College of Architecture, Art and Design, Ajman University, Ajman P.O. Box 346, United Arab Emirates
2 Independent Researcher, Ajman P.O. Box 346, United Arab Emirates
* Correspondence: c.jung@ajman.ac.ae

Abstract: This study investigates the tourism-oriented heritage districts in the United Arab Emirates (UAE) and explores a new approach to fostering heritage tourism. This is envisaged through conducting a post-occupancy evaluation (POE) to detect the characteristics of the tourism-oriented heritage districts from the residents and tourists. The systematic evaluation of opinions is based on indicators developed by UNESCO and ICOMOS international principles. According to the survey results, it is found that Dubai's approach was successful at reuniting people with their history via modern materials in the restoration works. This has preserved the heritage value and increased their culture and national identity. The social goal to reconnect residents to their cultural history has been realized despite the inconsistencies compared to the specification of the World Heritage Convention, which requires technical authenticity and integrity. People perceived the identities of tourism-oriented heritage districts in Dubai, Sharjah, and Ajman differently. Based on the POE survey, the Shindagha area and Al Fahidi Historical Area in Dubai were perceived as traditional destinations for tourist attractions. The Heart of Sharjah area was perceived as cultural and educational tourism, and the Ajman Heritage District area was perceived as community tourism.

Keywords: heritage district; tourism; conservation; post-occupancy evaluation (POE); United Arab Emirates (UAE)

Citation: Awad, J.; Arar, M.; Jung, C.; Boudiaf, B. The Comparative Analysis for the New Approach to Three Tourism-Oriented Heritage Districts in the United Arab Emirates. *Heritage* **2022**, *5*, 2464–2487. https://doi.org/10.3390/heritage5030128

Academic Editors: Daniela Fico and Daniela Rizzo

Received: 4 July 2022
Accepted: 9 August 2022
Published: 1 September 2022

Publisher's Note: MDPI stays neutral with regard to jurisdictional claims in published maps and institutional affiliations.

Copyright: © 2022 by the authors. Licensee MDPI, Basel, Switzerland. This article is an open access article distributed under the terms and conditions of the Creative Commons Attribution (CC BY) license (https://creativecommons.org/licenses/by/4.0/).

1. Introduction

Many European cities symbolize architectural heritage and aesthetics, such as Venice, Rome, and Paris [1,2]. Throughout history, these cities have succeeded in maintaining their value represented in the harmonious integration of historic districts with the contemporary urban context [3]. However, before the Industrial Revolution, these cities did not recognize the architectural heritage as an object for preservation. It became common to demolish the monuments of ancient Rome and build Gothic or Renaissance cathedrals from the ruins of ancient Rome [4]. There was not a solid act of preservation until the modern era, but the existing architectural heritage has suffered as much social fluctuation today [5,6]. From the end of the 18th century to the beginning of the 19th century, various European countries tried to confirm their identity through the historical remains [7,8]. Since then, historic buildings have become significant, admired, and protected by nations and governments [9,10]. During the 20th century, national and international institutions were established to protect the architectural heritage from the threat of destruction [11]. Likewise, historical districts were highly considered for preservation when their heritage value was threatened by negligence, misuse, and destruction [12].

In the case of the United States, there has been less attachment to history as the Puritan ancestors rejected the European colonial traditions and had their independence from Britain in 1776 [13,14]. However, since the country has a comparatively short history, it has shown great interest in its ancestors' achievements and historical relics [15]. At the Second United Nations Conference on Human Settlements (HABITAT II) in Istanbul in 1996, the United Nations Educational, Scientific and Cultural Organization (UNESCO) organized several international research projects on historical districts worldwide [16]. In 2008, at UN-HABITAT in Seville, UNESCO's Social and Human Sciences Sector identified significant indicators and parameters to emphasize social capital and cohesion in historic districts [17].

The United Arab Emirates was founded as a federation in 1971, including seven united Emirates [18]. Due to the unprecedented urbanization processes and rapid economic growth, most of the significant architectural legacy in some historic districts along the coast was swept away [19,20]. After the disappearance of most of the UAE's fragile urban heritage, the UAE began to search for its past and cultural heritage. During the late 1970s and early 1980s, all governments of the seven Emirates started to protect and develop the irretrievable historical districts towards achieving the shared national goal of boosting heritage tourism [21]. At least one historical district is present in each Emirate, usually a complex of original heritage buildings and a replica of traditional houses, markets, and public spaces [22].

The general strategy for the historic district in the UAE was an adaptive reuse of the restored structures from the start. Several municipalities believed that a harmonious mixture of different uses such as heritage, tourism, culture, and residential could be successful for urban revitalization [23]. As a result, this approach attracts many visitors and revitalizes the historic districts. The restoration of the building's original appearance via local building materials and basic construction techniques mixes adobe and coral stones. There is, however, little evidence to suggest that the proper materials and mixing methods are currently used. No studies have examined these issues in more detail to date.

Even though the heritage districts require the relevant consumers for survival, a balance should be maintained among cultural, social, and economic purposes. Without various support from the visitors, it could be neglected again.

Currently, there is no legislation guiding the conservation and protection of the architectural and urban heritage in the UAE. Some feedback was obtained from visiting experts, but no reference was made to international charters, which could have been a solid foundation for further actions. Moreover, foreign experts contributed to confusion among fast-growing but less experienced Emirates professionals regarding the best restoration methodologies.

This paper aims to examine the historic districts in the UAE and explore a new approach to fostering heritage tourism and reviving the culture and community. Additionally, the paper aims to highlight the significance of boosting heritage tourism in the UAE to support the plan of the government to increase the percentage of tourism from AED 68.5 billion (5.2%) to AED 116.1 billion (5.4%) by 2027 [24].

2. Materials and Methods

Initiatives for historic preservation took place in the UAE as part of the sustainable tourism plans to boost the hospitality sector [25]. These initiatives were considered after the success of the Al Bastakiya area, which includes the oldest building in Dubai that dates back to 1780, the Al Fahidi Fort, the Dubai Museum, and the Sheikh Mohammed Centre for Cultural Understanding [26,27]. The development of historic districts requires careful decision making since the resources are irreplaceable and affect the cultural and physical aspects of the future [28]. In 1999, during the 12th General Assembly of ICOMOS (The World Congress of the Conservation of Cultural Heritage), it was discussed to extend the definition of historical conservation to include cultural and physical properties [29]. Additionally, it was emphasized that cultural tourism and historic cities should be given careful attention to implement sustainable development since landmark resources are irreplaceable and belong to the nations [30]. Even though the rapid urbanization of Dubai and other Emirates has challenged the traditional city fabric, the government managed to harmonize the historical urban fabric with the contemporary city through a series of iconic building projects [31].

The city is a cultural phenomenon in its own right and tells its past and transfers its memory. Historic areas are the physical manifestation of memory that makes places significant. These areas contain visions of previous generations alongside urban and cultural life in a state of constant change, and urban development must be combined with the protection of cultural heritage. Preserving cultural and natural heritage, bringing it within reach of all, making cultures and civilizations better known, improving daily living conditions, and reducing poverty give meaning to tourism development's sustainability.

Historic areas in the UAE have been through metamorphoses [32]. This transformation usually involves updating the infrastructure and accommodating the historic buildings in commercial activities. Near them, the insertion of contemporary architecture is entirely different in terms of typology of buildings and materials of construction. This process has been noticed in most of the Gulf cities and particularly in Al Bastakiya (Dubai) and Al Kasbah (Sharjah), and there is nothing wrong with that; problems begin when this global change harms historic areas. Through some observations and surveys on three cases (Dubai, Sharjah, and Ajman), we will examine questions dealing with the perception and behaviors of the tourists towards the introduction of some commercial activities, signage control, and their repercussions on the preservation of historical heritage in terms of the built environment, and user preference and satisfaction with historic areas. The objective is to inform those factors that need to be considered in tourism development as a source for revitalizing historical sites. The historical regions use the same strategy and preservation objective for the selected cities (Dubai, Sharjah, and Ajman). Still, their implementations vary in many ways: tourist functions and a range of services as well as historical values and elements of the Conservatory Program. With attention to this differentiation, it seems possible to divide the historical areas into the following three groups:

- The first group consists of areas where the preserved buildings present an excellent technical state, considered invariable elements of the urban fabric and important touristic centers, used for nostalgic/prestigious and public programs, with rich cultural–educative programs (museums, exhibitions, souqs). They constitute a magnet for tourist attraction, contributing to the cultural life of the tourist and the local community (the case of Fahidi District and Shindaghah in Dubai [33], the souqs: Shinasiya, Saqr, and Mohammed along the Khor in Sharjah) [34].
- Traditional buildings included in the second group are those whose technical state or location does not permit the development of so many functions. Still, with the initiative of the local authorities, it is possible to organize occasionally varied open-air events: concerts, theatrical and traditional parade performances, knights' tournaments and feasts, and arrangements of the "light and sound" type. These programs are attracting both the local community and tourists; we may say that these buildings are acting as crucial centers of cultural life (the case of a Kasbah in Sharjah) [35].
- The third group (fortunately comparatively small) is constituted by some buildings in a poor technical state, and most are located in small touristic itinerary areas. Compared to the other facilities, their owners or managers might look passive in that their attitude towards the use of buildings is restrained to keep them open for sightseers, but they do not make an effort to overcome the stereotype programs and to create a valuable touristic attraction. It seems that in some of the historic buildings, particularly in the case of Souq Salah in Ajman, there is a chance to create an exciting and vital center influencing the cultural life of the historic area. In addition, there is an opportunity to implement some educational programs on the subject of tradition and history of the city (focusing on the local identity). The liveliness of the cultural life and the proposal of engaging activities would simultaneously promote tourism development, an opportunity to develop sustainable, ecological, and cultural tourism.

The methodology adopted for assessing the selected sites was based on several questionnaire surveys composed of four main sections: personal information; city history/culture; heritage resources; and the environment. These sections emphasize perceptual and social attributes. This paper explores the significance of the selected historical areas as heritage areas for cultural tourism and their impacts on revitalizing the historical sites. A systematic multi-level assessment method determines the tangible and intangible elements that highlight the area's characteristics.

In the case of the UAE and mainly the selected topics, the traditional buildings and historical and archeological sites as part of heritage are listed by the municipality or its representative (Department of the Historical Buildings and Archeology for the case of Dubai, the Museum of Art & Traditions for Ajman); we have to highlight that the Architectural Heritage Society is very dynamic and involved in all the actions taken at the level of the municipality or the property. The degrees, types, and priorities of intervention are identified by the Heritage Architectural Association in collaboration with the representative of the city. The Emiratis' traditional buildings, by their picturesqueness, secular history, and connection with the country's events and heroes, create an exciting magnet for tourism development [36]. Their values and preservation lead to the development of cultural tourism as a sustainable form of tourism [37]. According to Arch. Rashad Bukhash (Chairman of the Architectural Heritage Society and Chairman of ICOM-UAE), the strategy of the preservation of the historical area is based on the following:

- stop all demolitions of the properties (buildings) in the area delimited as historical or archeological sites;
- diversity of the users to encourage the involvement and participation of the community in the process of the preservation of the tangible and intangible objects reflecting the character and the identity of the area;
- one of the ways to avoid the proliferation of incompatible typologies will be achieved through the revitalization of the historical area;
- development of heritage tourism as one of the main activities for the revitalization of the area and as a source of income for the population and at the same time as an opportunity to sustain the know-how of the traditional art craft;
- promoting the awareness of the importance of preserving the urban heritage, mainly for the young people, in a way to let them be proud of what their predecessors had done and to highlight it as a part of the identity;
- use social media as a means for the promotion of the information and the awareness of the general public;
- the buildings of great value (architectural or symbolic) are to be listed as a heritage properties and open to the public.

2.1. *Tourism-Oriented Heritage Districts in Dubai*

Dubai Historic District is an ongoing project to preserve the UAE's traditional culture and heritage. Additionally, 60 other separate projects in a 1.5 km^2 area are being developed, serving the same goal [38]. This includes the Shindagha area, Al Fahidi Historical Area, and Saruq Al-Hadid Museum area, consisting of museums, exhibition centers, trading boats, restored trade stores, forts, restaurants, and performance centers [39] (Figure 1).

Shindagha Heritage District for tourism, known as Al Shindagha, is located on the shores of Dubai Creek and bordered by Bur Dubai in the south and Port Rashid in the west [40]. It was fully renovated to show the national heritage that is rich and significant. Dubai Municipality plans to qualify its historic districts for UNESCO World Heritage [41]. A museographic project is being used to implement restoration, rehabilitation, and creation of buildings to display the historical and archeological details of Dubai's traditional architecture and history [42]. The former ruler of Dubai, Sheikh Saeed Al Maktoum, lived in Al Shindagha from 1912 to 1958, and his house was restored as a public museum (Figure 2). Next to the house of Sheikh Saeed Al Maktoum, another 150 traditional houses were restored. Four were converted into museums, Craft Weaving and Textile House, Traditional Healthcare House, Jewelry House, and Traditional Fashion House [43]. Dubai Municipality will restore more historic houses and 23 secondary museums to create the largest open-air museum in the world that accommodates 12 million annual visitors [44]. Shindagha Heritage District is said to be well-crafted storytelling of the inhabitants with its buildings converted into museums [45]. Visitors navigate Shindagha Heritage District with narrow streets, "sikkah", that lead to the six-house complex, which displays the old sailors' sea navigation techniques, instruments, and skills (Figure 3). The exhibition areas show the various living environments in the UAE, such as living by the sea, desert, and oasis, which are all linked with a courtyard garden with vernacular plants, educational centers, cafés, and restaurants [46].

Figure 1. Map of Shindagha Heritage District.

Figure 2. The house of Sheikh Saeed Al Maktoum in Shindagha Heritage District (photo by authors).

Al Fahidi Historic District, also known as Al Bastakiya, is one of Dubai's oldest heritage districts for tourism in Bur Dubai along the Dubai Creek [47] (Figure 4). It holds the memory of pearl trading routes, connecting Persia, Rome, Portugal, Africa, Britain, India, and China with the Silk Road [48]. Historical markets for pearl and textile trading are preserved in Al Bastakiya, as it still has the old urban fabric with narrow traditional winding alleyways (sikkah) and wind catchers (barjeels) [49] (Figure 5). Al Bastakiya is nominated to be the UNESCO World Heritage Site in Dubai Municipality.

Figure 3. Historical Pictures exhibition in Sheikh Saeed Al Maktoum House (photo by authors).

Figure 4. Map of Al Fahidi Heritage District.

Al Bastakiya was initially built with sixty houses in the 1690s by wealthy Persian immigrants from the Bastak who escaped religious suppression. Most of the original inhabitants had emigrated to the new part of Dubai called Al Satwa after discovering oil in 1969 [50]. Unfortunately, more than 50% of Al Bastakiya was demolished in the 1970s due to new office complex development. At the same time, old houses were used as storage and labor camps for foreign workers [51]. Dubai Municipality had intended to deconstruct the remaining parts of Al Bastakiya in 1989 [52]. A British architect, Rayner Otter, had initiated the preservation campaign and written a letter to Prince Charles in the UK to visit the UAE. After his arrival, he explored the Al Bastakiya Heritage District for tourism with Rayner Otter. He proposed that this area be preserved, and Dubai Municipality canceled the demolition plan, and the restoration project for Al Fahidi Historical District started in 2005 [53].

Al Fahidi Historic District is composed mainly of 50 houses converted into museums, galleries, and cafés such as Sheikh Mohammed Centre for Cultural Understanding, Alserkal Cultural Foundation, Mawaheb Gallery, Majilis Gallery, Arabian Tea House Café, XVA Gallery, Coffee Museum, and Make Art Café [54]. Sheikh Mohammed Centre for Cultural Understanding (SMCCU) has a program to promote heritage awareness so visitors can learn more about UAE's heritage and tradition. Additionally, local homemade meals are offered to the visitors to experience the Bedouin hospitality (Figure 6).

Figure 5. Al Fahidi Heritage District (Al Bastakiya) in Dubai (photo by authors).

Figure 6. Courtyard of Al Bstakiya (photo by authors).

2.2. *Tourism-Oriented Heritage District in Sharjah*

The Heart of Sharjah is the largest heritage project in the UAE and the Gulf region, initiated to restore and revitalize the traditional souq area of Sharjah. The project covers 35,000 m^2 and is planned to be completed in 2025 in five phases [55] (Figure 7). Sharjah Investment and Development Authority (SHUROOQ) is leading this project along with the Sharjah Department of Culture. The Heart of Sharjah includes various cultural, commercial, and residential functions to attract tourists [56] (Figure 8). The traditional old houses within the site were converted into several museums such as Sharjah Heritage Museum, Sharjah Al Hisn Museum, and Sharjah Calligraphy Museum, an art gallery (Sharjah Art Foundation), boutique hotel (Al Bait Hotel) (Figure 9), traditional souq with retail shops (Souq Sager, Souq Al Shinasiyah, and Souq Al Arsah), cafés, restaurants, and offices. The land-use ratio is as follows: hospitality and entertainment (37.9%), mixed-use enterprises (19.5%), cultural (18.4%), residential (9.3%), commercial (7.6%), and community (7.3%) [57]. SHUROOQ plans to nominate the project for UNESCO World Heritage under "Sharjah, Gateway to the Trucial States" [58]. The project's first phase included the restoration of Al Hisn Fort, constructed in 1820 by the Ruler of Sharjah, Sheikh Sultan bin Saqr Al Qasimi, and was almost destroyed during the 1970s. Additionally, the Al Naboodah house, Al Midfa house, and Al Tawaweesh house were restored after being checked for structural augmentation [59,60]. The restoration of Bank Street, the first modern commercial street in Sharjah in 1978, was initiated with a non-intrusive radar survey of the ground to avoid conflict with the archeological excavation that T. Sasaki led at the Department of Archaeology at the University of Kanazawa, Japan [61,62].

A significant part of Bank Street, Souq Al Shinasiyah, has been renovated as it was a central area for trading in the 1950s. In addition, it has significant value as a cultural and heritage place as it connects the urban and social fabric of Souq Sagr in Al Shuwaiheen with Souq Al Arsah in Al Marijah. SHUROOQ would consider revitalizing this area as a heritage landmark to attract tourists and increase business activities and the overall economy [63].

Figure 7. Map of the Heart of Sharjah Heritage District.

Sharjah Art Foundation (SAF)'s galleries are also placed in the Heart of Sharjah to display the cultural identity (Figure 10). These art spaces are aimed to reactivate the Heart of Sharjah through contemporary art that enriches the socio-cultural context of Sharjah. SAF provides five galleries of restored buildings with minimal interior exhibition spaces and various exterior exhibition spaces such as courtyards, sikkahs, pocket parks, and an inter-connected roofscape [64]. It was a suitable reinterpretation of balance between the historical context and contemporary art to showcase the Bedouin style of living within the courtyard house typology [65]. This subtle design approach enables sustaining the past by reusing old houses and creating functional spaces for social and spatial functions while enjoying contemporary art [66].

Figure 8. The Heart of Sharjah heritage district in Sharjah (Source: Aga Khan Trust for Culture: https://archnet.org/collections/1662/sites/19005, accessed on 9 August 2020).

Figure 9. Courtyard of Al Bait Hotel in Sharjah.

Figure 10. Sharjah Art Museums (Al Mureijah Art Spaces) in Sharjah (photo by authors).

2.3. Tourism-Oriented Ajman Heritage District

In 2016, the Chairman of Ajman Municipality and Planning Department, Sheikh Rashid bin Humaid Al Nuaimi, approved the budget of AED 25 million for the renovation of Ajman Heritage District for tourism that is located adjacent to one of the oldest places in the Gulf, Ajman Museum [67] (Figure 11). Compared to Shindagha and Al Fahidi Heritage Districts in Dubai and the Heart of Sharjah, Ajman Heritage District (5300 m^2) is comparatively small. However, it is distinguished by a 1200 m long pedestrian walkway integrated with a children's playground, cafés, and restaurants, all of which exist within the traditional old buildings. Moreover, vernacular plants, such as sidra and ghaf, are planted to represent the rich heritage and environment of the UAE [68]. Ajman Municipality aims to boost the tourism and economy in Ajman through this project while serving the primary goal of preserving the cultural heritage of Ajman [69].

Figure 11. Ajman Heritage District (source: Ajman Municipality 2020).

The green canopies of trees distinguish Ajman Heritage District from the other heritage districts of Dubai and Sharjah [70]. The greenery results in a unique cultural experience within the valuable relics of the past. Ajman Heritage District (Figure 12) is a venue for preserved heritage and local and international events. While showcasing the history of Ajman, the project also aims to offer opportunities for investors and entrepreneurs to run businesses since there are 37 shops (40–100 m^2) [71].

Figure 12. Ajman Museum in Ajman Heritage District (photo by authors).

Three sites are selected for investigation based on their common characteristics subject to restoration and renovation: the Shindagha area and Al Fahidi Historical Area in Dubai, the Heart of Sharjah in Sharjah, and Ajman Heritage District in Ajman. The old houses within these heritage districts for tourism will be converted into museums to preserve and rejuvenate the historical site and support the economy [72].

The World Travel and Tourism Council report shows that the UAE ranked fourth in tourism investment among 181 countries in 2019 [73]. The UAE government emphasized the hospitality industry, which is why many heritage districts were developed to attract tourists and international investors. However, preserving place identity is not prioritized, leading to redundant investment and wasting many resources.

This research aims to find and preserve the identity of each heritage district in the three selected sites via a structured questionnaire survey. The survey is developed based on the categories extracted from several international principles such as the International Cultural Tourism Charter at the 12th General Assembly of the International Council on Monuments and Sites (ICOMOS) in Mexico in 1999, UNESCO Memorandum on World Heritage and Contemporary Architecture in Vienna in 2005, UNESCO Recommendation on the Historic Urban Landscape in Nanjing in 2011, and UNESCO Criteria for inscription on the lists established by the Convention for the Safeguarding of the Intangible Cultural Heritage in Paris in 2005 (Table 1).

Table 1. Extracted categories for the investigation of heritage districts based on the international principles.

	International Principles	Contents	Extracted Categories
1	12th General Assembly of International Council on Monuments and Sites (ICOMOS) in Mexico, October 1999	Major tourism attractions: the natural and cultural heritage as diversities and living cultures	(01) Diversity of heritage resources
			(02) Attractive heritage resources
			(03) Social, cultural, and economic vitality
2	UNESCO Memorandum on World Heritage and Contemporary Architecture in Vienna in 2005	8: The historic urban landscape is embedded with current and past social expressions and place-based developments	(04) Character-defining elements of the urban structure
		16: Consideration of the emotional connection between human beings and their environment. Contributing to a city's economic success and social and cultural vitality is fundamental	(05) Urban environmental quality
			(06) Social, cultural, and economic vitality
		23: Spatial structures around historic sites are to be enhanced via urban design and art	(07) Distinctive urban planning and art design
3	UNESCO Recommendation on the Historic Urban Landscape in Nanjing in 2011	9: This broader context includes the site's topography, geomorphology, hydrology, and natural features. (1) Built environment, both historic and contemporary, (2) open spaces and gardens, (3) land-use patterns and spatial organization, (4) perceptions and visual relationships, (5) social and cultural practices and values as related to diversity and identity	(08) Natural characteristics of ruins
			(09) Existing urban environment
			(10) Infrastructures
			(11) Land-use patterns and spatial organization
			(12) Perceptions and visual relationships
			(13) Character-defining elements of the urban structure
			(14) Social, cultural, and economic vitality
			(15) Diversity of heritage resources
4	UNESCO Criteria for inscription on the lists established by the Convention for the Safeguarding of the Intangible Cultural Heritage in Paris in 2005	Represents a masterpiece of creative genius	(16) Unique artistry or aesthetics
		Exhibits an important interchange of human values, over some time, or within a cultural area of the world, on developments in architecture or technology, monumental arts, town-planning, or landscape design	(17) Influence of heritage resources
		To bear exceptional testimony to a cultural tradition	(18) Preciousness of heritage resources
		An outstanding example of a type of building, architectural, technological ensemble, or landscape	(19) An outstanding example of a type
		An outstanding example of a traditional human settlement, land use, or sea use which is representative of a culture, or human interaction with the environment	(20) Traditional human settlement or human interaction
		Is directly or tangibly associated with events or living traditions, with ideas, or with beliefs, with artistic and literary works of outstanding universal significance	(21) Associated with events or living tradition

The selected sites depict valuable aspects such as urban history and culture, heritage resources, and environmental values. Sub-attributes for the investigation were chosen to include 21 elements used by the international principles as indicators for evaluation.

Based on the questionnaire survey, the expert survey was conducted on 100 experts via e-mail from 4 April to 4 May 2019, in which 82 respondents participated (82% response rate). The questionnaire was composed of four main sections: personal information such as academic background, specialization, occupation, and work experience in related fields; five questions about city history/culture; eight questions about heritage resources; and nine questions about the environment. The degree of respondents' agreement with the evaluation index was obtained using the five-step Likert scale, with 1 = very low, 2 = low, 3 = medium, 4 = high, and 5 = very high [74].

2.4. Validity Verification

The collected data in the first and second expert surveys were validated through content validity ratio (CVR), a method that C. H. Lawshe suggested in 1975 [75]. CVR proposes a minimum value based on the number of panels, and it is judged that the item has content validity when it exceeds the minimum value. The equation of CVR is as follows: CVR = $(ne \times -N/2)/(N/2)$, and it corresponds to the number of experts (in this paper, the frequency of respondents with 4 or 5 points on the Likert 5-point scale for the questionnaire), and N corresponds to the total number of experts who have responded.

The validity of the expert research techniques can be presented by analyzing the degree of expert opinion collection and consensus. The degree of convergence is 0 when opinions are all converged at one point, and the value increases when the deviation of the views is significant [76]. The degree of consensus is 1 when Q1 and Q3 coincide and are ultimately agreed upon and decreases when the difference of opinion is substantial. The closer the degree of convergence is to 0, and the closer the degree of agreement to 1, the more correct the item is. The equation follows degree of agreement = $1 - (Q3 - Q1/Mdn)$, degree of convergence = $(Q3 - Q1)/2$. (Q: quartile, Mdn: median). Data analysis was performed using the IBM SPSS Statistics 26 program. This research study judged that a CVR value of 0.40 or more could be a valid indicator. In addition, the criteria for accommodating the response result due to the expert panel's consensus were set to be 3.50 or more, consensus level 0.75 or more, and convergence level between 0 and 0.5.

2.5. Selection of Candidates for Evaluation Indicators

The significance of the heritage district for tourism lies in its influential characteristics or attributes as an attractive place for tourists [77]. The selected characteristics of the heritage districts for tourism are education from an open-air museum, history, culture, art of the Bedouin lifestyle, and diversity and uniqueness of the heritage resources. In this research study, the indicators were selected by identifying the characteristics of the heritage district and reviewing the international principles such as those of UNESCO and ICOMOS. The evaluation criteria are divided into three categories: history/culture of the city, heritage resources, and environment. According to the 12th General Assembly of the International Council on Monuments and Sites (ICOMOS) in Mexico, tourism mainly comprises natural and cultural heritage, diversity, and culture. Based on that, the evaluation indexes are selected: (1) Diversity of heritage resources, (2) Attractiveness of heritage resources, and (3) Vitality of society, culture, and economy. Since Articles 8, 16, and 23 of the UNESCO Memorandum on World Heritage and Contemporary Architecture in Vienna in 2005 satisfy the environmental domain of the city, the evaluation index is selected to include: (1) Distinct urban components, (2) The environment of a historic city, (3) The vitality of society, culture, and economy, and (4) Characteristic urban planning and art. In addition, the selected evaluation indicators from Article 9 of the UNESCO Recommendation on the Historic Urban Landscape in Nanjing in 2011 are: (1) Natural characteristics of the historic site, (2) Urban environment, (3) Basic facilities, (4) Land and space use, (5) Visual perception, (6) Distinct urban components, (7) Social, cultural, and economic vitality, and (8) Diversity of heritage resources. Lastly, the selected evaluation indicators from the UNESCO Criteria for inscription on the lists established for the Safeguarding of the Intangible Cultural Heritage in Paris in 2005 are: (1) Unique artistry or aesthetics, (2) Influence of heritage resources, (3) Preciousness of heritage resources, (4) An outstanding example of a type, (5) Traditional human settlement or human interaction, (6) Associated with events or living tradition (Table 1).

2.6. Derivation of Evaluation Indicators via Expert Survey

The expert survey was conducted to validate the indicators for the derived evaluation criteria. Table 2 shows the expert survey results in the heritage districts for tourism in the UAE. The importance of the three evaluation areas was generally high in the expert survey.

Table 2. The result of the expert perception survey.

Evaluation Criteria	Evaluation Indicators	Expert Perception Survey				
		Mean	Standard Deviation	Content Validity Ratio	Agreement	Convergence
Urban History and Culture	Historic	4.79	0.38	1.00	1.00	0.04
	Traditional	4.51	0.49	1.00	0.82	0.52
	Community	4.39	0.48	0.99	0.77	0.54
	Educational	4.67	0.46	1.00	0.82	0.54
	Cultural	4.42	0.50	1.00	0.77	0.52
Heritage Resources	Diversity of heritage resources	4.23	0.42	1.00	0.96	0.16
	The attractiveness of heritage resources	4.61	0.47	1.00	0.82	0.52
	Influence of heritage resources	4.14	0.43	0.92	0.98	0.10
	The uniqueness of heritage resources	4.29	0.45	1.00	0.77	0.52
	The preciousness of heritage resources	4.32	0.53	0.92	0.77	0.54
	Magic of heritage resources	4.04	0.42	0.90	0.99	0.02
	Natural characteristics of ruins	4.19	0.64	0.78	0.77	0.50
	Unique artistry or aesthetics	4.26	0.62	0.82	0.78	0.50
Urban Environment	Social, cultural, and economic vitality	4.56	0.49	1.00	0.77	0.54
	Characteristics of the urban structure	4.17	0.62	0.76	0.78	0.54
	Urban existing environmental quality	4.39	0.70	0.75	0.82	0.54
	Natural and historical environment	4.42	0.48	1.00	0.77	0.52
	Distinctive urban planning and art design	4.17	0.52	0.89	0.96	0.23
	Infrastructures	4.36	0.47	0.98	0.77	0.50
	Perceptions and visual relationships	4.56	0.48	0.99	0.82	0.52
	Interesting sites	4.48	0.50	0.98	0.81	0.52
	Regional characteristics	4.23	0.70	0.72	0.77	0.54

In the Urban History and Culture criterion, all indicators showed higher average values, indicating that history and culture are essential compared to the other standards. In the requirements of Heritage Resources, the average Attractiveness of heritage resources was the highest (4.61). In the Urban Environment conditions, Social, cultural, and economic vitality and Perceptions and visual relationships were the highest (4.56 on average). The expert survey results were distributed around an average of 4.04~4.79 in each index; the standard deviation was 0.38~0.70; the degree of convergence was 0.04~0.54; the content validity was 0.72 or more, and the agreement was 0.77 or more. Based on experts' opinions, the results of the suitability analysis of the two indicators in the three criteria were all above the reference value, so it was concluded as a valid evaluation index.

According to the expert perception survey, candidate indicators 5, 8, and 9 were selected for Urban History and Culture, Heritage Resources, and Urban Environment (Table 3).

Table 3. Evaluation criteria and indicators of heritage district.

Evaluation Criteria	#	Evaluation Indicators
Urban History and Culture	5	(1) Historical, (2) Traditional, (3) Community, (4) Educational, (5) Cultural
Heritage Resources	8	(1) Diversity of heritage resources, (2) Attractiveness of heritage resources, (3) Influence of heritage resources, (4) Uniqueness of heritage resources, (5) Preciousness of heritage resources, (6) Magic of heritage resources, (7) Natural characteristics of ruins, (8) Unique artistry or aesthetics
Urban Environment	9	(1) Social, cultural, and economic vitality, (2) Characteristic elements of the urban structure, (3) Urban existing environmental quality, (4) Natural and historical environment, (5) Distinctive urban planning and art design, (6) Infrastructures, (7) Perceptions and visual relationships, (8) Interesting sites, (9) Regional characteristics

3. Results

Based on the extracted evaluation indicators from the expert perception survey, a general survey was conducted on tourists and residents in the UAE to explore the reason behind their visit to the heritage districts (Shindagha area and Al Fahidi Historical Area in Dubai, the Heart of Sharjah in Sharjah, and Ajman Heritage District in Ajman). Based on the post-occupancy evaluation (POE) survey, the identities of each heritage district could be derived and compared with the original objectives of the municipalities of Dubai, Sharjah, and Ajman.

3.1. The Characteristics of Surveyors

The total number of participating respondents was 195 males (48.6%) and 206 females (51.4%). The participants were in their 40s (20.4%), 30s (19.3%), 50s (18.7%), and the rest in their 20s. The percentage of participating tourists was 57.6%, and 42.4% residents. Regarding the touristic value, the Shindagha area and Al Fahidi Historical Site in Dubai have gained 77.2% and 81.4%. This was higher than the Heart of Sharjah in Sharjah (36.7%) and Ajman Heritage District (34.0%). The occupation of the survey respondents was students (23.2%), self-employed (16.9%), housewives (16.5%), and governmental jobs (14.5%). As for the gender, there were no significant differences. As for the age group, the Shindagha area (23.8%) and Al Fahidi Historical Area (23.5%) in Dubai had the highest ratio of participants in their 40s, the Heart of Sharjah in Sharjah (22.4%) had the highest percentage of participants in their 20s, and Ajman Heritage District in Ajman (20.0%) had the highest ratio of participants in their 30s and 40s (Table 4).

Table 4. The characteristics of the respondents.

Classification		Dubai: Shindagha Area #	Dubai: Shindagha Area %	Dubai: Al Fahidi Historical Area #	Dubai: Al Fahidi Historical Area %	Sharjah: Heart of Sharjah #	Sharjah: Heart of Sharjah %	Ajman: Ajman Heritage District #	Ajman: Ajman Heritage District %	Total #	Total %
# of Respondents		101	100	102	100	98	100	100	100	401	100
Gender	Male	48	47.5	49	48.1	48	48.9	50	50	195	48.6
	Female	53	52.5	53	51.9	50	51.1	50	50	206	51.4
Type	Tourist	78	77.2	83	81.4	36	36.7	34	34	231	57.6
	Resident	23	22.8	19	18.6	62	63.3	66	66	170	42.4
Age Range	10s	12	11.8	15	14.7	20	20.4	12	12	59	14.7
	20s	14	13.8	12	11.8	22	22.4	19	19	67	16.7
	30s	19	18.9	18	17.7	20	20.4	20	20	77	19.3
	40s	24	23.8	24	23.5	14	14.3	20	20	82	20.4
	50s	21	20.8	23	22.5	13	13.3	18	18	75	18.7
	60s	11	10.9	10	9.8	9	9.2	11	11	41	10.2
Occupation	Office Worker	20	19.8	17	16.7	9	9.2	8	8	54	13.5
	Student	26	25.7	27	26.5	21	21.4	19	19	93	23.2
	Housewife	14	13.9	17	16.7	18	18.4	17	17	66	16.5
	Professional	19	18.9	16	15.7	9	9.2	12	12	56	13.9
	Government	9	8.9	11	10.7	18	18.4	20	20	58	14.5
	Unemployed	1	0.9	2	1.9	1	1	2	2	6	1.5
	Self-Employed	12	11.9	12	11.8	22	22.4	22	22	68	16.9
	Other	0	0	0	0	0	0	0	0	0	0

3.2. Transportation

Considering the means of transportation of respondents to each destination, the total walking activity is 13.4%, and the highest walking percent was for the Shindagha area (15.6%) and Al Fahidi Historical Area (12.2%) due to a large number of nearby hotels. Most respondents went to the heritage district by car (46.3%) and taxi/Careem (34.1%). In the Shindagha area and Al Fahidi Historical Area in Dubai, cars (38.6%) and taxis/Careem (39.4%) were the most common means of transportation. In Ajman Heritage District, a car was used by 54.1% of respondents, and a taxi/Careem was used by 31.7% of them. The Heart of Sharjah had more car dependence as a car was used by 54.8% of respondents compared to the other heritage districts. It is implied that Sharjah and Ajman have less public transportation infrastructure than Dubai for the heritage districts (Table 5).

3.3. Time Spent in Heritage Districts for Tourism

The overall duration of the time spent by respondents within the heritage districts was reasonably even. In the Shindagha area, 70.0% of respondents have spent time there and stayed between 1.5 and 2.5 h. In Al Fahidi Historical Area, 73.2% of respondents spent between 1.5 and 2.5 h there. In the Heart of Sharjah, 71.1% of respondents have spent time and stayed between 1.5 and 2.5 h. Finally, in Ajman Heritage District, 23.4% of respondents spent their time which lasted more than 2.5 h. It was concluded that people had spent more extended time in Ajman Heritage District (Table 5).

Table 5. The characteristics of user behavior of three heritage districts.

Categorization		Shindagha Area	Al Fahidi Historical Area	Heart of Sharjah	Ajman Heritage District	Total (401)
Transportation (%)	Walk	15.6	12.2	10.2	10.1	12.1
	Car	37.8	38.6	54.8	54.1	46.3
	Taxi/Careem	38.4	36.1	30.5	31.7	34.1
	Bus/Metro	8.2	13.1	4.5	4.1	7.5
	Total	100	100	100	100	100
The time spent in the project (%)	1.0–1.5 h	13.4	12.6	17.2	11.2	13.6
	1.5–2.0 h	34.8	33.4	32.7	24.6	31.3
	2.0–2.5 h	35.2	39.8	38.4	40.8	38.6
	Above 2.5 h	16.6	14.2	11.7	23.4	16.5
	Total	100	100	100	100	100
Companion Type (%)	Family	30.5	29.9	24.2	48.8	33.1
	Friends	12.6	13.2	42.4	23.2	22.9
	Colleagues	4.8	3.7	10.2	6.4	6.2
	Group Tourists	52.1	53.2	23.2	21.6	37.8
	Total	100	100	100	100	100

3.4. Companion Types

The most frequent type of companion was group tourists (37.8%), followed by family (33.1%). On average, it was found that the Shindagha area (52.1%) and Al Fahidi Historical Area (53.2%) had double the number of tourists as compared to the Heart of Sharjah and Ajman Heritage District. It was found that the percentage of visitors with friends (42.4%) in the Heart of Sharjah was much higher than in other heritage districts. This is due to the initiatives of the Sharjah Municipality to implement art programs and galleries to attract the younger generation [78]. However, in Ajman Heritage District, the percentage of families (48.8%) was higher than in the other heritage districts. Ajman visitors are assumed to approach the heritage district as a family during weekends (Table 5).

3.5. Evaluation Criteria Survey

To analyze and verify the survey results, the analysis of variance (ANOVA) test was used to ensure that the survey results of the four heritage districts were significantly meaningful [79]. Concerning the Urban History and Culture criteria, the indicators of Historic character (3.92) and Traditional character (3.83) were above 3.5, which means that these two indicators are relatively impressive to the visitors of the Shindagha area in Dubai. For Al Fahidi Historical Area in Dubai, the Traditional character (3.91) and Historic character (3.82) were above 3.5, reflecting their significance to the visitors. Since the survey included tourists, most considered this area a unique area of Bedouin history and tradition. In the Heart of Sharjah district, visitors replied that Cultural (3.93) and Educational (3.84) were the most critical indicators even though buildings were properly restored to remind them of the Bedouin history and tradition. It was found that Sharjah Art Foundation was well integrated with this area and has operated many art and culture-related programs in the Heart of Sharjah [80]. For Ajman Heritage District, Community character and Traditional character were above 3.5, which means these two indicators are essential to the visitors. It was comprehended that visitors used this heritage district as a family gathering place on weekends in Ajman. Compared to Dubai and Sharjah, Ajman Municipality put more effort into integrating the heritage district with more commercial programs (Table 6).

Table 6. The result of the analysis of variance (ANOVA) test for the selected heritage districts.

Survey Criteria	Indicators		Dubai		Sharjah	Ajman
	Heritage District		Shindagha	Al Fahidi	HOS	AHD
	Number		101	102	98	100
Urban History and Culture	Historic	M	3.92	3.82	3.31	3.28
		SD	0.76	0.71	0.71	0.64
	Traditional	M	3.83	3.91	3.42	3.41
		SD	0.72	0.78	0.68	0.69
	Community	M	2.86	2.92	2.82	3.98
		SD	0.69	0.69	0.73	0.75
	Educational	M	3.34	3.28	3.84	2.87
		SD	0.57	0.61	0.71	0.68
	Cultural	M	3.45	3.46	3.93	2.99
		SD	0.76	0.74	0.78	0.71
Heritage Resources	Diversity of heritage resources	M	3.64	3.74	3.93	2.91
		SD	0.69	0.73	0.75	0.68
	The attractiveness of heritage resources	M	3.91	3.93	3.81	3.80
		SD	0.76	0.71	0.74	0.72
	Influence of heritage resources	M	3.79	3.74	3.71	3.92
		SD	0.72	0.75	0.62	0.76
	The uniqueness of heritage resources	M	3.84	3.82	3.79	3.42
		SD	0.71	0.75	0.72	0.63
	The preciousness of heritage resources	M	3.33	3.32	3.28	3.36
		SD	0.72	0.71	0.69	0.75
	Magic of heritage resources	M	2.44	2.51	2.62	2.71
		SD	0.72	0.69	0.75	0.69
	Natural characteristics of ruins	M	2.21	2.31	2.26	3.08
		SD	0.68	0.65	0.67	0.71
	Unique artistry or aesthetics	M	3.62	3.71	3.63	3.53
		SD	0.77	0.74	0.74	0.71
Urban Environment	Social, cultural, and economic vitality	M	3.66	3.82	3.68	3.92
		SD	0.75	0.72	0.72	0.72
	Character elements of urban structure	M	2.53	3.02	3.12	2.64
		SD	0.71	0.69	0.72	0.72
	Urban existing environmental quality	M	3.87	3.92	3.81	3.69
		SD	0.72	0.76	0.73	0.72
	Natural and historical environment	M	3.34	3.28	3.21	3.22
		SD	0.74	0.69	0.72	0.64
	Distinctive urban planning and art design	M	2.84	3.03	3.91	2.97
		SD	0.72	0.68	0.72	0.71
	Infrastructures	M	3.44	3.52	3.02	2.78
		SD	0.63	0.71	0.63	0.64
	Perceptions and visual relationships	M	3.02	2.89	3.35	2.81
		SD	0.68	0.71	0.73	0.69
	Interesting sites	M	3.61	3.62	3.42	3.83
		SD	0.76	0.69	0.72	0.73
	Regional characteristics	M	3.93	3.78	3.59	3.71
		SD	0.77	0.71	0.73	0.65

In the Heritage Resources criteria, the Attractiveness of heritage resources (3.91) and Uniqueness of heritage resources (3.84) were relatively essential indicators to the visitors of the Shindagha area in Dubai. The Attractiveness of heritage resources (3.93) and Diversity of heritage resources (3.74) were above 3.5, which means that these two indicators are essential to the visitors to Al Fahidi Historical Area in Dubai. It was shown that Dubai Municipality's strategy has succeeded in attracting tourists based on the survey result of the high importance of attractiveness, uniqueness, and diversity [81]. In the Heart of Sharjah, the most critical indicators were Diversity of heritage resources (3.93) and Attractiveness of heritage resources (3.81). Due to many programs related to the community and young generation, Sharjah Art Foundation seems well integrated with the built environment. Sharjah Art Foundation is committed to renovating the existing environment; a good example is the Flying Saucer project [82]. For Ajman Heritage District, the Influence of heritage resources (3.92) and the Attractiveness of heritage resources (3.80) were essential indicators to the visitors. Based on our interview with the executive director of Ajman Municipality, Mohammed bin Omair Al Muhairi, the Ajman Heritage District was planned and developed from the masterplan phase as a hub for local and international events (Table 6).

In the criteria of Urban Environment, Regional characteristics (3.93) and Urban existing environmental quality (3.87) were relatively essential indicators to the visitors of the Shindagha area in Dubai. The values of Urban existing environmental quality (3.92) and Social, cultural, and economic vitality (3.82) were above 3.5. These two indicators are essential to the visitors to Al Fahidi Historical Area in Dubai. Based on this result, it can be explained that the strategy of Dubai Municipality concerning replica heritage is interwoven with the existing urban fabric of old markets such as the gold souqs in Bur Dubai and Deira [83]. In the Heart of Sharjah, Distinctive urban planning and art design (3.91) and Urban existing environmental quality (3.81) were the most critical indicators. Thanks to the Sharjah Art Foundation, this heritage area's museums and galleries were well integrated with the existing urban fabric [84]. For Ajman Heritage District, the indicators of Social, cultural, and economic vitality (3.92) and Interesting sites (3.83) were essential to the visitors (Table 6).

4. Discussion

In the twentieth century, urban restoration is usually initiated in cities due to the large-scale damages from civil revolutions, wars, or global vandalism. In the case of the UAE, many architects and scholars enquire about heritage restoration methods. The recurring question is: "Can subsequent restoration with modern materials disturb continuity and heritage value?" The Dubai Municipality explored the answer: "as a device to reunite people with their history and tradition."

The innovative approach of Dubai to reinterpret its heritage districts for tourism for historical authenticity can be represented in the cultural solidity. The cultural tool is believed to reconnect people with their history and traditions since Dubai is experiencing unprecedented modernization and globalization in history. Additionally, there is a social goal to reconnect the local people with their own Bedouin history, which is a simple approach compared to the Operational Guidelines of the World Heritage Convention, which require technical criteria of authenticity and integrity. Based on this research, the satisfaction level of the local people was very high, and many tourists have been visiting the Shindagha area and Al Fahidi Historical Area. This successful restoration approach was spread to other Emirates such as Abu Dhabi, Sharjah, Ajman, and Umm Al Quwain. The practice of historic restoration is required in a city with diverse cultures. The heritage preservation and continuity approaches are significant in supporting the city's urban identity.

5. Conclusions

This research study has examined three tourism-oriented heritage districts in the United Arab Emirates and explored a new approach to foster heritage tourism and revive the culture and community. This was envisaged through conducting a post-occupancy evaluation (POE) to detect the characteristics of the heritage districts as conceived by the residents and tourists. The systematic evaluation of opinions about the buildings in use was based on indicators developed by UNESCO and ICOMOS international principles and then evaluated by an expert survey to identify the characteristics of the UAE heritage districts. Urban History and Culture criteria included five indicators: Historical, Traditional, Community, Educational, and Cultural. The requirements of Heritage Resources included eight indicators: Diversity of heritage resources, Attractiveness of heritage resources, Influence of heritage resources, Uniqueness of heritage resources, Preciousness of heritage resources, Magic of heritage resources, Natural characteristics of ruins, and Unique artistry or aesthetics. The criteria of Urban Environment included nine indicators: Social, cultural, and economic vitality, Characteristic elements of the urban structure, Urban existing environmental quality, Natural and historical environment, Distinctive urban planning and art design, Infrastructure, Perceptions and visual relationships, Interesting sites, and regional characteristics.

Firstly, in the Urban History and Culture criteria, Historical (3.92) and Traditional (3.83) were essential indicators to the visitors of the Shindagha area in Dubai, and Traditional character (3.91) and Historic character (3.82) were crucial to the visitors of Al Fahidi Historical Area. On the other hand, the indicators of Cultural (3.93) and Educational (3.84) were essential to the visitors of the Heart of Sharjah district; and Community character (3.98) and Traditional character (3.41) were important in Ajman Heritage District.

Secondly, in the Heritage Resources criteria, Attractiveness of heritage resources (3.91) and Uniqueness of heritage resources (3.84) were essential indicators to the visitors of the Shindagha area in Dubai. The Attractiveness of heritage resources (3.93) and the Diversity of heritage resources (3.74) were critical to the Al Fahidi Historic Area visitors in Dubai. Furthermore, the Diversity of heritage resources (3.93) and Attractiveness of heritage resources (3.81) were essential indicators to the visitors of the Heart of Sharjah district, and the Influence of heritage resources (3.92) and Attractiveness of heritage resources (3.80) were important indicators in Ajman Heritage District.

Thirdly, in the Urban Environment criteria, Regional characteristics (3.93) and Urban existing environmental quality (3.87) were essential indicators to the visitors of the Shindagha area in Dubai. On the other hand, Urban existing environmental quality (3.92) and Social, cultural, and economic vitality (3.82) were essential indicators for Dubai's Al Fahidi Historical Area visitors. Distinctive urban planning and art design (3.91) and Urban existing environmental quality (3.81) were indicators critical to the visitors of the Heart of Sharjah district; and Social, cultural, and economic vitality (3.92) and Interesting sites (3.83) were important indicators in Ajman Heritage District.

Table 7 summarizes the evaluations of the questionnaire surveys. The attraction is the primary indicator for any action leading to the success of cultural heritage tourism.

Table 7. Summary of three selected heritage districts.

Criteria	Dubai	Sharjah	Ajman
Urban History and Culture	History and Tradition	Culture and Education	Community and Tradition
Heritage Resources	Attraction uniqueness	Attraction and Diversity	Attraction
Urban Environment	Social, Cultural, and Economic vitalities	Urban planning	Social, Cultural, and Economic Vitalities

In similar sites, keeping the area as residential only is not recommended since such sites constitute the city's center. It can support the development of one or a combination of the four following dimensions:

- Social Dimension: Diversity of the users to encourage the involvement and participation of the community in the process of the preservation of the tangible and intangible objects reflecting the character and the identity of the area.
- Economic Dimension: Tourism is one way of re-using heritage sites and buildings. On the one hand, it allows visitors to experience a living heritage; on the other, it generates income for the local community with which heritage resources can be conserved

and maintained and sustain the traditions and the know-how from one generation to another.

- Environmental Dimension: One of the ways to avoid sprawl development and the proliferation of incompatible typologies will be achieved through the revitalization of the historical area.
- Cultural Dimension: The sense of the place will be appreciated through the architectural language that will give character to the area and will be a source of attraction while simultaneously emphasizing the identity.

Author Contributions: All authors contributed significantly to this study. J.A., C.J. and B.B. identified and secured the example buildings used in the study. The data acquisition system was designed and installed by C.J. and J.A., M.A. and B.B. were responsible for data collection. C.J. and J.A. performed data analysis. The manuscript was compiled by C.J. and reviewed by J.A., B.B. and M.A. All authors have read and agreed to the published version of the manuscript.

Funding: This research received no external funding.

Institutional Review Board Statement: The study was conducted according to the guidelines of the Ajman University Research Ethics Committee.

Informed Consent Statement: Informed consent was obtained from all subjects involved in the study.

Data Availability Statement: New data were created or analyzed in this study. Data will be shared upon request and consideration of the authors.

Acknowledgments: The authors would like to express their gratitude to Ajman University for APC support and the Healthy & Sustainable Built Environment Research Center at Ajman University for providing an excellent research environment.

Conflicts of Interest: The authors declare no conflict of interest.

References

1. Yotsumoto, Y.; Vafadari, K. Comparing cultural world heritage sites and globally important agricultural heritage systems and their potential for tourism. *J. Herit. Tour.* **2021**, *16*, 43–61. [CrossRef]
2. Timothy, D.J.; Boyd, S.W. Heritage Tourism in the 21st Century: Valued Traditions and New Perspectives. *J. Herit. Tour.* **2006**, *1*, 1–16. [CrossRef]
3. Tan, P.L.; Noor, S.; Rasoolimanesh, S.M.; Mustafa, H. Communication and visitor factors contributing towards heritage visitors' mindfulness. *J. Herit. Tour.* **2020**, *15*, 27–43. [CrossRef]
4. Borri, A.; Corradi, M. Architectural Heritage: A Discussion on Conservation and Safety. *Heritage* **2019**, *2*, 631–647. [CrossRef]
5. Dewi, C. Rethinking architectural heritage conservation in post-disaster context. *Int. J. Herit. Stud.* **2017**, *23*, 587–600. [CrossRef]
6. Ung, A.; Vong, T.N. Tourist experience of heritage tourism in Macau SAR, China. *J. Herit. Tour.* **2010**, *5*, 157–168. [CrossRef]
7. Glendinning, M. *The Conservation Movement: A History of Architectural Preservation*; Routledge: London, UK, 2013; pp. 212–224. [CrossRef]
8. Krakover, S. A heritage site development model: Jewish heritage product formation in south-central Europe. *J. Herit. Tour.* **2017**, *12*, 81–101. [CrossRef]
9. Pickard, R. A Comparative Review of Policy for the Protection of the Architectural Heritage of Europe. *Int. J. Herit. Stud.* **2002**, *8*, 349–363. [CrossRef]
10. Del Barrio-García, S.; Prados-Peña, M.B. Do brand authenticity and brand credibility facilitate brand equity? The case of heritage destination brand extension. *J. Destin. Mark. Manag.* **2019**, *13*, 10–23. [CrossRef]
11. Little, C.; Bec, A.; Moyle, B.; Patterson, D. Innovative methods for heritage tourism experiences: Creating windows into the past. *J. Herit. Tour.* **2020**, *15*, 1–13. [CrossRef]
12. Carughi, U.; Visione, M. *Time Frames: Conservation Policies for Twentieth-Century Architectural Heritage*; Routledge: London, UK, 2017; pp. 224–236.
13. Tyler, N.; Tyler, I.R.; Ligibel, T.J. *Historic Preservation: An Introduction to Its History, Principles, and Practice*; W. W. Norton & Company: New York, NY, USA, 2018; pp. 38–46.
14. Hartmann, R. Dark tourism, thanatourism, and dissonance in heritage tourism management: New directions in contemporary tourism research. *J. Herit. Tour.* **2014**, *9*, 166–182. [CrossRef]
15. Mason, R.F. *Giving Preservation a History: Histories of Historic Preservation in the United States*; Routledge: London, UK, 2003; pp. 120–128.
16. Habitat, U.N. *Meeting the Urban Challenges*; 2008; pp. 4–10. Available online: https://mirror.unhabitat.org/downloads/docs/5883_42389_Meeting%20the%20Urban%20Challenges.pdf (accessed on 15 June 2020).

17. Iwamoto, W. *Historic Districts for All, a Social and Human Approach for Sustainable Revitalization*; 2008; pp. 8–16. Available online: https://unesdoc.unesco.org/ark:/48223/pf0000158331 (accessed on 15 June 2020).
18. Picton, O.J. Usage of the concept of culture and heritage in the United Arab Emirates—An analysis of Sharjah heritage area. *J. Herit. Tour.* **2010**, *5*, 69–84. [CrossRef]
19. Rai, B.S. Heritage Tourism in Focus in Dubai & Tourism Development. 2013. Available online: https://www.emirates247.com/news/emirates/heritage-tourism-in-focus-in-dubai-2013-05-31-1.508628 (accessed on 20 July 2020).
20. Hadjri, K.; Boussaa, D. Architectural and Urban Conservation in the United Arab Emirates. *Open House Int.* **2007**, *32*, 16–26. [CrossRef]
21. Boussaa, D. Urban Heritage and Tourism in the Gulf: The Case of Dubai in the UAE. *J. Tour. Hosp. Manag.* **2014**, *2*, 174–181. [CrossRef]
22. Al Kaabi, N. The UAE Has Made Great Strides in the Preservation of Its Rich Heritage. 2019. Available online: https://www.thenational.ae/opinion/editorial/the-uae-has-made-great-strides-in-the-preservation-of-its-rich-heritage-1.941523 (accessed on 22 July 2020).
23. McKercher, B.; Du Cros, H. *Cultural Tourism: The Partnership Between Tourism and Cultural Heritage Management*; Routledge: London, UK, 2002; pp. 80–86.
24. Michael, N.; Reisinger, Y.; Hayes, J.P. The UAE's tourism competitiveness: A business perspective. *Tour. Manag. Perspect.* **2019**, *30*, 53–64. [CrossRef]
25. Seraphim, J.; Haq, F. Challenges for promotion of heritage tourism: Case study of the UAE. In *Conservation and Promotion of Heritage Tourism*; IGI Global: Hershey, PA, USA, 2019; pp. 35–56.
26. Langham, E.; Barker, D. *Spectacle and Participation: A New Heritage Model from the UAE*; Routledge: London, UK, 2016; pp. 85–98. [CrossRef]
27. Sharfuddin, M. Al Bastakiya—Al Fahidi Historical Neighborhood. 2019. Available online: https://www.dubaiomg.com/al-bastakiya/ (accessed on 2 August 2020).
28. Gravari-Barbas, M. Tourism as a heritage producing machine. *Tour. Manag. Perspect.* **2018**, *25*, 173–176. [CrossRef]
29. Veldpaus, L.; Roders, A.R.P.; Colenbrander, B.J.F. Urban Heritage: Putting the Past into the Future. *Hist. Environ. Policy Pract.* **2013**, *4*, 3–18. [CrossRef]
30. Turnpenny, M. Cultural Heritage, an Ill-defined Concept? A Call for Joined-up Policy. *Int. J. Herit. Stud.* **2004**, *10*, 295–307. [CrossRef]
31. Seyfi, S.; Hall, C.M. *Cultural and Heritage Tourism in the Middle East and North Africa: Complexities, Management and Practices*; Routledge: London, UK, 2020; pp. 64–72.
32. Kyriazis, A.; Almehairi, A.; Al Seri, I. The Morphology of Abu Dhabi's Superblocks: A Modernist Vision. *Urban Morphol.* **2002**, *1*, 21–32. [CrossRef]
33. Prager, L. Displaying origins: Heritage museums, cultural festivals, and national imageries in the UAE. *Horiz. Humanit. Soc. Sci. Int. Refereed J.* **2015**, *1*. [CrossRef]
34. Madichie, N.; Madichie, L. City brand challenge 101: Sharjah in a globalised UAE context. *Int. J. Bus. Glob.* **2013**, *11*, 63–68. [CrossRef]
35. Melotti, M. Heritage and Tourism. Global Society and Shifting Values in the United Arab Emirates. *Middle East-Top. Argum.* **2014**, *3*, 71–91. [CrossRef]
36. Boussaa, D. Cultural heritage in the Gulf: Blight or blessing? A discussion of evidence from Dubai, Jeddah and Doha. *Middle East-Top. Argum.* **2014**, *3*, 55–70.
37. Richards, G. Cultural tourism: A review of recent research and trends. *J. Hosp. Tour. Manag.* **2018**, *36*, 12–21. [CrossRef]
38. Zaidan, E.; Abulibdeh, A. Master planning and the evolving urban model in the gulf cities: Principles, policies, and practices for the transition to sustainable urbanism. *Plan. Pract. Res.* **2021**, *36*, 193–215. [CrossRef]
39. Khobragade, S. 9 Things to Know about the Shindagha Heritage District Project. 2018. Available online: https://www.cntravellerme.com/content/12800-nine-things-you-didn%E2%80%99t-know-about-the-shindagha-heritage-district-development-project (accessed on 22 July 2020).
40. Henderson, J.C. Global Gulf Cities and Tourism: A Review of Abu Dhabi, Doha and Dubai. *Tour. Recreat. Res.* **2014**, *39*, 107–114. [CrossRef]
41. Abit, L. Dubai Cultural Tour at Al Shindagha Heritage District. 2014. Available online: https://www.ladyandhersweetescapes.com/2014/03/Dubai-Heritage-Tour-AlShindagha.html (accessed on 24 July 2020).
42. AlHasawi, M. The Policies and Process of Preserving the Historical and Heritage Buildings in Dubai City: Case Study of Historical District. *Acad. Res. Community Publ.* **2019**, *3*, 39–48. [CrossRef]
43. Stephenson, M.L. Tourism, development and 'destination Dubai': Cultural dilemmas and future challenges. *Curr. Issues Tour.* **2014**, *17*, 723–738. [CrossRef]
44. Saundalkar, A. Watch: Inside the Al Shindagha Museum in Old Dubai. 2019. Available online: https://whatson.ae/2019/02/watch-inside-the-al-shindagha-museum-in-old-dubai/ (accessed on 26 July 2020).
45. Thani, S.; Heenan, T. Theme park Arabism: Disneyfying the UAE's heritage for Western tourist consumption. In *Cultural and Heritage Tourism in the Middle East and North Africa*; Routledge: London, UK, 2020; pp. 157–169.

46. Rodrigues, J. Al Shindagha Days: New 10-Day Culture Festival Is Coming to Dubai. 2020. Available online: https://www.thenational.ae/arts-culture/al-shindagha-days-new-10-day-culture-festival-is-coming-to-dubai-1.960543 (accessed on 22 July 2020).
47. Shannon, B. THE 'DUBAI EFFECT'. In *The Emerging Asian City: Concomitant Urbanities & Urbanisms*; Routledge: London, UK, 2012; p. 254.
48. Farahani, F. Indications of Iran Bastak Architecture in the spatial Structure of Dubai Al Bastakiya Houses. *Tradit. Islamic Art* **2013**, *1*, 113–139.
49. Milek, J. Dubai's Heritage: A Tour Of Al Bastakiya. 2018. Available online: https://theculturetrip.com/middle-east/united-arab-emirates/articles/dubais-heritage-a-tour-of-al-bastakiya/ (accessed on 22 July 2020).
50. Manchuette, A.S. Al Bastakiya, The Old Iranian District of Dubai. 2020. Available online: https://www.odyfolioemirates.com/en/dubai/bastakiya--iranian (accessed on 28 July 2020).
51. Boussa, J. A Future to the Past: The Case of Fareej al-Bastakiya in Dubai. In *UAE. Proceedings of the Seminar for Arabian Studies*; Archaeopress: Oxford, UK, 2006; Volume 36, pp. 125–135.
52. Mazzetto, S. Heritage Restoration as a Tool to Promote Architectural Identity in the Gulf Regions. *Preserv. Digit. Technol. Cult.* **2018**, *47*, 3–11. [CrossRef]
53. Alawadi, K. Place attachment as a motivation for community preservation: The demise of an old, bustling, Dubai community. *Urban Stud.* **2017**, *54*, 2973–2997. [CrossRef]
54. Bastaki, S. Al Bastakiya Dubai: The Oldest Residential Area in Dubai. 2014. Available online: http://flashydubai.com/al-bastakiya-dubai-the-oldest-residential-area-in-dubai/ (accessed on 22 July 2020).
55. Khamis, J. 'Heart of Sharjah' Project under Way. 2013. Available online: https://gulfnews.com/business/tourism/heart-of-sharjah-project-under-way-1.1156359 (accessed on 24 July 2020).
56. Pallathucheril, V. New Hearts for Two Gulf Cities. *Arch. Des.* **2015**, *85*, 92–99. [CrossRef]
57. Ashill, N.J.; Williams, P.; Chathoth, P. 10 Examining the marketing opportunities of Sharjah as an Islamic tourism destination. In *International Tourism Development and the Gulf Cooperation Council States: Challenges and Opportunities*; Routledge: London, UK, 2017; p. 171.
58. Al Sarkal, M. Heart of Sharjah: Discover a Fascinating Past. 2019. Available online: https://shurooq.gov.ae/project/heart-of-sharjah/ (accessed on 22 July 2020).
59. Velegrinis, S.; Katodrytis, G. Drawing on Sand: Cities in the Making. *Arch. Des.* **2015**, *85*, 72–79. [CrossRef]
60. Rab, S. Sharjah: Seascape Urbanism in a Khaliji Port City. In Proceedings of the ACSA Association of Collegiate Schools of Architecture Annual Meeting, Montreal, Canada, 3–6 March 2011; pp. 3–7.
61. Chitty, G. *Heritage, Conservation and Communities: Engagement, Participation and Capacity Building*; Routledge: London, UK, 2016; pp. 108–116.
62. Wam, S. Visiting the Past in the Heart of Sharjah. 2016. Available online: https://www.khaleejtimes.com/20160714/1498/heart-of-sharjah-heritage-project (accessed on 26 July 2020).
63. Al Mutawa, Y. Shurooq Revives Historical Souq Al Shinasiyah in the Heart of Sharjah. 2014. Available online: https://shurooq.gov.ae/shurooq-revives-historical-souq-al-shinasiyah-in-the-heart-of-sharjah/ (accessed on 26 July 2020).
64. El-Mousfy, M. Sharjah Biennial 11, 2013: New SAF Art Spaces, Heart of Sharjah. 2013. Available online: https://universes.art/en/sharjah-biennial/2013/new-saf-art-spaces (accessed on 30 July 2020).
65. Ibrahim, I. ECO-Traditional Courtyard Houses in UAE: A Case Study of the Sharjah Museums. 2018. Available online: https://doi.org/10.2495/arc180021 (accessed on 22 July 2020). [CrossRef]
66. Abdel-Razzaq, J. 3 Must-Visit Cultural Destinations at the Heart of Sharjah. 2019. Available online: https://www.admiddleeast.com/heart-of-sharjah-places-to-visit (accessed on 24 July 2020).
67. Al-Oraibi, M. Ajman Ruler Briefed on Heritage Park Plans. 2016. Available online: https://www.thenational.ae/uae/ajman-ruler-briefed-on-heritage-park-plans-1.189170 (accessed on 24 July 2020).
68. Rix, J. The Rival to Dubai that Is on the Brink of a Tourism Boom. 2019. Available online: https://www.telegraph.co.uk/travel/destinations/middle-east/united-arab-emirates/articles/the-alternative-to-dubai-enjoying-a-tourism-boom/ (accessed on 25 July 2020).
69. Hussein, H. Ajman Urban Development Surges in Recent Years. 2019. Available online: https://www.wam.ae/en/details/1395302765499 (accessed on 29 July 2020).
70. MacLean, M. Suburbanization, National Space and Place, and the Geography of Heritage in the UAE. *J. Arab. Stud.* **2017**, *7*, 157–178. [CrossRef]
71. Abdullah, A. 25 Million AED Ajman Heritage District in the Works. 2019. Available online: https://www.khaleejtimes.com/nation/sharjah/dh25-million-ajman-heritage-district-in-the-works (accessed on 28 July 2020).
72. Kumar, A. Visit Cultural, Heritage Sites to Learn UAE's History. 2019. Available online: https://www.khaleejtimes.com/uae/abu-dhabi/visit-cultural-heritage-sites-to-learn-uaes-history- (accessed on 22 July 2020).
73. Grimaldi, E. World Travel & Tourism Council: Top 10 Fastest Growing Tourism Cities Are All in Asia. 2019. Available online: https://worldtourismwire.com/world-travel-tourism-council-top-10-fastest-growing-tourism-cities-are-all-in-asia-2392/#gsc.tab=0 (accessed on 30 July 2020).

74. Johnson, R.; Morgan, G. *Survey Scales: A Guide to Development, Analysis, and Reporting*; Guilford Publications: New York, NY, USA, 2016; pp. 134–148.
75. Ayre, C.; Scally, A.J. Critical values for Lawshe's content validity ratio: Revisiting the original methods of calculation. *Meas. Eval. Couns. Dev.* **2014**, *47*, 79–86. [CrossRef]
76. Litwin, M. *How to Measure Survey Reliability and Validity (Survey Kit)*; SAGE Publications, Inc.: Southend Oaks, CA, USA, 1995; pp. 42–54.
77. Sharpley, R. *Rural Tourism: An Introduction (Tourism and Hospitality Management Series)*; International Thomson Business Press: London, UK, 1997; pp. 87–94.
78. Gronlund, M. Sharjah's Rain Room: Four Things to Know before You Go to the Permanent Installation. 2018. Available online: https://www.thenational.ae/arts-culture/art/sharjah-s-rain-room-four-things-to-know-before-you-go-to-the-permanent-installation-1.753130 (accessed on 22 July 2020).
79. Strunk, K.; Mwavita, M. *Design and Analysis in Educational Research: ANOVA Designs in SPSS*; Routledge: London, UK, 2000; pp. 122–134.
80. Abdel-Razzaq, J. Sharjah Art Foundation, Smithsonian Institute Offer Training Program on the Arts. 2019. Available online: https://www.admiddleeast.com/sharjah-art-foundation-teams-up-with-smithsonian-institute (accessed on 27 July 2020).
81. Malcolm, C. Panoramic Vision: How the UAE Is Safeguarding Its Heritage. 2017. Available online: https://www.thenational.ae/lifestyle/family/panoramic-vision-how-the-uae-is-safeguarding-its-heritage-1.89076 (accessed on 28 July 2020).
82. Illankoon, K. Sharjah Art Foundation Completes Renovation of Iconic Brutalist Building, 'The Flying Saucer'. 2020. Available online: https://www.cbnme.com/news/sharjah-art-foundation-completes-renovation-of-iconic-brutalist-building-the-flying-saucer/ (accessed on 22 July 2020).
83. McArthur, R. A Guide to Local Favorites in Al Fahidi, Dubai. 2015. Available online: https://www.washingtonpost.com/travel/united-arab-emirates/al-fahidi-neighborhood-dubai/ (accessed on 25 July 2020).
84. Llbo-on, L. Heart of Sharjah to be Restored by 2025. 2014. Available online: https://www.khaleejtimes.com/nation/general/heart-of-sharjah-to-be-restored-by-2025 (accessed on 24 July 2020).

MDPI AG
Grosspeteranlage 5
4052 Basel
Switzerland
Tel.: +41 61 683 77 34

Heritage Editorial Office
E-mail: heritage@mdpi.com
www.mdpi.com/journal/heritage

Disclaimer/Publisher's Note: The title and front matter of this reprint are at the discretion of the Guest Editors. The publisher is not responsible for their content or any associated concerns. The statements, opinions and data contained in all individual articles are solely those of the individual Editors and contributors and not of MDPI. MDPI disclaims responsibility for any injury to people or property resulting from any ideas, methods, instructions or products referred to in the content.

www.ingramcontent.com/pod-product-compliance
Lightning Source LLC
Chambersburg PA
CBHW041410210926
51850CB00013B/3232
9783725827435